Verification of Autonomous Systems

Signe Redfield • Donald Sofge • Mae Seto •
John Sustersic
Editors

Verification of Autonomous Systems

Springer

Editors
Signe Redfield
U.S. Naval Research Laboratory
Washington, DC, USA

Donald Sofge
U.S. Naval Research Laboratory
Washington, DC, USA

Mae Seto
Dalhousie University
Halifax, NS, Canada

John Sustersic
RTX BBN Technologies
Portland, ME, USA

ISBN 978-3-031-88545-7 ISBN 978-3-031-88546-4 (eBook)
https://doi.org/10.1007/978-3-031-88546-4

This Springer imprint is published by the registered company Springer Nature Switzerland AG
The registered company address is: Gewerbestrasse 11, 6330 Cham, Switzerland

This book is dedicated to Dr. Susan Calvin and to all the autonomous systems that highlight why verification is necessary: those that end up operating outside their nominal parameters, those that operate in environments that go beyond their designers' and evaluators' expectations, and those that end up publicizing their inability to cope with the real world.

Foreword

Autonomous systems have become part of our daily lives whether we realize or not. We often hear about self-driving taxis operating on the streets of San Francisco or food delivery robots getting into trouble with the police or pedestrians. Moreover, the lucky travelers might have occasionally spotted cleaning robots with cute faces on their screens vacuuming the floors of airports like Incheon in Korea or Changi in Singapore.

Many other domains are building autonomous systems for a wider range of applications than we can even begin to imagine, ranging from home care to military. To deal with our world, which is chaotic, unstructured, and sometimes unpredictable, and tackle realistic applications in it, autonomous systems end up being quite complex.

Autonomous systems comprise the collection of many modules working together. These modules enact interactions of hardware and software that come from many sources. Furthermore, the modules and the systems interact with environments that can change, are difficult to describe and predict, and reflect the diversity of the people in them.

The existence of frameworks and best practices to build autonomous systems depends on the maturity of the individual technologies, along with the societal concerns of ensuring people's safety and property preservation. Nonetheless, there is a widespread lack of documentation and guidance on how to build autonomous systems that are safe and trustworthy, i.e., dedicated verification processes. Also, the available knowledge has not been widely transferred to the younger academic research community, startups, and end users.

Over the years, pioneering textbooks on verification methods have been published, with a strong focus on the theory and workings of a particular method or tool. For example, there is a plethora of books on the theory of model checking and formal verification methods including "Model Checking" by Edmund M. Clarke or "Principles of Model Checking" by Christel Baier and Joost-Pieter Katoen. Additionally, there is a vast range of literature on the verification of embedded systems and software through model-based methods and simulation, including "Practical Model-Based Testing: A Tools Approach" by Mark Utting and Bruno Legeard. But regarding the verification of autonomous systems, there are only a few very recent options, such as the book "Verifiable Autonomous Systems: Using Rational Agents to Provide Assurance about Decisions Made by Machines" by Louise A. Dennis and Michael Fisher. Furthermore, there is a lack of literature on the practicality of how to verify a specific autonomous system given its characteristics and interaction domain, i.e., choosing the most suitable method(s), reconciling the results, and what are the limitations of a particular system–method combination.

This book offers the first comprehensive guide to figure out how autonomous systems can be verified from a pragmatic perspective and according to the available methods and tools. Across a range of types of tools and methods, this book helps to understand what is available, how it works, how to use it for a particular system or application domain, and how to not use it or the limitations and gaps in the approaches. The authors come from a wide variety of domains, and this is reflected in the book, but their vast experience in building autonomous systems in practice provides invaluable insights onto what works best and what might not.

This book considers different personas and their viewpoints: the academics studying the construction of autonomous systems or the area of verification applied to autonomous systems; the advanced end users interested in knowing how their systems can be specified, modeled, analyzed, and certified; the regulators that need to grasp the best practices and the future focus areas; and the system developers who need to verify the systems that they are building. Note that some knowledge of mathematics and logic is required to dive into the details of each chapter.

Any of these personas should read this book to:

- Better understand the challenges of building autonomous systems in practice, i.e., what it takes to make a "good" useful system.
- Review the state of the art of many methods to verify autonomous systems in practice as of today and what are the gaps that need to be addressed in a timely manner to cover a wide range of autonomous systems and scenarios.
- Learn the practical aspects of verification and how to apply these methods to autonomous systems.

This book provides a great outline of the verification challenges for autonomous systems and provides reasonable answers when they exist, i.e., when research and development has enabled building methodologies, frameworks, tools, and practices to evaluate a system's performance. The chapters in the book nicely guide the reader through state-of-the-art work to describe (specify, model), measure, test (systematically and meaningfully), and analyze (in a mathematical or formal manner) autonomous systems.

About the Book Structure

As we have seen with novel technologies, such as artificial intelligence (e.g., large language models), but even more challenging when embodied like with robots or self-driving cars, the use of technologies leads to several questions regarding their performance. Whether we are end users, system builders (developers for hardware, software, or both), system evaluators and testers, theoretical researchers, government members and regulators, or even bystanders, we ponder: Is this system useful, i.e., does it do its intended task? Is it trustworthy? Am I able to interact with it safely and without harm? Why is it doing that? Is it always like that or could it change? How can I get any guarantee that the system is going to work well? Is it going to work in any situation?

All these are valid questions that, in this book, become verification challenges once posed in an adequate and systematic scientific manner in Chap. 2. These challenges require answers, to increase users' comfort and enable these technologies to succeed in the real world.

After the introduction of the verification challenges pertaining to autonomous systems, this book proceeds to describe how certain methods and tools have been used in practice, following the structure and flow laid out in Chap. 2. A plethora of examples are mentioned throughout each chapter to strengthen the claims.

Chapter 3 presents a comprehensive review of terminology on the concepts of "verification" and "autonomous systems" to lay ground for the rest of the book. This is followed by a proposal to use formal methods to formulate system requirements, runtime verification methods for the verification of systems that change over time, formal methods for the verification of systems that have machine learning components, and agile and DevOps frameworks and infrastructure derived from software engineering for development purposes, as opposed to traditional waterfall practices.

Chapter 4 complements the terminology from the previous chapter to define the concept of "specification." Then, the chapter focuses on a thorough review of capability analysis tables as a tool for requirement specification development, i.e., to describe what a system should and should not do. Contrary to first thoughts, formulating requirements is not a trivial task limited to writing a bulleted list, since it is easy to start designing a system before appropriately scoping its desired behaviors or intentions, leading to incongruent results: a system that is not fit for purpose. Having a systematic way and tools to formulate requirements helps to achieve their "correctness" and "completeness" in a measurable way, to ensure that the designs that come later can be evaluated adequately.

Chapter 5 dives into the realm of mathematically modeling autonomous systems (through an ontology) for their analysis in simulation. In this case, a "metamodel" or a hierarchical abstract modeling language is applied to a system's capabilities, and examples providing analysis of performance bounds are provided.

Chapter 6 deals with the verification of autonomous systems that learn, using multiple tools such as domain specification languages, auto-encoder models to represent or translate high-level requirements into neural networks and using formal logic constraints to guide neural network training. Although the motivation for this chapter is to enable the construction of autonomous systems that learn by demonstration from human behavior, the approach can be generalized to many learning-related use cases.

Chapter 7 continues with definitions of verification and validation, but this time in a testing- and metrics-driven context. Several performance metrics and key indicators that can be applied to autonomous systems are described, followed by metrics for experiment design in testing. Two use cases illustrate comprehensively the use of metrics and testing methods for industrial robots and human–robot teams.

Chapter 8 starts with a comprehensive review of maritime autonomous systems, both underwater and surface based, and their verification challenges. This is followed by a review of verification tools that can be applied to these systems, from modeling formalisms to formal methods (including those for runtime), testing hardware and software in the loop, and metrics. An interesting aspect mentioned is the need and methods for verification once the autonomous systems are in the water, beyond runtime verification.

Chapter 9 addresses the challenge of generating suitable tests to thoroughly examine a system in simulation. Generating a test suite that effectively and efficiently stresses the system is the ultimate goal when the system under test has almost infinitely many possible interesting stimuli and behaviors to examine. This chapter introduces the Range Adversarial Planning Tool and related framework, which uses adaptive sampling to explore the testing state space. Once the tests are executed in simulation, boundaries for the system's performance are computed, and metrics are defined and used to evaluate the results. A comprehensive example is provided to illustrate the use of the framework, based on an unmanned underwater vehicle use case.

Chapter 10 jumps straight into applying verification to spacecraft software, using some of the methods mentioned in the previous chapters. Three use cases are presented: an error detection system at rocket launch time, an autonomous navigation system that acts when communication is lost with the ground, and a system to compensate for crosswinds when on Mars. These systems sound simple in title, but the complexity required to deal with uncertainty in an autonomous manner and the cost and importance of ensuring that the systems keep operating call for comprehensive approaches from requirement solicitation to design and testing.

Chapter 11 closes the book with an overall review of all the tools and methods from verification in general that can possibly be used to evaluate autonomous systems. An overview of autonomous systems, architectures to build them, and their relation to machine learning and artificial intelligence is provided, from a perspective of highlighting the characteristics that make verification different from other systems and complementing Chap. 2. Then, the chapter moves on to outline all the different existing verification methods, drawing from domains such as embedded systems and software design, and how some of them have been used already for autonomous systems, complementing the examples from the previous chapters.

Overall, this book delves deeply into why the verification of autonomous systems reflects their complexity through numerous examples, provides a first glance on how to verify some of the existing autonomous systems with available tools and methodologies, and points out a myriad of challenges, gaps, and opportunities that need addressing. It is a great source of inspiration for those who strive to build better and useful autonomous systems for society.

Antwerp, Belgium Dejanira Araiza-Illan

Acknowledgements The editors would like to express their thanks. This book would not have been possible without our contributors, who provided a comprehensive view of the ways in which the challenges in this area are being addressed. It would not have been possible without our reviewers, who took time from their busy schedules to give us valuable feedback. It would also not have been possible without the ongoing support of our employers. And, most importantly, this book is the culmination of years of effort, years in which this book took attention from our other priorities—this book would not have been possible without the support of our families. We thank you all.

Contents

Contributors

Ali Bassam Mechanical Engineering Department, Dalhousie University, Halifax, NS, Canada

Andrew Bouchard NSWC PCD, Panama City, FL, USA

Jennifer Case NSWC PCD, Panama City, FL, USA

Daniel J. Dorney NASA, Langley Research Center, Hampton, VA, USA

Anthony Downs NSWC PCD, Panama City, FL, USA

Martin S. Feather Jet Propulsion Laboratory, California Institute of Technology, Pasadena, CA, USA

Benjamin Hartman NSWC PCD, Panama City, FL, USA

Constance L. Heitmeyer Naval Research Laboratory, Washington, DC, USA

Yordan Hristov School of Informatics, University of Edinburgh, Edinburgh, UK

Craig Innes School of Informatics, University of Edinburgh, Edinburgh, UK

Demetrious Kutzke NSWC PCD, Panama City, FL, USA

Craig Lennon Army Research Laboratory, United States Army, Bel Air, MD, USA

Jamie Lennon Naval Research Laboratory, Washington, DC, USA

Elizabeth I. Leonard Naval Research Laboratory, Washington, DC, USA

Jeremy A. Marvel NSWC PCD, Panama City, FL, USA

Galen Mullins The Johns Hopkins University Applied Physics Laboratory, Laurel, Maryland,

Lorraine E. Prokop NASA, Langley Research Center, Hampton, VA, USA

Subramanian Ramamoorthy School of Informatics, University of Edinburgh, Edinburgh, UK

Signe Redfield U.S. Naval Research Laboratory, Washington, DC, USA

David Scheidt Weather Gage Technologies, Annapolis, MD, USA

Craig Schlenoff NSWC PCD, Panama City, FL, USA

Mae Seto Mechanical Engineering Department, Dalhousie University, Halifax, NS, Canada

Donald Sofge Navy Center for Applied Research in Artificial Intelligence, U.S. Naval Research Laboratory, Washington, DC, USA

John Sustersic RTX BBN Technologies, Portland, ME, USA

Richard Tatum NSWC PCD, Panama City, FL, USA

Abbreviations

ACC-ORL-OPB	U.S. Army Contracting Command Orlando
ADCP	Acoustic Doppler current profiler
AI	Artificial intelligence
ALFUS	Autonomy levels for unmanned systems
ANOVA	Analysis of variance
ARE	Autonomy requirements engineering
ARIAC	Agile Robotics for Industrial Automation Competition
ASIS	Abort Sensing and Implementation System
ASV	Autonomous surface vehicle
ATEVV	Assurance, test, evaluation, verification, and validation
ATM	Army Technical Manual
AUT	Autonomy under test
AUV	Autonomous underwater vehicle
C2	Command and control
CAE	Claims, arguments, and evidence
CAT	Capability analysis table
CCD	Charge-coupled device
CCP	Commercial Crew Program
CDMA	Code-division multiple access
cFS	Core Flight System
CMU	Carnegie Mellon University
COLREGs	Convention on the International Regulations for Preventing Collisions at Sea
CODE	Collaborative Operations in Denied Environments
CPS	Cyber-physical system
CR	Capability representation
CRD	Command receiver decoder
DARPA	Defense Advanced Research Projects Agency
DART	Double Asteroid Redirection Test
DBSCAN	Density-based spatial clustering of applications with noise
DDS	Data distribution service

DevOps	Software development and information technology operations
DIMES	Descent Image Motion Estimation System
DMP	Dynamic movement primitive
DNN	Deep neural network
DOUG	Dynamic Onboard Ubiquitous Graphics
DSB	Defense Science Board
DVL	Doppler velocity log
EDGE	Engineering DOUG Graphics for Exploration
EDL	Entry, descent, and landing
EDS	Emergency Detection System
EM	Expectation maximization
EO-1	Earth Observing One
EPSRC	Engineering and Physical Sciences Research Council
ESA	European Space Agency
FDIR	Fault detection, identification, and remedy/recovery
FORMS	FOrmal Reference Model for Self-adaptation
FRET	Formal Requirement Elicitation Tool
FSD	Finite state diagram
FVM	Four-variable model
Auto GCAS	Automatic Ground Collision Avoidance System
GM	General Motors
GNC	Guidance, navigation, and control
GNSS	Global navigation satellite system (GPS (USA), Galileo (EU), GLONASS (Russia), BeiDou (China))
GPS	Global positioning system (USA)
GSN	Goal structuring notation
HCI	Human–computer interaction
HRI	Human–robot interaction
HITL	Hardware-in-the-loop
IAW	In accordance with
IEEE	Institute of Electrical and Electronics Engineers
IFAS	Integrity and fault assessment system
IMU	Inertial measurement unit/inertial management unit
INCA	Inertial navigation and control assembly
INS	Inertial navigation system
ISO	International Organization of Standardization
ITL	Integrated Test Laboratory
JHU/APL	Johns Hopkins University Applied Physics Laboratory
JLTV	Joint Light Tactical Vehicle
JSC	Johnson Space Center
KPI	Key performance indicator
LfD	Learning from demonstration
LIDAR	Light detection and ranging
LRO	Lunar Reconnaissance Orbiter
LTL	Linear temporal logic

LVC	Live-virtual-constructive
MAS	Marine autonomous system (of which AUS and ASV are examples)
MBSE	Model-based systems engineering
MC	Model checking
MER	Mars Exploration Rover
ML	Machine learning
MLOs	Mine-like objects
MOOS-IVP	Mission Oriented Operating Suite-Interval Programming
MSSF	Mission success/safety failure
MSSS	Mission success/safety success
MTBF	Mean time between failure
MTTF	Mean time to failure
MTTR	Mean time to repair
NASA	National Aeronautics and Space Administration
NASA-TLX	NASA Task Load Index
NATO	North Atlantic Treaty Organization
NIST	National Institute of Standards and Technology
NITRD	Networking and Information Technology Research and Development
NMCM	Naval mine countermeasures (mission)
NN	Neural network
NOAA	National Oceanic and Atmospheric Administration
NOC	National Oceanography Center in Southampton, UK
NUWC	Naval Undersea Warfare Center
OEE	Overall equipment effectiveness
OFDM	Orthogonal frequency division multiplexing
OFT	Orbital flight test
OODA	Observe, orient, decide, act
ONR	Office of Naval Research
OpNav	Orion Optical Navigation
PAA	Project assurance action
RAPT	Range Adversarial Planning Tool
RCTA	Robotics Collaborative Technology Alliance
ReLU	Rectified linear unit
RE	Requirements engineering
RF	Radio frequency (communication)
RGB	Red, green, blue
RGBD/RGB-D	Red, green, blue, depth
RL	Reinforcement learning
ROI	Return on investment
ROS	Robot operating system
RQV	Runtime quantitative verification
RV	Runtime verification
SAL	Symbolic Analysis Laboratory

SAT	Boolean satisfiability
SCR	Software cost reduction
SIL	Systems Integration Laboratory
SITL	Software-in-the-loop
SLAM	Simultaneous localization and mapping
SLOC	Source lines of code
SMT	Satisfiability modulo theory
SNR	Signal-to-noise ratio
SRM	Solid rocket motor
STL	Spatial temporal logic
SUT	System under test
SysML	Systems modeling language
T&E/S&T	Test and evaluation/science and technology
TOP	Test operation procedure
TRL	Technology readiness level
TRMC	Test Resource Management Center
UAV	Unmanned air vehicle
UKRI	United Kingdom Research and Innovation
ULA	United Launch Alliance
UML	Unified modeling language
UMS	Unmanned system
USV	Uncrewed (or unmanned) surface vehicle
UUV	Uncrewed (or unmanned) underwater vehicle
V&V	Verification and validation
V3D	Three-dimensional vector
VCC	Vehicle control computer
VAE	Variational autoencoder
XUV	Experimental unmanned vehicle

Part I
Introduction

Chapter 1
Introduction

Mae Seto, John Sustersic, Signe Redfield, and Donald Sofge

> *"As machine-learning systems grow not just increasingly pervasive but increasingly powerful, we will find ourselves more and more often in the position of the 'sorcerer's apprentice': we conjure a force, autonomous but totally compliant, give it a set of instructions, then scramble like mad to stop it once we realize our instructions are imprecise or incomplete—lest we get, in some clever, horrible way, precisely what we asked for."*
>
> —Brian Christian
> *The Alignment Problem: Machine Learning and Human Values*

In 2014, the first meeting of the Verification of Autonomous Systems Working Group was held in the naïve hope that if we assembled roboticists, computer scientists, electrical and mechanical engineers, mathematicians, and verification specialists (along with at least one translator), we could explain our verification problem to the mathematicians and either they would identify a mathematical space or representation that would help us handle our complexity problem, or they would determine that it is theoretically impossible for a mathematical space or representation to help handle our complexity problem, or they would determine that

M. Seto
Mechanical Engineering Department, Dalhousie University, Halifax, NS, Canada
e-mail: mae.seto@dal.ca

J. Sustersic
RTX BBN Technologies, Portland, ME, USA
e-mail: john.sustersic@rtx.com

S. Redfield (✉)
U.S. Naval Research Laboratory, Washington, DC, USA
e-mail: signe@ieee.org

D. Sofge
Navy Center for Applied Research in Artificial Intelligence, U.S. Naval Research Laboratory, Washington, DC, USA
e-mail: donsofge@gmail.com

S. Redfield et al. (eds.), *Verification of Autonomous Systems*,
https://doi.org/10.1007/978-3-031-88546-4_1

further work was necessary to figure out whether a mathematical representation that could help manage our complexity was possible.

Instead, what actually came out of the meeting was a list of 27 open research challenges, grouped into four main categories. Over the next few years, we met with one another, shared funding opportunities, identified new challenges, and made progress in addressing them. We established an IEEE Robotics and Automation Society Technical Committee and we're developing a guide to best practices under the auspices of the IEEE Standards Association. This book captures the state of research into these challenges as the initial research began to bear fruit.

1.1 Relevance

Before we get into the meat of our topic, we need to address two reciprocal questions: why is verification an important topic in autonomous systems, and why are autonomous systems an important and special area in verification?

Verification is the process of gaining confidence in the correctness of a design or ensuring that it has certain properties. These properties are often related to safety, security, or ethical considerations. We are interested in the middle step in the *verification and validation process*:

1. Validate the match between the purpose for which a system is designed and the system requirements. This may include development of a design or model of the system.
2. Once the system is designed, verify that the design (or a model of the design) meets the defined requirements.
3. Validate the system to ensure it matches the end user's needs.

This process creates an assurance case for the functional, security, or safety properties of a system. Autonomous systems are an important topic in verification because autonomous systems are typically poorly specified or underspecified and lack theoretical tools to constrain the region of the performance space that must be evaluated. Verification rests on the assumptions that an adequate specification exists against which the system can be evaluated, that it is possible to generalize from the verification evaluation process to operational results, and that the evaluation process can be limited in time and cost and still achieve the desired performance guarantees and reduce risk to an acceptable level.

Autonomous systems, for the purposes of this book, include any system that makes decisions based on local environmental information and **has an intractably complex interaction with its environment**. The purpose of the bold-faced clause is to restrict our focus to the systems where verification is particularly complex and poorly understood. They include physically and virtually instantiated agents that maneuver through their world (robots) and physically and virtually instantiated agents that are stationary but can modify their internal or external environment in some way (immobots). Autonomous systems include homogeneous and heterogeneous teams of robots and immobots that may also include humans. They may be centralized or decentralized in their sensing and decision-making. A *system* includes the hardware, software, perception, actuation, communications, and decision-making that are abstracted as a unit and act in the world. A robot that picks up a cup is a system, but the cup remains part of the environment. A robot with a tool changer includes both the compatible tool and the robot in the system regardless of whether the tool is attached or not.

Users and *operators* monitor the autonomous system as it operates, providing guidance and information as the system attempts to perform its task. A *user* controls the system without understanding the underlying technology or decision-making capability, while an *operator* understands the underlying complexity and utilizes it. Users may become operators as they define new tasks for the system to perform that the original designers did not envision—in that case, the users are operators for the purposes of that task.

Autonomous systems contain and exhibit *behavior*. Actions a system chooses to take in the world as it attempts to accomplish a task are the behavior it exhibits. An algorithm that defines what actions should be taken in a given situation is a designed behavior. For an autonomous system, these behaviors are designed to operate in *unstructured* environments, the correct action for the system to take at any given time is *uncertain*, and the interaction between the environment and the system is intractably *complex*.

These systems are poorly specified because their raison d'être is to select an appropriate action given the current circumstances they find themselves in. Their entire existence is predicated on the designer and developer *not* knowing the details of the precise circumstances they will face. Autonomous systems are an important topic in verification because the challenges they pose are beyond the scope of existing tools.

But why is verification an important topic in autonomous systems?

1.2 Context

In the larger picture of autonomous systems research, verification is a small corner of the test and evaluation ecosystem. But it is implicit in every concrete robotics research topic—every kind of robot, every autonomous behavior, every perception algorithm, control algorithm, decision-making process, and application domain each require understanding of how to verify that component, that platform, and that behavior. Every systems engineering decision, every design methodology choice, and every transition from design to development to operation require us to understand how we will evaluate whether each of these pieces can do their job or not, individually and as a whole. Everything involved in accomplishing a task using autonomy is impacted by our lack of understanding of how to verify these systems.

1.2.1 Industrial Context

In the industrial community, on both the product creation and sales sides, verification is an essential component of the process that ensures that a product is going to meet customers' expectations and not trigger lawsuits and loss of capital. It enables the company to identify the scope of the market niche they are attempting to fill and know that the product they put on the market fills that niche. On the production side, verification is the foundation of the buyer's confidence that the factory equipment they purchase will do the job they need it to.

The applications of autonomous systems to daily life are starting to become economically viable. Simple household systems like the Roomba®, commercial systems like the Delaval VMS™ milking machine, and industrial systems like the KUKA™ robot arms have all been extremely successful and profitable. As the technology matures, interest in applying it for profit increases. Venture capital investment in autonomy is only increasing, and this investment drives a need for a return. That return is only possible when we can reduce the risk associated with using that system, through verification and validation.

1.2.2 Academic Context

Beyond economic motives, however, we also have scientific motives. Verification of autonomous systems requires that we improve our scientific and technical understanding of these systems and how they operate in the world. It requires that we develop improved metrics and approaches to both specification and evaluation. And it requires that we effectively address some of the most fundamental challenges associated with establishing robotics as a genuine scientific field. In order to demonstrate that a system meets its requirements, its performance must be reproducible and generalizable. But any science requires the ability to reproduce experiments and develop performance predictions that generalize from a single, anecdotal experiment to system performance in a broad range of contexts. Verification is important to researchers, as a scientific topic, in part because it forces us to look at our existing research with a critical eye. It drives us to become more scientific in our approach to the development of autonomy algorithms and autonomous systems.

1.2.3 Public Interest Context

In addition to economic and scientific motives, there are public interests in verification of autonomous systems. Entities serving the public interest may include both governments and organizations like accreditation groups, non-profits, foundations, and government-affiliated agencies and companies. Governments and other entities serving the public interest design autonomous systems for government and public use when those uses are not served by systems developed for industrial or personal needs. Entities serving the public interest test, evaluate, verify, validate, certify, and deploy autonomous systems for use in public service. Governments regulate the production and function of autonomous systems across all use cases through standards and certification agencies. And governments make policy that drives acceptable and prohibited use cases for these systems. Governments and entities serving the public interest need the problem of verification of autonomous systems solved because they need to design, deploy, and regulate these systems, and because they need to craft policy around them. The degree to which verification of these systems is possible effectively defines the degree to which they can be deployed and regulated.

1.2.3.1 End Users

Governments serve many different kinds of end users. They develop new autonomous systems and behaviors for applications that are not served by the commercial realm. They purchase autonomous systems from commercial providers to accomplish dirty, dull, and dangerous jobs, and they evaluate autonomous systems both for government use and to develop regulations regarding their commercial and private use. These end users include military users, other government agencies, and both individual and industrial civilian consumers.

Military

Military users have a driving need to solve the verification of autonomous systems problem, since this type of assurance is necessary to transition technologies from research laboratories to operational availability.

The core challenges they face depend on how they break up the verification challenge across the various components of their autonomous systems. The OODA (observe, orient, decide, act) loop is a decision-making approach that filters available information (*observe*), puts it in context (*orient*), and makes a timely decision (*decide*) with which to *act* upon. This type of strategy is suited for circumstances where the ability to adapt and react to evolving circumstances faster than the competitor confers an advantage. It is one of many decision-making approaches used by the military and other organizations.

In a military context, robotic autonomy has utility if it can get one to close their own OODA loop faster than the adversary's. The autonomous robot can, in a timely manner, support the *observe* and *orient* pieces to provide the situational awareness to inform the *decide* or decision. For some scenarios, the *decide* could be implemented in the autonomy, as, for example, a calculation based on in situ measurements that compare two competing probabilities and that factors in the tolerance for false positives or negatives. The *act* may not be difficult to implement in the autonomy either. However, there are other considerations that go into the decision.

Autonomy for military robots must consider the roles of international regulations, doctrine, and Rules of Engagement. These could be viewed as another set of requirements that are superimposed on the technical OODA loop ones if the autonomy is to integrate them. If the autonomy goes to the point of engagement with other entities (e.g., other robots, manned vessels, humans, other humans (nations), etc.), how that autonomy will integrate into the larger group must also be considered. Rules of Engagement vary with nation, doctrine can vary with groups of nations (e.g., NATO), and regulations (e.g., COLREGs) depend on whether the nation that deploys the robot observes them. This makes for unique autonomy implementations and subsequently, much more complex verification.

Additionally, military deployment of robots also has to consider the intent of the adversary, and any information, knowledge, or intelligence available to inform that. In a circumstance where the uncertainty may be greater than a threshold (e.g., as defined by say a decision theoretic that the autonomy can implement), the autonomy may be "taken offline" in deference to the operator's broader experience, perspective, context, and beliefs, i.e., the operator's knowledge. The operator becomes a vital autonomy component to the system, and the adversary becomes a significantly more complex environmental factor.

Verification of military systems must include systems that are, at some point, a hybrid of robot and human operator, where intelligent adversarial situations must be considered, and where the consequence of decision errors has a very high cost.

Other Government Agencies

Other government agencies may also use autonomous robots. For these agencies, verification is essential to ensure that the systems they procure meet both the specification for the desired functionality and the specifications for the required safety and security.

The agencies responsible for oversight of power plants, for example, have a vested interest in ensuring the functionality, safety, and security of autonomous systems used to investigate and assess dams, power plants, and nuclear reactors. As with military systems, the consequence of decision error may have a very high cost depending on the application in which the system is used. For these systems, however, the environmental factors are adversarial in the sense of being difficult to overcome and work around rather than being deliberately and intelligently attempting to thwart the autonomous system's ability to accomplish the task. As with military systems, verifying that the system provides the correct functionality is just as important as ensuring it meets safety and security guidelines.

Consumers

The government also has a role in ensuring that systems designed for individual and industrial consumer use meet existing legal guidance. Agencies responsible for oversight of transportation systems and commerce have a vested interest in ensuring the functionality, safety, and security of autonomous systems used for transit by individual citizens like personal automobiles and public transit systems, and of systems used for shipping by companies like cargo ships and truck convoys. These groups are also responsible for assessing autonomous systems used in industrial settings like robot arms used on factory floors. Verification is an essential element of that process. Government agencies like the US National Institute of Standards and Technology (NIST) are actively engaged in understanding these systems with a view to developing standards for verification of autonomy.

1.2.3.2 Public Good

When an autonomous system is developed for the public good, specifically to support publicly funded activities, the government's concerns include not only functionality, safety, and security, but also the ethical and moral implications of the use of the system to a much greater degree. Robots used in war are subject to the rules of war, robots used in factories are greatly constrained by both the factory environment and the company's need for them to be cost-effective, and robots used in the home may be purchased or avoided depending on whether the consumer feels that the ethics and morals of the company and its products are acceptable to them. Robots used by an entity serving the public interest, performing tasks in the public interest, are subject to the scrutiny and ethical and moral demands of their actions as they relate to the public interest. The verification of these systems must address these ethical and moral implications.

Regulatory and Policy Implications

Verification is an essential component of developing regulations and policy for autonomous systems. The degree to which a system's performance can be verified, the situations in which a system can be verified to function acceptably or not, the uses to which a system is put, all affect what regulations are needed, what policy decisions need to be taken, and how these systems are allowed into the commercial and service industries.

Standards

The government may also develop standards to support commerce and public safety. In some cases, these standards provide a metric against which commercially developed autonomous systems must be measured. In others, they define the specifications of the environments in which they are expected to operate. Verification enables autonomous system developers to determine whether they respond correctly to the anticipated environment and meet the needs of the industrial or commercial setting for which they are intended.

There are many ways in which public safety can be affected by autonomous systems. Beyond the obvious aspects, like ensuring the system does not injure bystanders, there are more subtle and complex interactions. Even when any given robot is able to avoid people, are bystanders safe when a large group of these robots assembles? Are there any requirements associated with damage to property, especially in cases where localized damage to a building or structure causes it to become unsafe?

What about safety issues when the system is in a low-power mode? What degree of sensing is necessary in order to ensure the safety of not only the system but its surroundings?

What about safety issues beyond the individual bystander? Verification provides a point at which the sufficiency of the specification can be evaluated. Does the interaction between the robot and its environment trigger catastrophic cascades of events that lead to unsafe conditions? The development of standards that address autonomous system behavior will inevitably rely on the ability of individuals and corporations to verify that systems meet those standards.

Verification of Autonomous Systems encompasses the processes, tools, and fundamental understanding of system and behavior design, operation, and evaluation. It enables the creation of assurance that an autonomous system or any of its cognitive elements will perform the task it is required to accomplish, safely, securely, and reliably. And verification enables the reduction of risk associated both with incorrect assumptions about the unstructured, complex environment in which the task is set and with uncertainty about the correct required action.

1.3 Impact

Each application domain for autonomous systems has its own risks and concerns that must be addressed during the verification process. Systems that operate in terrestrial environments experience different verification challenges than systems that operate in aerial, maritime, or space domains. Further, specific considerations associated with effective operation of autonomous systems drive their own constraints and challenges. In particular, the interfaces between humans and autonomous systems require complex considerations related to both human and machine safety, and all autonomous systems require extensive consideration of how verification can support or limit both their trust and their trustworthiness. This section addresses the risks and verification considerations associated with these domains and elements of autonomous system operation.

1.3.1 Terrestrial Impact

In terrestrial systems, the risk of accidentally falling on someone or suddenly appearing underneath someone is significantly lower than it is for aerial systems or underwater systems. A terrestrial robot is not going to suddenly sink in the middle of the ocean. In many cases, the neutral state of a terrestrial robot is "stopped"—removing power defaults it to a safe state. And in many situations, this state is in fact safe—an autonomous car stopped by the side of the road is not going to *initiate* actions that make someone unsafe; an autonomous vacuum cleaner stopped in the middle of the floor is at worst an obvious tripping hazard much more obvious and therefore much less dangerous than a stray piece of Lego®; an autonomous security robot face down in a mall pool is more of a tourist attraction than a danger.

But in practice, simply showing that the system is safe in the sense of "not active" is insufficient. An autonomous car that shuts down in the fast lane of a highway is not operating safely and will endanger its passengers. A robot engaged in walking or running cannot simply stop mid-stride and assume it will not hit anything important when it falls.

Furthermore, terrestrial robots are more likely than other robots to be performing tasks that involve interacting with humans. From robots that provide nursing support like the Moxi robot[1] to autonomous automobiles,[2] to delivery robots,[3] to security robots,[4] to vacuum cleaners like the Roomba® (which shipped 3.9 million robots in the first 9 months of 2021[5]), to medical robots like the DaVinci,[6] terrestrial robots are more likely to be involved in tasks that are taking place in the same spaces as humans.

The risk of being in a situation for which stopping is the wrong answer is lower, but the risk of inadvertently injuring a human while operating is higher. The impact of being able to verify these systems is reduction of risk to humans in their environment.

1.3.2 Space Impact

Space systems operate in a completely different environment. Once deployed, it is virtually impossible for them to be retrieved or repaired, and a malfunctioning

[1] https://www.cedars-sinai.org/newsroom/robots-help-nurses-get-the-job-donewith-smiles-and-beeps/

[2] https://www.kbb.com/car-advice/self-driving-cars/

[3] https://www.gpsworld.com/delivery-robots-begin-to-look-real/

[4] https://www.engadget.com/2017-07-17-dc-security-robot-throws-itself-into-pool.html

[5] https://investor.irobot.com/news-releases/news-release-details/irobot-reports-third-quarter-2021-financial-results

[6] https://www.davincisurgery.com/

system can cause incalculable damage even if it's incapable of anything except following an unpredictable trajectory. These systems are horrendously expensive to develop (due in part to the hostility of their operating environment and in part to the verification and validation efforts that are required to overcome the risk associated with their operation). Even though they typically operate at great distances from the nearest person, they can have staggering impacts on humanity (due to interference with cellular communications or GPS navigation satellites or due to unplanned crashes to earth). Fault management and verification of functionality, intended to ensure that they don't lose contact with their operator during available windows or function unpredictably while out of contact, are critical aspects of the development of these systems.

Some relatively small or hobbyist space systems (like microsatellites and StarLink[7] and the "Humanity Star"[8]) are much less risky. They do not have the time or resources to support the level of risk reduction associated with larger communications satellites or long duration NASA missions through the solar system, and the consequences of their failure are typically small. But even these systems can produce unintended consequences, affecting astronomers[9] and space agencies as they attempt to ensure that their larger, more expensive systems do not collide with these smaller units and cause catastrophic damage due to debris circling the Earth.

Comprehensive verification of functionality, safety, and security is *essential* for space systems to address, document, and reduce this risk.

1.3.3 Maritime Impact

Maritime systems can be just as remote from human control, but have their own unique challenges driving specific difficulties in verification. The difference in environment fundamentally changes the autonomy and with it, its verification requirements. The autonomous underwater vehicle (AUV) and the autonomous surface vehicle (ASV) will serve here as illustrative examples.

For most of the history of maritime autonomy, the militaries have led the development of these systems. The military has had the developmental funding, access to costly infrastructure (jetties, barges, boats/ships, etc. and their maintenance) and are driven by compelling security and safety requirements.

[7] https://www.space.com/spacex-starlink-satellites.html

[8] https://www.theatlantic.com/science/archive/2018/03/humanity-star-satellite-crash/555965/

[9] https://www.nzherald.co.nz/business/astronomer-ian-griffin-says-rocket-lab-and-peter-beck-have-vandalised-the-night-with-satellite/YUKI4YVPB56VHHM7UDCZ32TFYQ/ Sergal, Michael (25 January 2018). "Astronomer Ian Griffin says Rocket Lab and Peter Beck have 'vandalised the night' with satellite". The New Zealand Herald. Retrieved 2 June 2021.

The underwater environment has many unique challenges, but the ones addressed here, due to their impact on verification, are (1) sustained long range and high bandwidth communications; (2) a complex and generally poor sensing environment (that is, the signal-to-noise ratio (SNR) is low); and (3) the requirement for persistence and endurance in a harsh environment.

Unique challenge (1) means the robot operator cannot communicate well, if at all, with the deployed AUV. The operator cannot easily intervene to assist much when an under-ice AUV experiences an unanticipated or critical event like a loss of navigation capability. In such circumstances, there is a greater onus on the AUV to "figure it out"; thus, on-board autonomy becomes paramount on AUVs. As in spacecraft, the verification of critical systems like autonomous navigation is essential to ensure the system's operation is predictable enough to satisfy operators that may be out of contact for extended times.

Unique challenge (2) is especially relevant to underwater (military) robots that can be in a position to acquire sensor measurements to inform decisions that can have high consequences for errors. For example, an anti-submarine warfare mission may be entirely reliant on determining whether an in-water contact is friendly or not. The example is further compounded by the greater uncertainty (due to the generally poor signal-to-noise ratio) that is part of sensor measurements in underwater environments. If the autonomy includes operator decision aids to recommend a course of action, determining how the verification for this would be performed given the large uncertainties (in terms of both high false positives and high false negatives) is a significant challenge. This challenge reduces to how to verify the accuracy of a sensor measurement from the autonomous fusion of multiple robotic sensors that may have high uncertainty, but yet a decision, or even a suggestion, is a required outcome that entails high consequences for error.

Unique challenge (3) addresses robotic military or scientific measurements with a requirement for persistence and endurance despite the remote and harsh conditions. Residential AUVs and intelligent landers are deployed for days to weeks. Anti-submarine warfare occurs over the course of weeks and sometimes longer. Under-ice missions can go for weeks. AUVs that are hibernating on the seabed to be awakened by a triggering event which transitions them to a higher level of activity to address threats or intruders for maritime archeological sites[10] and protected reserves may need to operate for years without maintenance or updates. For these systems, it is not enough to verify functionality in the short term and to prove that the system has the capacity to function properly after being resident in an environment for a day or two. These systems must show that they will work when they are encrusted with biological organisms and debris, and that they will work after a storm has pushed them around on the seabed. All the environmental persistence assumptions that come into play when evaluating terrestrial and aerial systems are violated—landmarks can change, local populations can change, and the system may not be able to identify how it has been moved.

[10] https://en.unesco.org/underwater-heritage/threats-and-protection

Fault management on these longer term systems[11] is likewise both more essential and more complex. Verifying the efficacy and appropriateness of fault detection, identification, and recovery mechanisms is essential, especially for systems that will lie motionless on the sea floor for extended periods of time. The space community has similarly inaccessible systems and their expertise can be drawn on. However, even within the space community verification of autonomous systems is a time and labor intensive activity requiring significant financial investment with uncertain rewards.

1.3.4 Aerial Impact

Beyond ground, space, and maritime domains, the aerial domain poses many challenges of its own. Commercial passenger travel, small quadcopters flying in formation during air shows, package delivery, fixed-wing agricultural aircraft, and rotary-wing military aircraft all involve different levels of trust and different levels of risk. They each drive different considerations in terms of how much risk is acceptable, what kind of risk is acceptable, and how the system's operation, safety, and security should be verified.

While in the terrestrial robotics world, the fundamental safe response is to slow down and stop, aerial systems must execute more or less complex landing maneuvers to reach a safe state. If they simply turned off their motors like a ground vehicle, they would fall out of the sky and potentially seriously injure someone, something, or themselves. But an aircraft that operates over largely empty fields is subject to different safety analysis and requirements than a commercial jet full of passengers.

The commercial aircraft domain poses unique risks, since it marries a typically non-risky behavior (move from point A to point B) with the risks associated with human operator interaction and complex transitions of responsibility and the risks of carrying human passengers in the context of this more complex mode of travel. Underwater systems need only rise to the surface to provide a safe environment for passengers. Ground vehicles need merely stop away from traffic. But safety considerations for aircraft include not only when to land and where to land but how to land so that the passengers remain safe during the landing operation, even when there is no air traffic control to monitor and direct the situation or when the plane is experiencing a major equipment failure. Pilots are trained to make trade-offs related to how to manage critical situations but assessing responsibility and determining whether a system will make the "right" decision in extremely rare cases that have traditionally been left to human judgment is extraordinarily difficult.

[11] Seto and Svendsen, "Advanced AUV fault management,", in Autonomous Underwater Vehicles: Design and practice, F. Ehlers, ed., pp. 419–445, IET 2020, https://doi.org/10.1049/SBRA525E_ch

1.3.5 Human–Robot Interaction Impact

Regulating times and places where people and autonomous systems interact is hugely complicated and expensive. Factories explicitly eliminate the problem by physically deconflicting people and robots—the robots are placed in areas that humans are not allowed to enter while the robot is operating. Where that kind of physical separation is not possible, robots are usually physically designed to minimize opportunities for problematic interaction. A robot vacuum moves too slowly to injure and consists entirely of rounded edges.

User and operator safety is not just a matter of not being physically harmed by the robot. Certifying that a robot will not hurt a human involves more than just ensuring it won't cut or bruise someone. A robot designed to travel ahead of a convoy on dangerous, poorly maintained roads requires an operator to travel in the convoy and watch its video feed. The combination of the instability of the video feed as the robot bounces around on the road and the bouncing experienced by the operator in the convoy can combine to produce extreme nausea in the operators. In order to ensure a system is safe and effective, the full experience of typical use cases must be explored.

Interface design is critically important, as it can engender or undermine trust by sharing the wrong information, or sharing the right information in the wrong way, or sharing the right information at the wrong time. Risk is increased when the user cannot make the decisions the robot is assuming they're making because the interface cannot provide adequate situational awareness or information. Defining the constraints within which the autonomous system will operate correctly and provide the desired capability is critical. In some cases, like the DART mission's final approach,[12] the interface is less important because once the user hands off control to the robot, the human will not regain control until the robot hands it back (or, in this case, until the robot has destroyed itself).

Beyond co-workers and users or operators, bystanders can also interact with robots—a robot hitchhiking around the world was first supported and sent on its way by helpful bystanders[13] and then destroyed by malicious bystanders.[14]

In the entertainment domain, the competing needs of many stakeholders must be considered. For a system performing on stage,[15] the task itself is poorly specified—how can we define "projecting the right emotion" in a quantifiable way? How do you verify the accuracy and function of a museum tour guide, of an artistic exhibition, or of an actor? Especially in cases where the robot is expected to interact with the public, but also in cases where the robot is effectively on display, the safety

---

[12] https://dart.jhuapl.edu/News-and-Resources/article.php?id=20220920

[13] https://hbr.org/2014/12/what-a-hitchhiking-robot-can-teach-us-about-automated-coworkers

[14] https://www.usatoday.com/story/news/nation-now/2015/08/03/hitchhiking-robot-destroyed-philadelphia-ending-cross-country-trek/31051589/

[15] https://www.youtube.com/watch?v=jAJtfb-OFXI "Robot actress appears alongside human actor in stage play".

considerations are considerable. For a robot actor, not only are we concerned for the safety of the system, we must address the safety of the stage crew, other cast members, and the risk to the audience if the robot should misjudge the edge of the stage. And these risks multiply when robots are effectively used as sideshow entertainment. Even the most inadequate risk assessment should have identified that a robot capable of breaking a child's finger and a child required to follow strict turn-taking rules without reminders or enforcement should require significant checks and balances before being allowed to interact.[16]

The medical domain encompasses both surgical and hospital robots and in-home care robots. For systems used in hospitals, environmental accommodations can ensure safety and support functionality. For in-home systems, few assumptions can be made about the environment or the capacity of the user. These systems require extensive safety and functionality testing, both of the system and of its interface.

In the delivery domain, system risk and functionality assessment must include not only consideration of the delivery process itself, but also consideration of bystander interactions. Within the delivery process, ensuring the ability of the robot to return from its delivery task unscathed, ensuring the cargo arrives intact, and ensuring the correct recipient can access it are critical. But *during* the delivery process, it is essential that the robots obey traffic and pedestrian laws or maneuver safely around drunk patrons,[17] that it act predictably in an extremely uncertain environment so that people in its environment can maneuver safely around it, and that it both be able to protect itself from theft and simultaneously not harm anyone.

In the military domain, it is essential to track both the safety of the operator and the desired function of the system. In some cases, such as demilitarized zones where anyone present is assumed to be a combatant, it is more important to accomplish the task than it is to ensure the safety of a bystander who by definition ought not to be present. For military systems, knowing the boundaries on safe operation and establishing rules of operation that enables users to accomplish the task is more important than minimizing risk, since the people working in these jobs go into them knowing they are risky. The environment in which these systems operate and the potential actions are much less predictable than civil applications. Military systems operate in extremely unstructured environments. These systems require verification of operations, verification of safety, and verification of security.

In the agricultural domain, tolerances are extremely important. Costs like seed and fertilizer and pesticides add up quickly, so systems that can apply them precisely are much more cost effective. Interactions with people are minimized in agricultural settings, since in most cases, the farmer will be operating the vehicle from a safe space and the fields are unlikely to contain people whose safety must be considered.

---

[16] https://www.theguardian.com/sport/2022/jul/24/chess-robot-grabs-and-breaks-finger-of-seven-year-old-opponent-moscow ("Chess robot grabs and breaks finger of seven-year-old opponent", Jon Henley, The Guardian, 24 July, 2022 08:37 am EDT).

[17] https://www.worldsbestbars.com/cocktail-revolution-drinks-and-drones/ ("Cocktail revolution: drones that deliver drinks" World's Best Bars, 25 Sept., 2018).

The safety of this expensive equipment, however, is a concern—hazards like tipping over on steep slopes and faults that could cause damage to the crops or wasted resources must be effectively and explicitly managed.

While the factory domain is similar to the agricultural domain in terms of the tolerances and precision of operation required, the users have significantly more control over the environment and can ensure both that system safety is addressed and that humans and robots do not interact by establishing spatially separate work areas and operational constraints like ensuring the robot is powered down before a human can enter its work area. This reduces the impact of safety constraints on the robot's behavior and work process design while increasing the burden on the users to ensure that external safety requirements are defined and operational rules are adhered to.

There are similarities between the human–robot interaction assumptions behind agricultural and factory robots and the chess playing robot described above. In both cases, the developers constructed a system that relied on physical separation between a robot that had the potential to be dangerous and its human users. But in the case of the chess playing robot, the human was expected to *continually interact* with the robot as part of the task. In the agricultural and factory cases, the humans are expected to *get out of the way* when the robot is operational. In the agricultural and factory cases, the robots must be precise. In the chess playing case, one reason the robot was dangerous was because it was imprecise—a precise robot would be designed to use the minimum force required to pick up a chess piece and would have been incapable of breaking a child's finger even if it had grasped it by mistake. The fundamental flaw in the chess playing robot was that it was designed and tested using factory-appropriate assumptions, when it should have been designed and tested using assumptions suited for a system providing in-home care.

From a verification standpoint, safety (in the form of human interaction and risk reduction) is paramount in the entertainment domain, both functionality (precision) and safety (risk reduction) are essential in the delivery domain (especially the aerial delivery domain), functional requirements are critical and safety requirements depend on the situation in the military domain, and precision operation is essential while safety is either accepted or managed primarily through environmental constraints in the agricultural and factory domains.

1.3.6 Trust Impact

But fundamentally, while it may be most obvious in human-centered tasks, the critical element in all of these domains is trust. Verification enables us to begin to assess whether the system is trustworthy, what domains it should be trusted in, and when it should not be trusted.

Trust evolves naturally over time. As users become familiar with a system and its strengths and limitations, they use it more confidently and they trust it to accomplish jobs for which it has demonstrated competence. This familiarity and trust are developed over time, through extensive observation of system performance under different circumstances. Trust requires understanding of system limitations. The system does not need to be perfect. Rather, the user's understanding of the system's imperfections drives their justified trust when it is attempting a task it is capable of and their justified distrust when it is attempting a task it is incapable of accomplishing well.

Verification provides a shortcut to establishing that kind of trust. Instead of each user needing to develop their own set of observations, the verification process provides a sufficient set of observations from which conclusions about the trustworthiness of the system can be drawn. The system inherits trustworthiness from the verification process, so the users know how and when (and when not) to use the system based on the assurance provided by the evidence gathered during verification.

An effective verification process supports and engenders trust. The process of verification establishes necessary limitations on the context in which the system can be trusted and defines the degree to which the system is trustworthy in specific combinations of circumstances. It describes the parts of the performance space where the system can and should be trusted, the parts of the performance space where the system is fundamentally untrustworthy, and the parts of the performance space where its performance is dependent on external factors like the willingness of the user to trust it.

In the airline industry, lessons learned are tracked from every design flaw and operational incident. They are made available to the community and folded into the design process to ensure they are considered and tested for in future systems. In the space industry, lessons learned from every test, every failure, every piece of knowledge we have been able to discover from prior missions are incorporated into future designs so that they may avoid these problems. Because they take the time and effort to track and address problems over time, because they do comprehensive risk assessment as part of their design and evaluation process, systems in these domains are trusted.

This is fundamentally a book about verification, not trust or risk assessment. It is about how we prove whether a system does what we claim it will, not about the implications of that proof or lack of proof or whether it will continue to do what we claim once it goes into the field for operational use. But verification requires an assessment of what counts as sufficient proof of the desired functionality, and "sufficient" is often determined implicitly or explicitly by the degree to which trust is either necessary or allowable.

Verification is essentially an attempt to establish trustworthiness. To build a baseline for performance expectations so that users will give the autonomous system a chance to gain a user's trust. To give the user enough information so that they understand when the system is trustworthy and when it is not.

The tools we use to achieve these goals are similar across the different domains of autonomous robotics, as are the challenges. But because autonomous systems require fundamentally different levels of trust than non-autonomous software systems and hardware components, new challenges exist and new tools must be developed. The level of trust associated with a wrench or a word processing application is entirely different from the level of trust required to let an autonomous robot loose in the world to deliver a pizza in 30 min or less.

The verification tools and processes captured in this book are a first step towards the creation of that trust.

1.4 Goals

The core problem this book addresses is how to verify autonomous systems. There are three different ways in which we can describe it:

- to be able to mathematically (or otherwise) express the boundary conditions on the predictability of the system's behavior—even when the system's specific actions are mediated by the environment and cannot be predicted in advance
- to be able to verify that a system has implicit understanding (possibly even the ability to generate and understand the implicit goals that must be established and accomplished in order to achieve the larger task)
- to be able to verify, at run time, that the system will not perform actions outside some expectation for safe operation

By merging the perspectives of the mathematicians and formal methods specialists, the testing and evaluation specialists, the behavior design specialists, and the verification agents, we conclude that it is not possible to define a single process for verification of autonomous behavior. The tools available to each practitioner will differ and the assurance that must be developed will change not just as a function of the requirements or the task description, but as a function of the designed and implemented behavior and the environment in which it is expected to operate.

These goals are addressed in the context of assumptions that any processes used to verify and validate autonomous systems cannot take years to perform (they must be relatively quick and flexible), and that while it may be appropriate to require system developers to adhere to design practices that make verification easier (or possible), in practice these kinds of design constraints are impractical. Developers will prefer systems engineering approaches that prioritize effectiveness and performance first, and verification to the extent necessary but no further.

Where some groups may have access to a digital twin and be able to perform many tests in software but only be able to support very limited hardware testing, others may have access to a complex suite of hardware testbeds and limited simulation capacity. Some groups may need to use formal analysis to bound the scenarios that must be tested while others will need formal analysis to verify that specific failure modes are not present.

Because the tools available vary, the processes must also vary. Some people will use simulation to connect the physical tests to the formal results while others will use simulation to define the physical tests that must be run. There is no one way to solve this problem, only a set of tools that can be put together as needed to accomplish the desired goal.

So the purpose of this book is to take a first look at the tools and processes that people are developing to solve this extremely challenging problem of how to verify autonomous systems.

1.5 How to Use This Book

This book connects specific challenges to specific tools and processes that people have developed. The technical challenges are captured in Chap. 2, Verification Challenges. Each technical challenge is linked to the specific contributed chapters that address it in a table at the end of Chap. 2, and each contributed chapter is prefaced by an editorial note that discusses how that chapter connects to each of the relevant challenges.

The book is intended both for researchers and for practitioners, for those who are responsible for verifying autonomous systems, those who are responsible for providing autonomous systems to be verified, and those who are responsible for defining what the autonomous system ought to do. We hope this book will help researchers and developers to better understand our current progress against each of these challenges and to better understand the challenges themselves. For practitioners, we hope that this book will help them better understand which challenges remain to be solved and must be worked around or accommodated in their verification processes, how these challenges affect their ability to verify these systems, and what new tools and processes are being developed that they can use to accomplish their goals. Because the book is structured to connect challenges to the solutions, we anticipate that practitioners will use this book as a reference, identifying the challenge they are currently trying to overcome and finding the chapters that have tools that they will be able to use. In addition to the technical challenges defined in Chap. 2, we provide stakeholder questions and connect them to chapters.

As shown in Fig. 1.1, consumers and end users are primarily interested in whether the system performs the desired task as advertised, to a desired degree of safety and security (*Use*). Consumers often include industrial entities who are concerned with the reliability and certifiability (*Commercialization*) of both the autonomous systems they buy and the autonomous systems they produce and entities that will use the system for the public good and therefore need additional assurance of its ethical, legal, and moral impact (*Public Interest*).

Regulators and certification agents must work in a context that is dominated by the need to both satisfy the agencies that oversee certification and define regulatory frameworks (including customers in the *Commercial* and *Public Interest* spaces) and to *Understand* the technology sufficiently well to be able to regulate and certify it effectively.

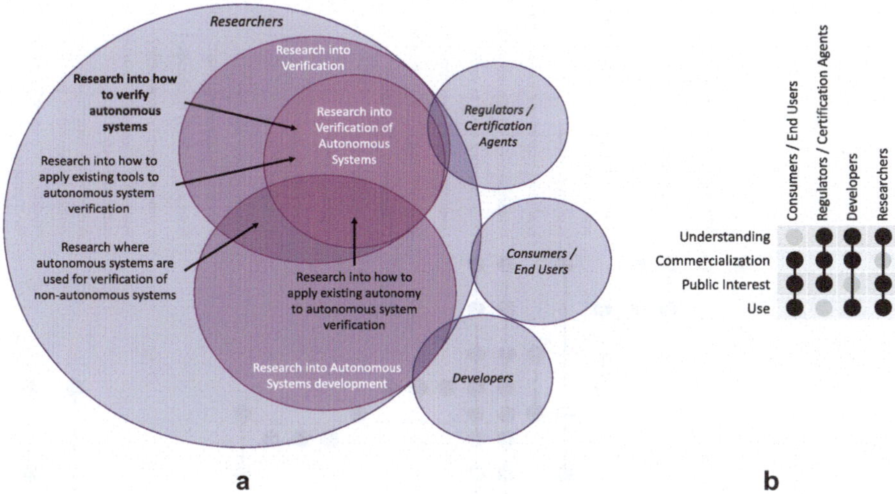

Fig. 1.1 Relationship between (**a**) Stakeholders and the work collected in this book and (**b**) Stakeholders and stakeholder goals. (UpSet diagram derived from Lex, A. et al., "UpSet: Visualization of Intersecting Sets" IEEE Trans. on Vis. and CG (InfoVis), 20(12): 1983–1992, https://doi.org/10.1109/TVCG.2014.2346248, 2014)

Developers are primarily interested in applying the technology to achieve specific goals. They must have *Understanding* of the technology, the *Use* case, and the *Commercial* production and certification implications of their design choices.

Researchers are concerned not only with *Understanding* the technology under test, but also with understanding its *Use* and its impact on the *Public Interest*. In addition to understanding autonomous systems themselves, some researchers are also focused on developing improved *Understanding* of how to verify existing systems, how to apply existing verification tools to the verification of autonomous systems, and how to apply autonomy to improve verification processes.

Each of these four goals, *Understanding*, *Use*, *Commercialization*, and *Public Interest*, reflects different questions that need to be answered. These questions overlap as shown in Fig. 1.2, so for example, knowing whether or not the product is useful is important to both the *Commercialization* and *Use* spheres.

The two fundamental questions at the core of these goals are (1) understanding why people verify systems and (2) understanding what needs to be verified.

People verify systems because their customers want assurance that they are getting what they pay for, because they need to satisfy legal regulations and policy guidance, because they want to use their systems to accomplish a task. Each of these reasons requires understanding of the system being verified, and each drives a different assessment of what needs to be verified. For some systems, verifying that they are incapable of causing a particular type of damage is sufficient. For others, verifying their invulnerability to security threats is essential. For autonomous

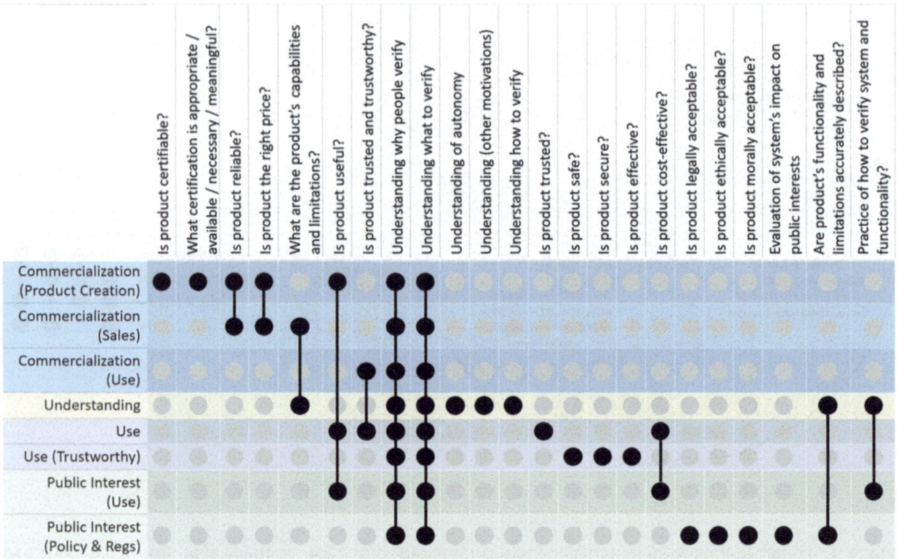

Fig. 1.2 Relationships between each interest area and the questions arising from them

systems, security is often important, but safety and functionality are critical. Saleable systems must be safe and reliable, and they must be able to be designed, developed, produced, and certified at a reasonable price point. Their capabilities must be known—they must be able accomplish the task effectively. Their limitations must be known so that the legal liability of the company that sells them is understood and accounted for. Each of the chapters in this book addresses at least one of these questions, as shown in Fig. 1.3.

Chapter 2 identifies the core underlying technical challenges associated with verifying autonomous systems from a research perspective—the fundamental conceptual problems that must be addressed in order to answer these questions. Each of these conceptual problems is then mapped to specific chapters. If, as a reader, you have a particular interest in a specific technical challenge, Chap. 2 will help you identify the chapters most relevant to your concern. If, however, you are motivated by the questions raised here, Fig. 1.3 illustrates the connection between these questions and the chapters in this book.

The book itself is organized into three parts.

Part I, Overview, contains this Introduction, which focuses on the different ways in which the material in this book is relevant to different readers, and the chapter on Verification Challenges.

Part II, Theoretical and Formal Approaches, contains chapters that focus on defining the theoretical underpinnings behind solutions to the fundamental challenges identified here and in Chap. 2, and case studies for theoretical approaches to verification.

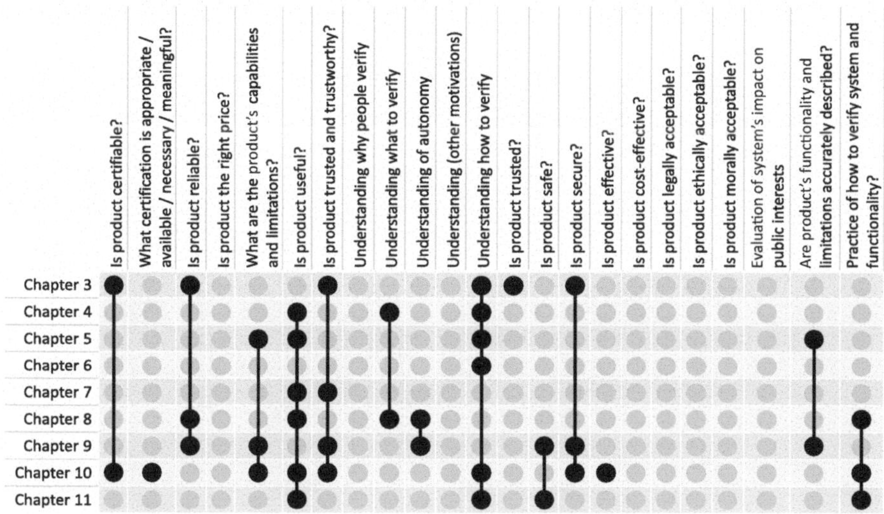

Fig. 1.3 Illustration of how the chapters in this book address the different questions posed by the various interested parties

Part III, Practical Considerations, contains chapters that focus on case studies and practical approaches to verification in the context of specific applications or technological contexts.

Each technical chapter is preceded by a short editorial note describing exactly how that specific chapter addresses both the questions identified here in Fig. 1.3 and the fundamental challenges identified in Chap. 2, to help the reader quickly determine whether a given chapter addresses their specific need.

Chapter 2
Verification Challenges

Mae Seto, John Sustersic, Signe Redfield, and Donald Sofge

Abstract The double-V diagram for testers provides a starting point for discussion of the technical challenges associated with verification when autonomy is introduced into the system. This diagram illustrates many different ways in which autonomy violates the underlying assumptions behind the diagram and existing tools. These assumption violations are aggregated into a list of underlying conceptual technical challenges. Each chapter in this book is mapped to the conceptual technical challenges it addresses.

2.1 Introduction

The core technical challenges associated with verification of autonomous systems are generally not materially different from the challenges associated with verification of any other type of system. Obtaining adequate, simple, testable requirements from users is challenging for any application. Meaningfully connecting them to form a structure where all the pieces work together is difficult. Defining tests and procedures that will gather the evidence needed to evaluate whether the actual system meets its requirements is expensive and time-consuming. And building a

M. Seto
Mechanical Engineering Department, Dalhousie University, Halifax, NS, Canada
e-mail: mae.seto@dal.ca

J. Sustersic
RTX BBN Technologies, Portland, ME, USA
e-mail: john.sustersic@rtx.com

S. Redfield (✉)
U.S. Naval Research Laboratory, Washington, DC, USA
e-mail: signe@ieee.org

D. Sofge
Navy Center for Applied Research in Artificial Intelligence, U.S. Naval Research Laboratory, Washington, DC, USA

S. Redfield et al. (eds.), *Verification of Autonomous Systems*,
https://doi.org/10.1007/978-3-031-88546-4_2

coherent design from those specifications and maintaining the specifications and the design in tandem as constraints change is always a problem.

So what makes autonomous systems special? What are the technical verification challenges associated with autonomy that are so much more serious and difficult than they are for other systems that we need to completely rethink how we approach the problem? What is it about autonomy that causes our existing tools to be inadequate and that drives the need for this book?

2.2 Why Is This So Hard?

The double-V diagram for testers from the CMU Systems Engineering Institute[1] shown in Fig. 2.1 captures the relationship between the models of the system under test used in the design phase (the downward stroke of the V), the system itself in the integration phase (the upward stroke of the V), and the various tests required at each stage.

This diagram is predicated on a set of assumptions about how the tests relate to the models and the system elements, how the models are created and how the system elements relate to each other, and how the tests relate to the evidence and assurance they provide in terms of reducing the risk of using the system in an operational context.

Verification of autonomous systems is challenging because the construction and action of autonomous behaviors break many of these assumptions.

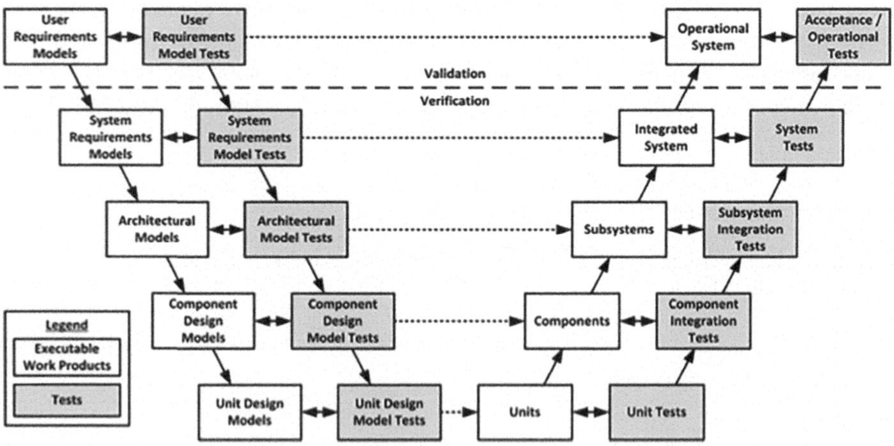

Fig. 2.1 CMU double-V diagram for testers

[1] Link to source: https://insights.sei.cmu.edu/blog/using-v-models-for-testing/

2.2.1 Specification Assumptions

Autonomous behavior is extremely difficult to specify. Physical objects are specified in terms of their properties such as dimensions, weight, color, hardness, strength, smoothness, electrical resistance, and heat tolerance. These are all things that are both quantifiable and measurable. A numeric threshold can be set and it is possible to measure the actual value and check whether the object is above or below the threshold. We have tools and methods to measure certain aspects of predictable processes. We can define the frequency at which an event occurs and its statistical properties. But we lack quantifiable descriptors of desired *behavior*. The point of an autonomous system is that it selects the correct action for the current circumstances when we do not have a priori knowledge of what the current circumstances might be. We are literally using this control methodology *because* we do not have enough information to write a clear and well-defined specification.

In Fig. 2.2, the filled x marks illustrate how our difficulty in defining adequate specifications for autonomous behavior affects the development of system requirements models and tests. Without a clear specification, we cannot define the goals and metrics against which our system will be evaluated. We cannot define rigorous and specific tests at the system requirements level that will adequately capture the desired behavior and prove that the system will do what we expect because we cannot define "what is expected."

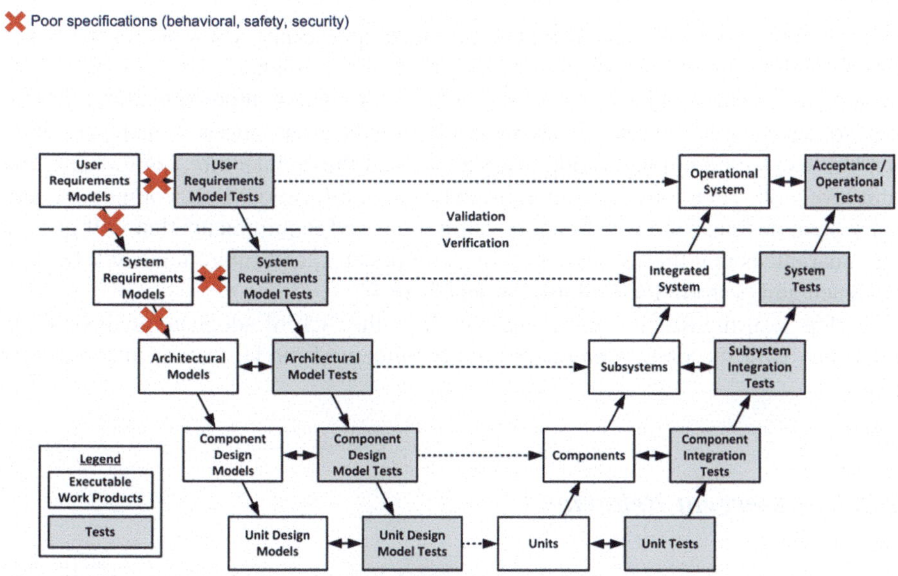

Fig. 2.2 Impact of poor specification on double-V diagram

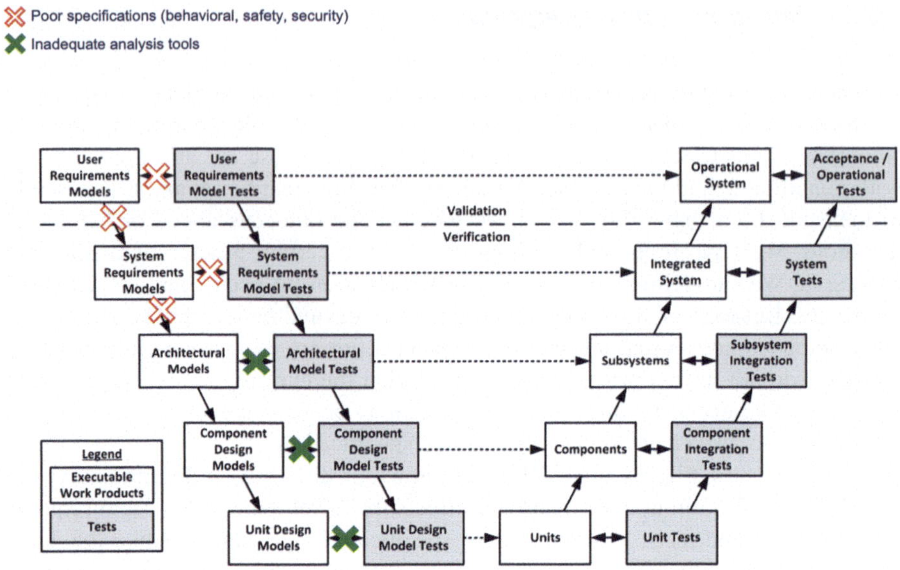

Fig. 2.3 Impact of inadequate analysis tools on double-V diagram

2.2.2 *Analysis Tools*

Autonomous system models also generally lack appropriate test tools for the lower levels of the system. As we dive deeper into the components of the system, we do not have a clear understanding of how the individual components affect system performance at higher levels. Each piece relies on the assumptions at the level above, but because we do not have solid analysis tools at the architecture, component, and unit levels, we cannot use testing at these levels to inform our unit, component, and subsystem designs, and we cannot develop improved requirements that will give us any guarantees that the subsequent units, components, and subsystems will be well designed and perform their allotted tasks correctly.

In Fig. 2.3, the filled x marks indicate how the lack of adequate analysis tools interferes with the unit, component, and architectural model testing in the design phase.

2.2.3 *Emergent Behavior*

Autonomous systems are prone to emergent behavior, where some unexpected interaction between the platform, the environment, and the designed behavior causes some change in the observed behavior that was not planned by the behavior designer.

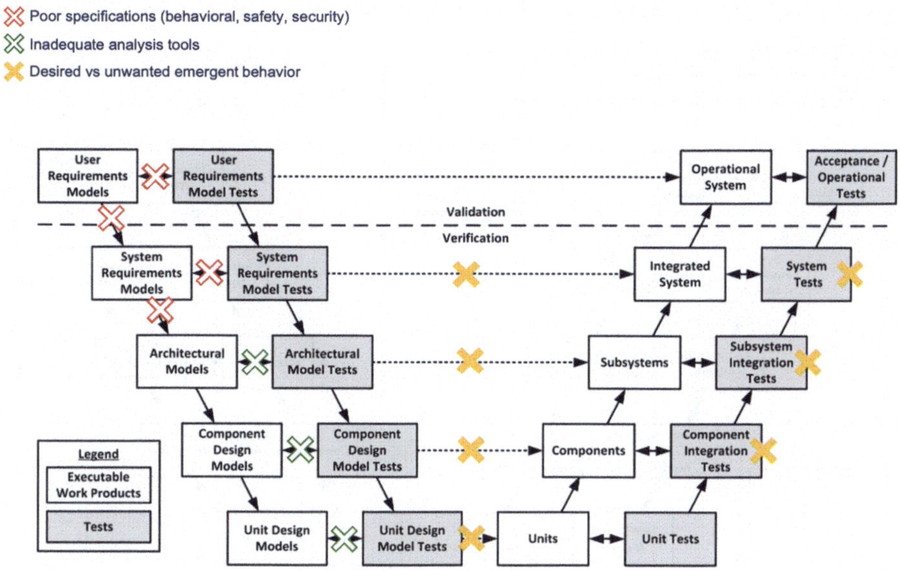

Fig. 2.4 Impact of emergent behavior on double-V diagram

In Fig. 2.4, the filled x marks show how our lack of understanding of the causes of emergent behavior affects the verification process. Because we cannot predict when the system will engage in unpredictable behavior, we cannot make the leap from the results of the design phase testing to the adequacy of the implemented elements, and we cannot determine how and when the tests on the implemented elements will prove to be invalid or inaccurate.

2.2.4 Unknown Unknowns

But emergent behavior is just one of the many aspects of the system that are difficult to identify. Autonomous systems are extremely complex, because what they do is fundamentally determined by the properties of the environment in which they operate. A system that performs perfectly in an environment with a clean concrete floor may fail catastrophically in a gravel parking lot. In order to specify the desired behavior, we must specify the driving factors in the environment. But we don't have tools or methods or rules of thumb to determine which factors in the environment are the driving factors, so every simulation and every test comes down to the "unknown unknowns"— the factors in the environment that we have assumed are not driving factors and have therefore abstracted from explicit consideration in our models, our designs, our implementations, our simulations, and our tests.

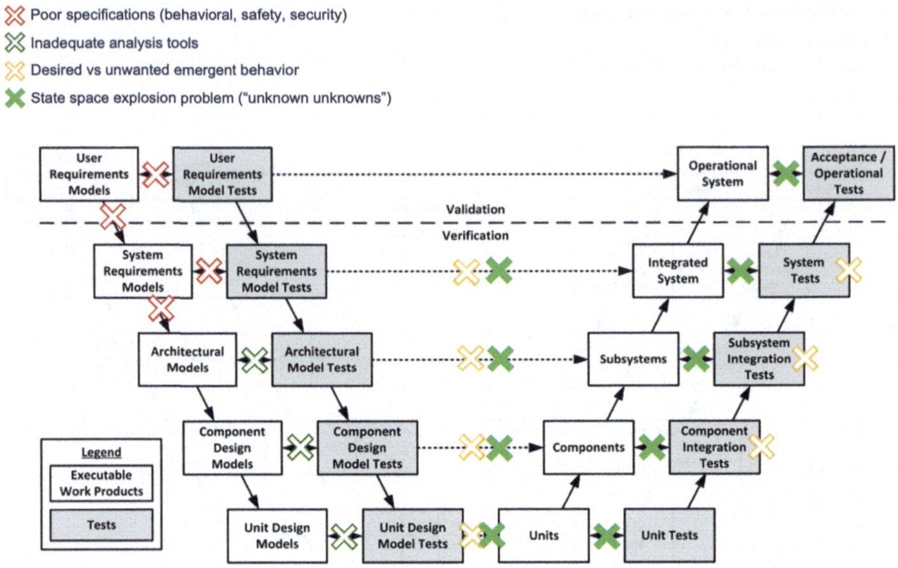

Fig. 2.5 Impact of state space explosion problem on double-V diagram

We can't include all of them. Even only addressing the things we know to be driving factors, we run into the "state space explosion problem," where the number of elements of the system and its environment that must be considered grow so large, and the different combinations of possible situations become so complex, that it's simply not tractable to test or evaluate an adequate fraction of them.

In Fig. 2.5, the filled x marks illustrate how not being able to define the driving factors and having to simulate or emulate too many scenarios affects the verification process. Again, design phase testing does not allow us to extrapolate that certain behavioral properties of the system will hold outside of those tests. But now we also lack tools that will enable us to effectively test and analyze the behavior of the individual components. We can't define adequate unit tests that will tell us useful information about whether the unit actually supports the components' needs, just as we can't define adequate acceptance and operational tests that will be sufficient to predict what the operational system will actually do in the field.

2.2.5 *Reproducibility*

Reproducing an experiment designed to assess autonomous behavior is extremely challenging. Every minor change in the environment or the system may or may not impact system performance. It is sometimes possible to control the environment and the system to the point where it is possible to reproduce a given experiment, and

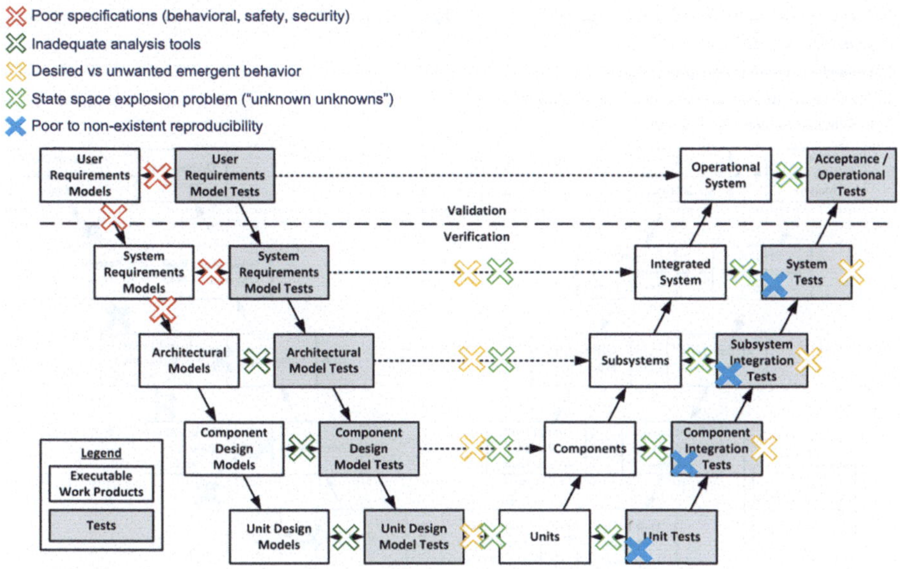

Fig. 2.6 Impact of reproducibility challenges on double-V diagram

sometimes a behavior is sufficiently robust that it will perform well under a range of environmental and system conditions. But in practice, demonstrating via test that a bug in the behavior has been fixed is extremely challenging. In many cases, the bug will not manifest again even when the environment and the system conditions have been replicated as accurately as possible, since its manifestation may depend on internal race conditions, on differences in timing of events, or on differences in how the environment or the platform responded to actions selected by the system.

In Fig. 2.6, the filled x marks indicate that reproducibility affects our ability to test the instantiated components at all. Without reproducibility, we cannot guarantee that what we see in a given test is actually true, or whether it is even predictive of future identical tests, much less whether it is predictive of performance in the field.

2.2.6 *Ground Truth*

Ground truth, or the measurement of what actually happened as opposed to what the system reports happened, is essential for verification tests to be meaningful. But in many autonomous systems, obtaining data for the actual actions of the system in its environment can be difficult to impossible to obtain. For ground vehicles, we have GPS and accurate localization systems, and if all else fails, we can pull out a tape measure or instrument the environment and know precisely where the vehicle actually was instead of having to rely on where it said it was. For maritime,

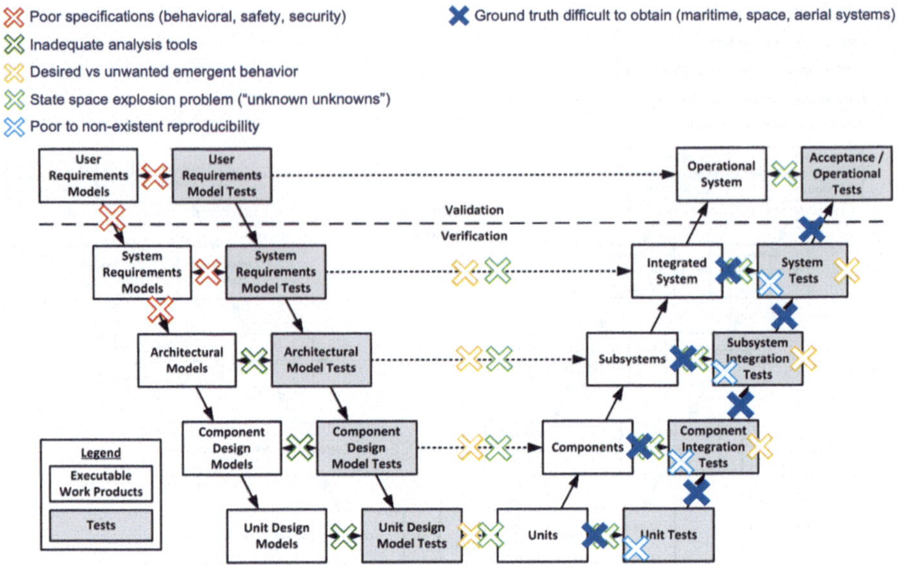

Fig. 2.7 Impact of lack of ground truth on double-V diagram

space, and aerial systems, however, we cannot just pause them indefinitely and have them passively hold position while the measuring process occurs. We must have either a heavily instrumented environment or we must trust the system's report of its pose and state. Systems exist for gathering data about aerial system positioning, but even the most sophisticated system has blind spots, regions where the calibration or physical configuration of the localization system is inadequate to capture detailed, accurate information, or limitations on the types or configurations of agents that can be detected and placed. In environments without those systems, like maritime and space environments, our best guesses involve limited external sensing and the reports that the vehicle itself provides about where it is.

In Fig. 2.7, the filled x marks show how the violation of the assumption that ground truth for the desired behavior is available affects the verification process. If we have no ground truth and if we cannot determine the difference between what the system is actually doing and what it says it's doing, we cannot assume that the test results reflect desired performance, and we have no way to move from one level of abstraction to the next.

2.2.7 Communications Latency

Variability in the delay associated with communications also affects the assumptions that underlie this process. In maritime, space, and urban systems in particular, there can be significant delays in communication both within teams of agents and between the autonomous system and its operator(s). These delays can be predictable or unpredictable, and they can range from a second or less to (in the case of distant space systems) indefinitely long. In the underwater environment, communication is complicated by the distance and unpredictable path of acoustic transmission, which causes correspondingly unpredictable latency and dropouts in addition to the complexity of ensuring that communications do not overlap in time. In space, while distance-based delays and dropouts due to occlusion of physical bodies can be predicted, the processing of the information as it travels from the receiving station to the operator can introduce unpredictable latency. Incorporating these communication challenges significantly increases the complexity of the tests that must be performed in order to determine whether the system is robust to these problems.

The filled x marks in Fig. 2.8 show how latency in communications affects our ability to test instantiated systems. Race conditions and decisions based on information arriving with different latencies affect unit tests, component integration tests, subsystem integration tests, and system tests.

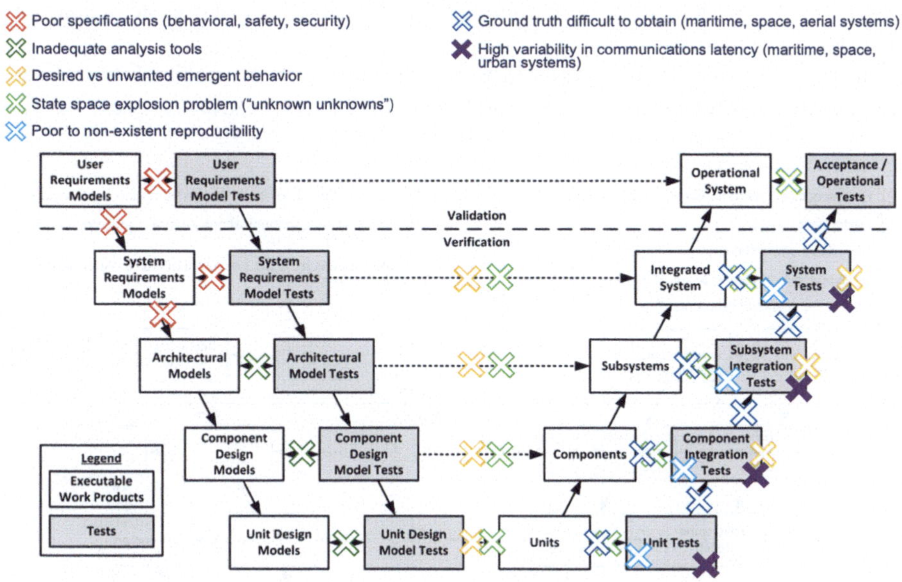

Fig. 2.8 Impact of communications latency and unpredictability on double-V diagram

2.2.8 Learning Systems

Beyond these challenges, we have one final problem. Some autonomous systems incorporate learning into their decision-making process. They may learn perceptual mappings from incoming sensor data to models of the world, or they may learn that certain actions are better in some situations than others. This goes beyond the normal responsiveness to the environment that all autonomous systems exhibit. In these systems, the behavior itself changes over time as the system interacts with its environment. The process of moving from the integrated system to the operational system, the process of moving from system tests to acceptance tests, and the ability to define acceptance and operational tests are all predicated on three assumptions. First, this process assumes that the thing that completes the verification process is the same thing that is being accepted for operational use. Second, this process assumes that the results of the system tests can be used as evidence to support the selection and definition of the acceptance/operational tests. And third, this process assumes that the results of the acceptance and operational tests are predictive of future performance. Because these systems learn after they are deployed into their operational environment, these assumptions aren't valid—the system that is operating in the field may be fundamentally different from the system that was originally accepted.

In Fig. 2.9, the filled x marks illustrate that this lack of predictability affects the transition from verification to validation. In the verification process, we can

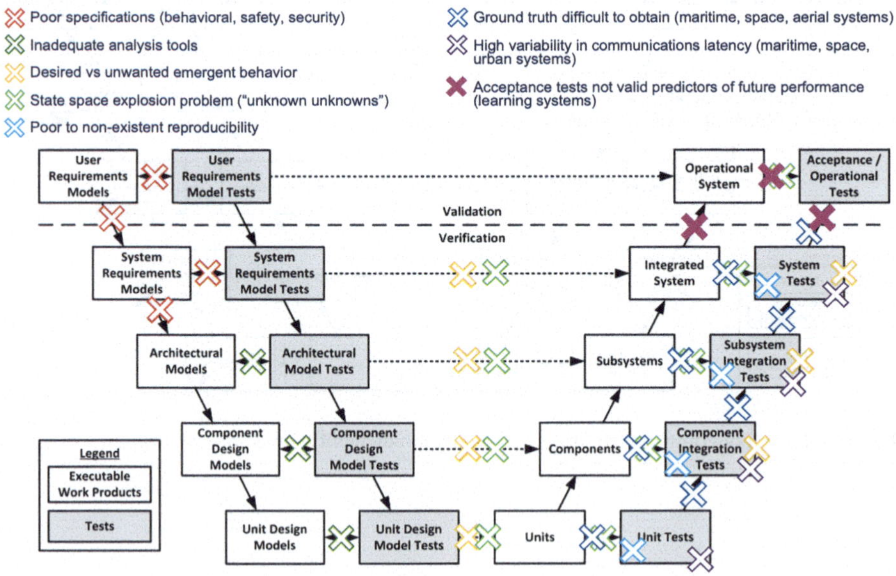

Fig. 2.9 Impact of learning on double-V diagram

theoretically prove that the system meets its requirements. We cannot say anything about whether it will continue to meet those requirements moving forwards or not.

Almost every aspect of the verification process must be assessed for the implications of these assumption violations and modified to support verification of autonomy.

2.3 Research Challenge Organization

Because these assumptions are violated, we cannot blindly follow a process like this and group the work in this book around a traditional verification structure. Instead, we identify commonalities across the different problems and define our challenges based on four main research areas. Because there are overlaps between these areas, some specific research challenges simultaneously fall into more than one of these research areas.

2.3.1 Challenge Categories

The four main research areas that we group our fundamental research challenges into are *Models*, containing challenges associated with development and type of models, *Abstraction*, containing challenges related to the level of fidelity with which a given system under test and its objectives are captured, *Testing*, containing challenges related to how to test and what information can be gained from testing, and *Tools*, containing challenges related to the identification and development of new tools to support verification of autonomous systems.

2.3.1.1 Models

The challenges in this category are concerned with the development of models representing the system, including both the development of models of the platform, the environment and any relevant objects within it, and the development of models of the behavior. The models may be specification-related models or models of the designed behavior or system. They include physical, virtual, and logical or event-based models.

The challenges address how the models are created, what kinds of models can or should be created, properties the models should have, and what information the models should contain.

Challenge 1: How is an adequate model of the system created?

This challenge requires that we understand the context in which the verification will be performed sufficiently well to define how "adequate" should be interpreted. The goal of this challenge is to identify what factors determine whether a model is adequate or not and to develop tools, processes, metric thresholds, and rules of thumb that will reliably produce an "adequate" model of the system for the purposes of the verification agent.

This challenge identifies a common problem underlying three of the previously defined broken assumptions:

- Inadequate analysis tools (in the sense of an adequate model forming one of the necessary analysis tools)
- Emergent behavior (in the sense that an adequate model should exhibit the same emergent behavior as the system, since detection of emergent behavior is a key functionality of a useful autonomous behavior model)
- The state space explosion problem (since by definition the purpose of a model is to limit the complexity of the state space required to represent the system being modeled)

While research resulting in the construction of models for these purposes is often undertaken, assessment of what constitutes an adequate model is still an active challenge. Chapter 10 (Feather, et al.) illustrates the degree to which synthetic data modeling the sensory environment of a system must be assured to support verification of a specific system.

Challenge 2: Common models and frameworks are needed that describe autonomous systems broadly enough so they can be used to standardize evaluation efforts and interfaces to the system.

This challenge identifies one particular analysis tool that is necessary to make appropriate connections between the specifications and the tests used to evaluate them. Chapter 5 (Tatum, et al.) introduces one approach to addressing this challenge.

Challenge 3: How should debug models of black box autonomous systems be developed? How is a mathematical and/or logical model suitable for formal analysis produced from empirical observations?

The core of this challenge is the connection between *observed* behavior of the system (since the design details of the system under test are not available) and model development. The double-V model assumes that the design process is under the same auspices as the testing—that the design and the testing are happening in tandem throughout the design and development lifecycle. This challenge addresses the case where the design occurs independently from the verification activities and where the design is hidden from the verification team. Chapter 6 (Ramamoorthy, et al.) develops an approach to producing a formal model from empirical observations.

Challenge 4: How should one identify and model components that are not captured yet (and what are their properties)?

This challenge encompasses both the components of the autonomous system that were originally considered to be outside the scope of the necessary modeling process or that were developed after the original modeling process was complete, and components of the environmental and task-related context that are not expected to affect the system's performance but, upon later reflection, are determined to have a significant impact.

This challenge identifies the research challenge at the core of the unknown unknowns problem and the emergent behavior problem and also feeds into our lack of adequate analysis tools. Since any given model is by definition an abstraction of the actual thing, something must be left out. Because of the limitations on our current theoretical understanding of which aspects of the system and the environment drive task performance, it is extremely difficult to ensure that the relevant components and properties are captured in our models. The core of this challenge is that we cannot predict whether the elements we failed to include are actually important, and therefore, we must have some mechanism or process to add those elements to the model. This is especially important where those elements are poorly understood, like how the color of ambient light changes as a function of time of day and time of year, or how contact dynamics in spacecraft interactions changes as a function of the materials involved. Identifying and modeling those components are essential if we are able to adequately model emergent behavior and the relevant portions of the system's state space, but this is an incredibly difficult problem, without an existing solution.

2.3.1.2 Abstraction

This topic area is concerned with defining the level of abstraction at which the autonomous system's behavior should be specified, at which its design should be tested and at which the assurance case should be built. The challenges in this topic area encompass the abstraction required or desired for models and simulations (the *Fidelity* category) and the level of abstraction defined in specifications (the *Requirements* category).

Fidelity

The Fidelity category contains challenges concerned with determining what level of abstraction is allowed in order for the assurance case to be valid.

Challenge 5: How do we define the appropriate level of fidelity required to extract the information we are interested in?

This challenge reflects challenge 4, but explicitly focuses not on what components are represented but on how accurately they are represented. Again, this challenge is one of several addressing the problem of inadequate analysis tools and emergent behavior—if the analysis or model is too abstract, it will be unable to provide the necessary information, but if it is too detailed, it will become unusably complex. Chapter 10 (Feather, et al.) discusses various scenarios in which the level of abstraction is shown to be adequate for the necessary verification.

Challenge 6: How is the adequate level of abstraction determined for the model of a robot, its behaviors, the simulation that tests the model, or for any given tool or system? How many environmental characteristics need to be specified? What are the aspects of the environment, the robot, and the autonomy algorithms that cannot be abstracted away without undermining the verification?

This challenge is the opposite problem from challenge 4, but addresses the same broken assumptions in the double-V model—instead of focusing on how to add components to an existing simulation or analysis, this challenge focuses on pruning the complexity until only the necessary elements exist. Chapter 10 (Feather, et al.) describes an approach to complexity pruning in the context of space systems.

Requirements

The Requirements category is focused on the relationship between the requirements development process and the design process, and on how implicit and explicit goals are identified and captured.

Challenge 7: How do we determine at what level we are transitioning from specifying requirements to designing the system and develop principled requirements that do not devolve into designing the solution?

This challenge does not explicitly appear in the double-V model because the model itself assumes that the separation between requirements and design is already established. In general, the interactions between the units and the components must be established in order to settle on a functional higher-level design. Instead of following a clearly linear path from most abstract to most detailed, autonomous system behavior specification typically follows a winding path with a lower-level unit specification relying on higher-level design choices while at the same time the higher-level architectural choices and system requirements are based on the selection of specific lower-level units in a component design. Chapter 4 (Redfield and Lennon) explores the tension between the defining the requirements and designing the behavior and proposes a potential solution.

Challenge 8: How does one ensure both the implicit and the explicit goals of the system are captured? How is a model of the system's goals from a human understanding of the task goals, the system, and the environment created?

This challenge addresses the specification assumption, since the implicit and explicit goals of the system are typically the most challenging elements to specify. Chapter 3 (Heitmeyer and Leonard) presents a method for developing a set of autonomous system behavior requirements using formal representations.

Challenge 9: How are performance, safety, and security considerations integrated?

This challenge also cuts across all the aspects of the double-V approach. At every stage, all three of these considerations must be included and addressed. But autonomous behaviors are designed to integrate these elements rather than keeping them separate. The performance of the avoidance algorithm, when isolated from its context, is defined in terms of system safety. The performance of the avoidance algorithm in the context in which it is used is evaluated in terms of its impact on the larger system's goals. Separating these elements into distinct requirements and identifying how they will be evaluated are challenging verification problems that are distinct from the double-V verification process assumption problem. In Chap. 3 (Heitmeyer and Leonard), tools to support assessment of performance, safety, and security properties are presented. Chapter 11 (Scheidt, et al.) discusses safety in the context of assuring and assessing the safety of autonomous systems through hardware testing and evaluation.

2.3.1.3 Testing

In this topic area, we capture the development of test scenarios, metrics, and performance evaluation mechanisms as well as the extension of simulations, test environments, and tools to enable generalization from tests to conclusions about system performance. This topic area also includes challenges related to replication of results and our ability to establish the predictive properties of a given test. It consists of four categories of challenges:

- *Test* challenges, which focus on evidentiary limitations and test activities
- *Scenarios* challenges, which focus on predictive test environments
- *Metrics* challenges, which focus on how tests are evaluated
- *Assurance* challenges, which focus on generalizability of test results and assurance case construction

Test

Challenge 10: At what point is there enough evidence to determine that a given autonomous system or behavior has been verified?

This is a common challenge across many types of systems and is not limited to autonomy, but it is particularly challenging in the case of autonomous systems due to the state space explosion problem. Chapter 10 (Feather, et al.) illustrates case studies where different methods for estimating the level of evidence required are used.

Challenge 11: How does one ensure it is possible, in the physical world, to test simulated situations that result in boundary cases?

One of the advantages of simulation is that it enables us to test scenarios that are impractical or impossible to test with physical systems. This is particularly problematic for autonomous systems. One purpose of verifying autonomous systems is to ensure that they select correct actions as they operate. These actions are typically triggered by specific combinations of environmental properties, system state, and task goals. Often, the same environmental circumstances should cause different actions to be selected within the confines of a single attempt to perform a single task. The selected action may depend on factors like the internal optimization metric or the battery level of the vehicle, or it may depend on probabilistic factors like the current estimated chance that something in the environment will move in a particular way. In some cases, the selected action will depend on the outcome of a race condition—if one piece of information arrives first, the vehicle will choose one action, while if another piece of information arrives first, the vehicle will choose a different action. If we run across circumstances where this causes a poor outcome in simulation, we may not believe that the simulation is an accurate reflection of performance in the physical world. In these cases, replicating those specific simulated circumstances in the physical world may be impossible. Chapter 11 (Scheidt, et al.) addresses best practices and pitfalls associated with this transition from simulation to reality.

The core of this challenge is determining which simulated scenarios can be replicated in the physical world. In order to address it, we need a fundamental understanding of what aspects of the simulation can be accurately replicated and what aspects cannot, and of what elements of the boundary cases can be driven to occur by specific environmental or system configurations, what elements cannot be driven and must occur spontaneously, and what elements cannot be replicated.

This is one particularly critical aspect of the ground truth assumption problem—when a simulated situation arises, how do you determine whether the simulated results are valid or not? In this case, the complexity of the interweaving of actions and environmental results of actions that defines autonomous behavior can make it impossible to definitively state that a given scenario identified in simulation has actually played out in the physical world. The high variability in communications latency in the physical world also requires that we address this challenge, but this challenge is still very much an unsolved problem.

Scenarios

The Scenarios category focuses on the development of specific predictive test environments or principles.

Challenge 12: How to design tests so that passing them indicates a more general capability?

This gets to the heart of the problem that our acceptance tests are not valid predictors of future performance. Even in non-learning systems, if we cannot generalize from our acceptance tests to different operating conditions, we cannot conclude that we have adequately assessed the system's operational performance. Chapter 9 (Mullins) proposes a tool that identifies the boundary regions of the state space, indicating under what conditions the system's performance is likely to change and under what conditions it is likely to stay the same.

Challenge 13: How are challenging design reference missions selected so that performing well against them indicates a more general capability for the system rather than for specific system components?

This challenge focuses on a slightly different aspect of the acceptance testing problem. Not just how to design tests that predict operational performance, but what kinds of missions predict performance at the system level rather than at the component level. Chapter 9 (Mullins) focuses on directed testing at the system level to identify where generalization is and is not likely.

Challenge 14: How can test sets be produced to yield the data required to create mathematical/logical models or find the boundary locations and fault locations in the larger state space?

This challenge addresses testing to develop models and analyses rather than testing for assurance. It is core to our inadequate analysis tools and our state space explosion problems—in order to develop adequate models and find boundary locations in the state space, we must be able to identify the test sets that will give us the most salient information. The tool presented in Chap. 9 (Mullins) identifies boundary locations and fault locations in the larger state space.

Metrics

The Metrics and Evaluation category focuses on how we measure the outcomes of the tests in the context of the desired system behavior.

Challenge 15: Once a reasonable model is created how would one determine whether all resulting emergent behaviors were captured and what are appropriate performance measurement tools?

This challenge is focused on the problem of identifying system–environment interactions that were not predicted during the design stage. In order to make progress on this, we need to understand how to describe system–environmental interactions in general, and how those interactions trigger specific behaviors. In the double-V model, this captures one of the challenges associated with the emergent behavior problem. Predicting emergent behavior remains one of the most challenging problems in autonomous systems.

Challenge 16: Measurement and evaluation are generally poorly understood—operators can describe the tasks for the robot but lack tools to quantitatively evaluate them. How should autonomous behaviors be measured in a way that consistently and accurately describes the capability embodied by a robot?

Defining how to measure the capability a particular autonomous behavior provides when it is instantiated in a particular system, in the context of a particular task and a particular environment in which that task must be accomplished can be done if the specifications for that particular task are extremely well-defined. In practice, however, we do not understand how to define performance such that all the relevant aspects of the behavior are captured. This impacts both our specification development and the tools used to evaluate the system, since the measurement method will determine how the tests are structured and whether the system is competent or not. Chapter 7 (Marvel, et al.) focuses on how to map the measurement to the embodied capability, and Chap. 11 (Scheidt, et al.) addresses capability representation and metric development. Chapter 5 (Bouchard, et al.) establishes a performance-based method for comparing measurement and enabling evaluation across capabilities, and Chapter 8 (Seto and Bassam) reviews verification of functionality across abstraction levels in the marine autonomous systems context.

Challenge 17: How is a metric defined for comparing solutions?

When different approaches are used to accomplish the same task, they will generally use different internal optimization functions to drive the system's behavior. This challenge is concerned with how higher-level metrics can be developed that apply independent of the system or approach used to accomplish it. Furthermore, in many cases, the evaluation metric itself is binary, simply capturing success or failure. In these cases, how do we define a comparative metric that encapsulates when one solution is better at the task than another? The *tatum* introduced in Chap. 5 (Bouchard, et al.) provides a way of representing the performance information such that it can be compared across implementations.

Challenge 18: How is the "optimal" defined against which the verification is performed? How is it shown that the solution is in fact, optimal? How is the performance of the system measured?

When we define the required performance of an autonomous system, in some cases, we cannot define in advance what a specific solution will look like. Instead, we may attempt to constrain the system behavior to optimize its actions in cases whose precise solution will vary depending on the environmental circumstances. In cases where no mathematical "optimal" solution can be defined, we are side-stepping the question of what constitutes "optimal" actions and placing an impossible burden on the shoulders of those assigned to verify the system's behavior. If our goal is to show that a given obstacle avoidance/path planning algorithm plans its path to "optimally" avoid objects, how do we measure that? How do we test whether it satisfies that requirement or not? These questions, while important, do not yet have answers.

Assurance

The Assurance and Generalization sub-category focuses on the predictiveness and generalizability of a given test and on mechanisms for building assurance cases for autonomy frameworks.

Challenge 19: How is the performance from a finite set of samples of the performance state space generalized across several variables and parameters?

In this challenge, we are concerned with being able to predict performance given a limited sample set. System performance is inherently complex, driven by the complexity of the possible combinations of environmental status and system state. Because so many factors play into the performance of the system, it is essential to determine which factors are having an outsize impact on performance in any given situation and which ones are not substantially affecting performance. The core of this challenge is how to simplify the set of factors driving the performance space function shape at any given time, so that predictions of performance can be made from the smallest possible set of samples. The challenge addresses the core of the state space explosion problem, and while a general approach has not yet been developed, Chap. 5 (Bouchard, et al.) provides a mechanism to compare systems and system components that enables assessment of performance generalization across specific system parameters.

Challenge 20: Autonomy frameworks are unable to determine whether all the resulting emergent behaviors have been captured or to supply performance measurement tools.

Existing autonomy frameworks focus primarily on the structure of the design of the system. We lack frameworks that capture not only the design components but also the results of the interactions of those design components. This challenge focuses on how we could go about predicting those results, both at the level of design inconsistencies and assumption violations, and at the level of interactions between the design and the environment. Even when we have the ability to generate these predictions, we will still struggle to compare solutions because the kinds of performance measurement needed will have to capture consistent performance metrics that meaningfully express relative performance by different systems.

This challenge addresses another way in which our analysis tools are inadequate. Existing frameworks do not yet address this problem.

2.3.1.4 Tools

New tools or techniques that must be developed in order to support autonomous system verification are identified in these challenges. These also address evaluation of a given system's fitness to accomplish a given task.

Challenge 21: What new tools or techniques need to be developed?

Different verification processes require different tools. A verification process that relies on the establishment of a behavior oracle that defines the correct behavior for each situation against which the system's behavior is checked during the design process will require different tools and techniques than a verification process that occurs entirely after the design is complete. This challenge focuses on the new and different ways we may need to develop to verify these systems and the new tools that will be required to support them. Chapter 7 (Marvel, et al.) identifies specific tools and techniques that are needed to support these processes, while Chap. 11 (Scheidt, et al.) focuses on tools and techniques in the hardware testing domain. Chapter 6 (Ramamoorthy, et al.) defines a new technique connecting formal analysis and learning systems.

Challenge 22: In general, how do we verify the fitness of a given physical robot structure for a given task or environment (obviously, a robot that cannot sense color or is operating in the dark with an infrared sensor is unfit to sort objects on the basis of color)?

Verifying that a specific platform does or does not have a specific component is straightforward. Verifying that a given system can or cannot perform a task is more complicated, especially since combinations of existing sensors or actuators with novel behaviors can sometimes replace more traditional technologies and succeed at performing the task. For example, if a robot is required to mix the separated liquid in a container, that could be accomplished by stirring the liquid with a gripper holding a tool, by shaking the container with an arm, or by vibrating the container by driving over a bumpy surface while holding it. All three options will succeed at mixing the liquid, but a human generating the specification may not think of all the options. Instead of specifying the job to be done in terms of the desired end state of the liquid, they may indirectly end up specifying how the robot should do the job by defining unnecessary platform details. Without evaluating both the platform *and the behavior*, it is difficult to state definitively that a given robot can or cannot perform a task without running the experiment in an appropriate environment. This challenge is concerned with developing predictive tools that enable us to generate these predictions without running the experiment. It captures a common thread across the specification, analysis tools, and emergent behavior problems. Chapter 4 (Redfield and Lennon) proposes a new tool addressing this challenge, while Chap. 8 (Seto and Bassam) provides a review of existing tools as they have been used in verification of marine autonomous systems.

Challenge 23: Descriptive frameworks are either too specific and constrain the developer to specific tools when designing the autonomous elements of the system or too broad and difficult to apply to specific cases. Tools are needed to analyze systems at both the specific and the broad levels.

This challenge gets at the key aspect of the inadequate analysis tools problem. We have two core problems that drive the need to address this problem. First, we lack understanding of how the lower-level components of a system affect its overall performance. Because we cannot predict how a change in the component will drive change in performance, we need to be able to define models and analyze the system at the level of the overall system/environment interaction and simultaneously understand the system at the level of individual algorithms. This need to capture the details of the system means that it is very difficult to apply the broad, generalizable frameworks to specific cases.

We also do not have tools that let us identify how a specific design approach will affect system effectiveness. This means that we can't simplify the analysis by assuming a design methodology and allowing the design methodology to drive the degree of simplification we can use when we analyze the system. Any descriptive framework that constrains the designer will exclude potentially more effective behaviors, so we need tools that can analyze systems generally, across different design paradigms. We need the ability to represent system elements at both the broad and specific levels. The specification tool proposed in Chap. 4 (Redfield and Lennon) provides this kind of descriptive framework. The hierarchical technique described in Chap. 5 (Bouchard, et al.) supports comparative analysis at an arbitrary level of abstraction, supporting broad and general descriptive functions.

Challenge 24: How is a structured process that allows feedback between the physical/ground truth layer and the formal methods verification tools developed?

The core challenge behind the specification problem is how to define a level of abstraction and a logical structure that accurately captures the potential flaws in the system and the desired properties of the system. There must be a feedback link between the physical system and the formal methods. Without that link we cannot ensure that the conclusions the formal methods draw about the system are valid, and we cannot develop tests of the physical system that can enable the formal methods to generalize across operational ambiguity. The key benefits of a formal approach to verification are the predictive identification of potential emergent behavior and management of the state space explosion problem. But without proper connectivity between the formal model and the physical system, we cannot utilize these formal tools effectively to reduce the verification burden. Although further work is necessary in this area, Chap. 4 (Redfield and Lennon) provides a starting point for the development of this kind of structured process.

Challenge 25: How to disambiguate between cases where the specification was incorrect (task description abstraction failed to capture some required system action) and those where the environmental model was incorrect (environmental abstraction failed to capture some critical system-environment interaction)? How to identify not just individual situations but classes of situations where the vehicle fails to be safe or to achieve safe operation (e.g., *a front wheel often falls off the cliff but the back wheels never do). How should unanticipated unknowns be accommodated?*

This challenge addresses the specification and the state space explosion unknown unknowns problems and highlights the difficulties inherent in capturing the source of a performance deficit. For a piece of simple hardware, properties like hardness define fixed limits for operational effectiveness. A screwdriver whose tip is too soft will erode away without unscrewing the screw; a screwdriver whose tip is too hard will strip the screw head and make it impossible to unscrew. The designer sets the hardness and defines the materials that the screwdriver will be effective against. Fundamentally, the screwdriver design drives the acceptable operational context. But while an autonomous system designer must make similar assumptions about the context in which it will be used, the massive size of the operational state space means that it is effectively impossible both to express those constraints in a way that is meaningful to the end user and to ensure the robot is able to identify when its design assumptions have been violated. Determining how to capture those assumptions and embed them into the system and its verification and validation testing is non-trivial. Chapter 9 (Mullins) provides an initial approach both to identifying classes of situations that are particularly safe or unsafe and to enabling disambiguation between environmental assumption and specification failure (where those environmental factors can be represented in the simulated environment). Chapter 11 (Scheidt, et al.) explores how the physical testing process addresses unsafe and unexpected scenarios.

Challenge 26: If an algorithm, or patch to an existing algorithm, was replaced, can it be proven that no new failure modes were introduced without re-doing the entire verification process?

This challenge gets to the core of the problem with modular autonomous behaviors and their design and development cycle. Ideally, the goal of a modular approach to system design is that new approaches to problems can be substituted for the old without incurring a significant verification and assurance burden—verifying the old complete system and verifying the new module means that only a very small subset of cases need to be considered in order to assure the new combination. In practice, however, autonomous system components are heavily interdependent. Each individual component relies on a complex web of interactions between the other components in order to provide the necessary function. These interdependencies mean that in order to ensure that no new failure modes are introduced, a more complex analysis is necessary to determine to what degree the original verification process must be replicated. This challenge is still unsolved—those analysis tools still need to be developed.

Challenge 27: How can we verify systems that are designed to learn as they operate?

This challenge specifically addresses the problem of ensuring that the operational system in the field retains the desirable properties and functionality assessed during the verification process. It highlights the challenges associated with systems whose acceptance tests are most unlikely to reflect their operational performance. Chapters 3 (Heitmeyer and Leonard), 6 (Ramamoorthy, et al.), and 11 (Scheidt, et al.) each address different aspects of the problem of verifying autonomous systems that learn.

2.4 Conclusion

There are many challenges associated with verifying autonomous systems. Approaching the problem from the standpoint of a practitioner, the double-V model is often used in verification. But this model relies on many assumptions that are violated by the nature and use of autonomous systems. The double-V model assumes that it is possible to write adequate, complete specifications, appropriate analysis tools exist, emergent behavior can be modeled and managed, relevant information about the system's operation and performance can be known, experiments are reproducible, ground truth is available, communications can be effectively modeled, and the system does not learn as it runs.

In practice, however, fundamental scientific questions like "How can test sets be produced to yield the data required to create mathematical/logical models or find the boundary locations and fault locations in the larger state space?" lie underneath these violated assumptions. Understanding how to produce test sets whose results illuminate performance boundaries and faults in the larger state space implies the creation of analysis tools and the ability to select the relevant information necessary to capture the system's operation and performance. The intersections of these challenges are shown in Fig. 2.10.

Answering these fundamental questions improves our understanding of the fundamental properties of autonomous systems. When we connect these fundamental questions and their answers to the challenges highlighted by the double-V model, our improved understanding enables us to determine how and when to apply our existing tools and how and when to use new tools to accomplish our verification goals. Verification of autonomous systems is a challenge, but it is not an unsolvable one. The chapters in this book provide improved understanding of specific fundamental questions as cross-referenced in Table 2.1 and point the way to more comprehensive solutions.

Category	Subcategory	Challenge	Specifications	Analysis Tools	Emergent Behavior	Unknown Unknowns	Reproducibility	Ground Truth	Communications Latency	Learning
Models		1. How is an adequate model of the system created?		●	●	●				
Models		2. Common models and frameworks are needed that describe autonomous systems broadly enough so they can be used to standardize evaluation efforts and inter-faces to the system.		●						
Models		3. How should debug models of black box autonomous systems be developed? How is a mathematical and/or logical model suitable for formal analysis produced from empirical observations?	●	●	●	●	●	●	●	●
Models		4. How should one identify and model components that are not captured yet (and what are their properties)?		●	●	●				
Abstraction	Fidelity	5. How do we define the appropriate level of fidelity required to extract the information we are interested in?		●						
Abstraction	Fidelity	6. How is the adequate level of abstraction determined for the model of a robot, its behaviors, the simulation that tests the model, or for any given tool or system? How many environmental characteristics need to be specified? What are the aspects of the environment, the robot, and the autonomy algorithms that cannot be abstracted away without undermining the verification?		●						
Abstraction	Requirements	7. How do we determine at what level we are transitioning from specifying requirements to designing the system and develop principled requirements that do not devolve into designing the solution?	●	●						
Abstraction	Requirements	8. How does one ensure both the implicit and the explicit goals of the system are captured? How is a model of the system's goals from a human understanding of the task goals, the system, and the environment created?	●							
Abstraction	Requirements	9. How are performance, safety, and security considerations integrated?	●	●	●	●	●	●	●	●
Testing	Test	10. At what point is there enough evidence to determine that a given autonomous system or behavior has been verified?				●				
Testing	Test	11. How does one ensure it is possible, in the physical world, to test simulated situations that result in boundary cases?					●	●		
Testing	Scenarios	12. How to design tests so that passing them indicates a more general capability?		●		●				
Testing	Scenarios	13. How are challenging design reference missions selected so that performing well against them indicates a more general capability for the system rather than for specific system components?		●	●					
Testing	Scenarios	14. How can test sets be produced to yield the data required to create mathematical/logical models or find the boundary locations and fault locations in the larger state space?		●		●				
Testing	Metrics & Evaluation	15. Once a reasonable model is created how would one determine whether all resulting emergent behaviors were captured and what are appropriate performance measurement tools?			●					
Testing	Metrics & Evaluation	16. Measurement and evaluation are generally poorly understood – operators can de-scribe the tasks for the robot but lack tools to quantitatively evaluate them. How should autonomous behaviors be measured in a way that consistently and accurately describes the capability embodied by a robot?	●	●						
Testing	Metrics & Evaluation	17. How is a metric defined for comparing solutions?	●							
Testing	Metrics & Evaluation	18. How is the "optimal" defined against which the verification is performed? How is it shown that the solution is in fact, optimal? How is the performance of the system measured?	●				●			
Testing	Assurance	19. How is the performance from a finite set of samples of the performance state space generalized across several variables and parameters?				●				
Testing	Assurance	20. Autonomy frameworks are unable to determine whether all the resulting emergent behaviors have been captured or to supply performance measurement tools.		●						
Tools		21. What new tools or techniques need to be developed?		●						
Tools		22. In general, how do we verify the fitness of a given physical robot structure for a given task or environment (obviously, a robot that cannot sense color or is operating in the dark with an infrared sensor is unfit to sort objects on the basis of color)?	●	●	●					
Tools		23. Descriptive frameworks are either too specific and constrain the developer to specific tools when designing the autonomous elements of the system or, too broad and difficult to apply to specific cases. Tools are needed to analyze systems at both the specific and the broad levels.		●						
Tools		24. How is a structured process that allows feedback between the physical/ground truth layer and the formal methods verification tools developed?	●							
Tools		25. How to disambiguate between cases where the specification was incorrect (task description abstraction failed to capture some required system action) and those where the environmental model was incorrect (environmental abstraction failed to capture some critical system-environment interaction)? How to identify not just individual situations but classes of situations where the vehicle fails to be safe or to achieve safe operation (e.g. a front wheel often falls off the cliff but the back wheels never do). How should unanticipated unknowns be accommodated?	●			●				
Tools		26. If an algorithm, or patch to an existing algorithm, was replaced can it be proven that no new failure modes were introduced without re-doing the entire verification process?		●						
Tools		27. How can we verify systems that are designed to learn as they operate?								●

Fig. 2.10 Summary of relationships between "double-V" challenges and fundamental verification challenges

Table 2.1 Cross-referencing the research challenges and the chapters in which progress toward addressing them is presented

	Challenge	Chapter
1	How is an adequate model of the system created?	Chapter 10
2	Common models and frameworks are needed to describe autonomous systems broadly enough so they can be used to standardize evaluation efforts and interfaces to the system	Chapter 5
3	How should debug models of black box autonomous systems be developed? How is a mathematical and/or logical model suitable for formal analysis produced from empirical observations?	Chapter 6
4	How should one identify and model components that are not captured yet (and what are their properties)?	–
5	How do we define the appropriate level of fidelity required to extract the information we are interested in?	Chapter 10
6	How is the adequate level of abstraction determined for the model of a robot, its behaviors, the simulation that tests the model, or for any given tool or system? How many environmental characteristics need to be specified? What are the aspects of the environment, the robot, and the autonomy algorithms that cannot be abstracted away without undermining the verification?	Chapter 10
7	How do we determine at what level we are transitioning from specifying requirements to designing the system and develop principled requirements that do not devolve into designing the solution?	Chapter 4
8	How does one ensure both the implicit and the explicit goals of the system are captured? How is a model of the system's goals from a human understanding of the task goals, the system, and the environment created?	Chapter 3
9	How are performance, safety, and security considerations integrated?	Chapters 3 and 11
10	At what point is there enough evidence to determine that a given autonomous system or behavior has been verified?	Chapter 10
11	How does one ensure it is possible, in the physical world, to test simulated situations that result in boundary cases?	Chapter 11
12	How to design tests so that passing them indicates a more general capability?	Chapter 9
13	How are challenging design reference missions selected so that performing well against them indicates a more general capability for the system rather than for specific system components?	Chapter 9
14	How can test sets be produced to yield the data required to create mathematical/logical models or find the boundary locations and fault locations in the larger state space?	Chapters 9
15	Once a reasonable model is created how would one determine whether all resulting emergent behaviors were captured and what are the appropriate performance measurement tools?	–

(continued)

Table 2.1 (continued)

	Challenge	Chapter
16	Measurement and evaluation are generally poorly understood—operators can describe the tasks for the robot but lack tools to quantitatively evaluate them. How should autonomous behaviors be measured in a way that consistently and accurately describes the capability embodied by a robot?	Chapters 5, 7, 8, and 11
17	How is a metric defined for comparing solutions?	Chapters 5
18	How is the "optimal" defined against which the verification is performed? How is it shown that the solution is in fact, optimal? How is the performance of the system measured?	–
19	How is the performance from a finite set of samples of the performance state space generalized across several variables and parameters?	Chapter 5
20	Autonomy frameworks are unable to determine whether all the resulting emergent behaviors have been captured or to supply performance measurement tools	–
21	What new tools or techniques need to be developed?	Chapters 6, 7, and 11
22	In general, how do we verify the fitness of a given physical robot structure for a given task or environment (obviously, a robot that cannot sense color or is operating in the dark with an infrared sensor is unfit to sort objects on the basis of color)?	Chapters 4 and 8
23	Descriptive frameworks are either too specific and constrain the developer to specific tools when designing the autonomous elements of the system or, too broad and difficult to apply to specific cases. Tools are needed to analyze systems at both the specific and the broad levels	Chapters 4 and 5
24	How is a structured process that allows feedback between the physical/ground truth layer and the formal methods verification tools developed?	Chapter 4
25	How to disambiguate between cases where the specification was incorrect (task description abstraction failed to capture some required system action) and those where the environmental model was incorrect (environmental abstraction failed to capture some critical system-environment interaction)? How to identify not just individual situations but classes of situations where the vehicle fails to be safe or to achieve safe operation (e.g., a front wheel often falls off the cliff but the back wheels never do). How should unanticipated unknowns be accommodated?	Chapters 9 and 11
26	If an algorithm, or patch to an existing algorithm, was replaced can it be proven that no new failure modes were introduced without re-doing the entire verification process?	–
27	How can we verify systems that are designed to learn as they operate?	Chapters 3, 6, and 11

Part II
Theoretical Approaches

Chapter 3
Assuring Autonomy Using MBSE, Formal Methods, and the Best Agile/DevOps Practices

Constance L. Heitmeyer and Elizabeth I. Leonard

Chapter 3 Editorial Note

This chapter focuses on mechanisms for assuring autonomy software based on mathematical and logical proofs and systems engineering practices. Through that lens, it answers stakeholder questions related to three aspects of autonomous system verification: certifiability and reliability of autonomous systems, trustworthiness and security of autonomous systems, and understanding the impact of formal representation and processes on how to verify these systems. Technical challenge 8 (capturing and modeling the system and the task's goals) is addressed by providing guidance for how to formally capture both the implicit and explicit goals of the system. This chapter's assessment of how formal tools and systems engineering processes can aid in integrating performance, safety, and security considerations, even for learning systems, provides context and techniques addressing challenges 9 (integration of performance, safety, and security considerations) and 27 (verification of learning systems). Stakeholder questions are presented in the Chap. 1, Introduction. Technical challenges are introduced in Chap. 2, Verification Challenges.

Abstract Increasingly, autonomous systems, including robots, are being used in both commercial and military applications. These systems span a wide range, from unmanned ground vehicles, e.g., "smart cars," to unmanned weapon systems. Autonomous systems can have major advantages—they perform some tasks faster and more efficiently than humans and can operate in dangerous environments where humans might be killed. Given the unsafe environments in which many autonomous systems operate, assurance is needed that they behave as intended. However, assuring autonomy is complex partly because autonomous systems often operate in rapidly changing environments, making predicting their behavior difficult. This

C. L. Heitmeyer (✉) · E. I. Leonard
Naval Research Laboratory, Washington, DC, USA
e-mail: constance.heitmeyer@nrl.navy.mil; elizabeth.leonard@nrl.navy.mil

S. Redfield et al. (eds.), *Verification of Autonomous Systems*,
https://doi.org/10.1007/978-3-031-88546-4_3

unpredictability means that existing assurance techniques, which typically assume the system behavior is known in advance, will often not work in assuring autonomy. This chapter describes four major challenges that arise in assuring autonomous systems and how to address each challenge. We conclude with recommendations describing how formal methods, agile, and DevOps practices can be combined to produce safe autonomous systems.

3.1 Introduction

In future years, autonomous systems, such as driverless vehicles, are expected to play an increasingly important role in society [14]. Recent advances in autonomous systems, such as autonomous delivery at Amazon [55], intelligent air and land transport at the Australian Centre for Field Robotics [59], and Lincoln Lab's autonomous steering and mine detection ground vehicle [82], demonstrate the huge potential of autonomous systems to increase system efficiency, reduce system costs, and reduce system reaction times. These systems should also reduce the number of humans operating critical systems, as well as human risk by autonomously performing dangerous tasks in complex environments, thus removing humans from harm's way. Autonomous systems will also perform some operations that humans cannot perform; for example, they can repair satellites traveling in geosynchronous Earth orbit [29]. However, several factors, such as the high potential in autonomous systems for unintended emergent behavior, make the development of autonomous systems, assured to safely and correctly perform their assigned tasks, uniquely challenging.

Below, we describe four major challenges that arise when humans and human intelligence are replaced by machines and machine intelligence, namely, how to elicit and precisely document autonomous system requirements, how to assure autonomous systems given the dynamic, rapidly changing environment in which they operate, how to assure autonomous systems which rely on machine learning (ML), and which process to use in designing, developing, and assuring autonomous systems. Addressing these challenges and assuring that autonomous systems satisfy critical requirements are critical. As stated in a recent report [21], "No good can come from autonomy without proper assurance."

This chapter is organized as follows. Section 3.2 defines our terminology and explains our focus on software, Sect. 3.3 describes the four major challenges introduced above, Sect. 3.4 suggests how each challenge can be addressed, and Sect. 3.5 describes how a process based on formal methods combined with effective practices of agile software development can assure the safety and correctness of future autonomous systems and shorten system development time. This process was inspired by a process proposed by Ghezzi in 2018 [37]. Finally, Sect. 3.6 describes our conclusions and open research problems.

3.2 Background

This section defines terminology used in this chapter. It also presents an overview of the substantial literature available to guide the design of autonomous systems and justifies our focus on assuring the *software* components of autonomous systems.

3.2.1 Terminology

This section describes what we mean by critical systems and three different kinds of systems that are often referred to as autonomous. Although, in the remainder of this chapter, we use the terms "autonomous system" and "autonomy" in a general sense, it is valuable to review the distinctions among various autonomous systems. The section concludes by distinguishing system validation from formal verification.

3.2.1.1 Different Classes of Systems

We define a *critical system*, also called a *dependable system* [5], as a system in which a system malfunction could have catastrophic consequences, such as human injury or death, or serious damage or loss of costly equipment. Many autonomous systems, such as robots, can be classified as critical systems [87]. Examples of noncritical systems include payroll systems and drawing programs, where errors rarely have catastrophic consequences.

Our interest is in three kinds of critical systems: tele-operated, automated, and autonomous. Although most of these systems will be unmanned, some, such as the Tesla vehicle described below, may be manned. The systems of interest in this chapter may also employ behaviors in more than a single category; for example, an unmanned air vehicle may be directly controlled by an operator using a joystick, or alternatively it may navigate on its own using a set of waypoints provided in advance.

A *tele-operated system* is a system whose behavior is directly controlled by a human operator, e.g., by entering commands at a control terminal or with a joystick. Examples of tele-operated systems include surgical robots and remotely controlled unmanned vehicles. Tele-operated systems rely on an operator to select the appropriate behavior when operational conditions change.

In contrast to tele-operated systems, an *automated system* performs its assigned tasks largely without human supervision. However, an automated system cannot usually adapt its behavior when operational or environmental conditions change. For this reason, industrial systems may be subject to differing levels of human supervision based on the task to be performed and the system environment. Some may be completely preprogrammed to perform their assigned task, while others may require a human to explicitly grant the system permission to continue unmanned operation at different points during the system's execution. Many industrial robots are automated and generally preprogrammed. Although they may perform complex tasks, such as product assembly and packaging, automated systems typically operate in controlled, well-understood environments.

An important and early example of an automated system is Tesla's Autopilot, which, upon the driver's command, takes over certain driving functions. Autopilot is capable of driving a vehicle without human guidance, keeping the vehicle in its lanes and adjusting the vehicle's speed to keep it a safe distance from other vehicles. However, because Tesla's ability to respond to environmental changes is limited, the driver of a Tesla vehicle is instructed to keep his/her hands on the wheel when using Autopilot, so the driver can respond when unexpected events occur (e.g., sunlight blocks the sensors' view of a road obstacle). Partly because humans are poor monitors and easily distracted, a number of accidents involving Tesla vehicles have occurred, some resulting in human death. In 2020, for example, a Tesla, traveling at high speed on a California freeway, ran a red light and hit a Honda Civic, killing the two people inside the vehicle; on the same day, another Tesla struck a parked firetruck on an Indiana highway, killing a passenger inside the vehicle [66]. Fatal crashes involving similar driver assistance systems from GM and other auto manufacturers have not been reported. To prevent serious accidents, GM monitors drivers with cameras and shuts down the automated driving system if a driver fails to watch the road [66].

According to analysts at the Rand Corporation, *autonomous systems* will not only perform complex tasks without human intervention, but, in contrast to both tele-operated and automated systems, they will also adapt independently and without human guidance to changes in their operational and environmental conditions [70]. While many tele-operated and automated systems have already been built and are currently used in practice, this is not the case for autonomous systems as defined by Rand. Although many advances in autonomy technology have been made in recent years and a number of autonomous systems are planned, autonomous systems capable of responding successfully to unexpected environmental changes are rare. In fact, a 2019 report by Rand predicts that autonomous systems that can handle unexpected events safely may not be ready for practical application for a decade or more [70].

In the remainder of this chapter, we use the terms "autonomous" and "autonomy" in a general sense: By an autonomous system, we mean any system, manned or unmanned, that can perform certain operations largely without human intervention. Hence, we use the term autonomous to refer to both automated systems as defined above and autonomous systems as defined by Rand. Given this general definition, note that most moderate to large commercial aircraft, while manned, can be viewed as operating autonomously—for most of a flight, the pilot monitors the behavior of the aircraft and remains passive unless unexpected events occur.

3.2.1.2 Requirements, Properties, Assurance, Validation, and Formal Verification

Requirements The *system requirements* are defined as the set of all acceptable system implementations [75]. Based on this definition, the system requirements should not be over specified—that is, they should not exclude any acceptable implementations, nor should they be underspecified—that is, they should not include any unacceptable implementations. In assuring a critical software system, a precise, unambiguous statement of the system requirements is essential. Precise requirements provide a solid foundation for design and testing and for future system updates when new versions of the system, for example, versions that support new functions, are needed. The requirements of a system are extremely important because they form the basis for all other activities in the development process-design, coding, testing, etc. A problem is that, currently, the vast majority of requirements documents are written in natural language which is inherently ambiguous and imprecise. Section 3.3.1 addresses the problem of imprecise, ambiguous requirements.

Property We define a *property* as a predicate that holds for all executions of a system. A synonym for property is *assertion*. The system requirements are often described as a set of properties and are expressed in natural language.

Assurance We define *assurance* as confidence that the system satisfies its requirements. This definition is equivalent to McDermid's definition [71] that an assured system behaves as intended in its environment of use. We note that many use the term "verification" as a synonym for assurance. In contrast, we use the term "verification" to mean formal verification as defined below.

Formal Methods *Formal methods* are mathematics-based techniques, such as formal models, formal verification tools, and validation tools, for representing and reasoning about systems. *Formal models* are models expressed in a language with an explicit formal semantics. In this chapter, our interest is in *formal system models*, formal models whose objective is to precisely capture the required system behavior. *Formal verification* is a mathematical process whose objective is to prove formally that a formal system model, or the system code, satisfies specified system properties, such as consistency, completeness, and safety properties. Many formal verification tools are available for checking formal models for properties of interest. One such tool, a *consistency checker*, checks a formal model for *coverage* and *disjointness*, where *coverage* means that each function in the model is total (no missing cases) and *disjointness* means that each function is well defined (no non-determinism) [47]. Consistency checking is automatic and an efficient technique for detecting missing cases and unwanted non-determinism in a formal model. Other examples of formal verification tools are model checkers [18], interactive theorem provers (see, e.g., [4]), and SMT (Satisfiability Modulo Theories) solvers [74].

In contrast to formal verification, which is designed to show that a formal model has the right properties, validation is designed to show that a formal model is right. More precisely, *validation* provides evidence that a system satisfies its requirements, that is, that the system behavior specified by a formal model captures the intended system behavior. In current software practice, testing is commonly used to validate that the system code captures the intended system behavior

Runtime Verification Currently, formal methods are rarely used in software practice. A major problem is that formulating the initial formal model can be very difficult. A promising technique that overcomes this and other problems that hinder the adoption of formal methods is *runtime verification* [31], a combination of testing and formal methods that monitors the runtime behavior of both the system and its environment, and when it detects the potential for a safety violation, either reports the potential violation or takes action to avoid or mitigate the violation. A major advantage is that, during system execution, runtime verification can detect and prevent safety violations that could not be predicted during system development.

Model-Based System Engineering A key aspect of our recommended approach (see Sect. 3.5) is Model-Based System Engineering (MBSE). We define *MBSE* as a systematic approach to developing critical systems which uses models as fundamental components throughout the system life cycle. The models used in MBSE may range from informal sketches to formal models.

3.2.2 Existing Literature

Many autonomy experts (see, e.g., [11, 79]) have pointed to the large body of literature in software engineering, formal methods, human computer interaction, and other areas, which has been published to guide the development of nonautonomous systems. They argue that many of these same techniques can help assure the behavior of autonomous systems. For example, on the topic of software system engineering, see the paper by Landwehr et al. which describes a set of capabilities that universities should teach to engineers who will design and build software systems [68]. For the problem of human-autonomy interaction, the 2012 report on assured autonomy by the Defense Science Board (DSB) points to "established human-centered [and system] design principles" that can lead to "reduced manpower, lower human workload, better performance, fewer errors, and greater user acceptance" of autonomous systems. In addition, there is a large body of literature on formal methods that could prove useful in assuring autonomous systems. See [87] for a survey of many of these formal methods and a description of how they have proven useful in assuring the behavior of critical systems that are not autonomous.

3.2.3 *Central Role of Software*

According to the 2012 Defense Science Board (DSB) report, the critical capabilities provided by autonomous systems will reside in the system software and further that the system software will be complex [22]. Although the report's focus is military systems, we expect that the software in most nonmilitary autonomous systems will also be complex and perform critical system functions. A major problem for both military and commercial development of autonomous systems is that the traditional approach to developing unmanned systems focuses on hardware. As stated in the DSB report [22], the result is that autonomy software may be treated as an afterthought added to a hardware platform later in the process. Unfortunately, autonomous systems developers may not have sufficient expertise in autonomy software nor have "testbeds for validation and verification" of autonomous systems whose primary component is software [22]. This is problematic because we expect the software to manage the sensors, processing power, communications, and "intelligent" components of autonomous systems and to use this information to make the critical decisions. Hence, assuring that the autonomy software behaves correctly and safely is extremely important. That is why this article focuses on assuring the *software* component of autonomous systems.

3.3 Challenges

This section describes four major challenges that arise in assuring the behavior of autonomous systems.

3.3.1 *Eliciting and Documenting the System Requirements*

It has long been recognized that formulating the system and software requirements precisely is extremely difficult. In 1987, in a widely cited article [15], Brooks described the challenge of writing requirements and the serious consequences of poorly written requirements:

> The hardest single part of building a software system is deciding precisely what to build. No other part of the conceptual work is as difficult as establishing the detailed technical requirements.... No other part of the work so cripples the resulting system if done wrong. No other part is as difficult to rectify later.

The difficulty of formulating precise requirements has also been recognized by the autonomy community. At an aerospace forum in 2016, Rozier stated that understanding the system requirements and representing the requirements formally was the "biggest bottleneck" in autonomy [80]. Similarly, a 2020 report sponsored by the NITRD (Networking and Information Technology Research and Development) program [21] concludes that the "lack of actionable and unambiguous requirements is a significant bottleneck" in assuring autonomous systems. More recently, in his 2021 testimony to Congress, Admiral Michael Gilday emphasized the importance of autonomous system requirements, stating that "...a growing consensus is forming that the Navy needs to get its requirements and systems right" before making a big investment in unmanned surface and undersea vehicles [38].

As stated above, in practice, the system requirements are most often described in natural language. Natural language descriptions of requirements are useful because they provide us with intuition about the required system behavior. The most common approach in practice is to describe the system requirements in a list of "shall" statements, such as "The system shall not maneuver if its remaining battery power is below an operator-specified threshold," and "The system shall not begin a maneuver unless a specified time has passed since its last maneuver." Use cases, user scenarios, and user stories are other informal techniques that describe requirements in natural language. In contrast to single shall statements, these techniques describe how one may use the system to perform a given task.

While natural language requirements are useful, they are inherently imprecise and ambiguous. To improve precision and remove ambiguity, a number of alternatives to natural language requirements, such as tables and finite state diagrams (FSDs), have been proposed for specifying *requirements models*. When these models have an explicit formal semantics, that is, a mathematical description of their meaning, they are called *formal requirements models*. What distinguishes requirements models from requirements described as shall statements is that requirements models can be checked, either manually or mechanically, for completeness and consistency. Moreover, when they are expressed in user-friendly notations, such as tables and FSDs, these models are easily understood by practitioners.

The A-7 requirements document, the first notable example of a requirements model [3, 54], specified the requirements of the flight program for the A-7, an existing U.S. Navy aircraft. Using a tabular format, the document defined a set of mathematical functions. These definitions were designed to be manually checked to ensure that each function was well-formed, i.e., *consistent*, and total, i.e., complete. The A-7 requirements document was not only an important research contribution—its tabular notation specified the requirements unambiguously and precisely, but also it was useful at the practical level. The engineers who maintained the flight A-7 software relied on the A-7 requirements document, rather than the earlier natural language requirements documents, when changes to the flight software were needed. In addition, the document proved useful in developing flight software for the A-6 aircraft, whose flight program was similar to the A-7's.

At the beginning of system development, understanding of the system requirements is poor. This is especially true when the system is brand new. As was demonstrated in the A-7 project, understanding the requirements is easier when the task is to build a new version of an already existing system. To understand the system requirements, developers can study the existing requirements documentation and then consult maintenance engineers to answer questions that remain.

3.3.2 How to Address Change During System Operation

An important difference between more traditional systems and autonomous systems is the nature of their environments. In many traditional systems, the behavior of the environment in which the system functions is known in advance. Thus, most testing and other analysis of traditional software assumes that the system environment is fixed. Such an assumption is often invalid for an autonomous system. As the 2019 Rand report notes [70], in contrast to the environment of most traditional systems, the environment of many autonomous systems will be dynamic and changing. Thus, developers of autonomous systems must plan for change in the behavior of the system environment after deployment and the resultant change that must occur in the required behavior of the system itself. Not as frequent is a second class of change: a change in the system requirements. How to assure autonomous systems whose requirements change after the system has been deployed is also a major challenge.

3.3.3 How to Assure Machine Learning Modules

Machine learning (ML) is a key enabling technology for autonomy. According to the 2012 DSB report [22], "Machine Learning has become one of the most effective approaches to developing intelligent, autonomous systems." ML can be used to perform tasks critical to the performance of many autonomous systems including obstacle detection and avoidance, navigation, path planning, action selection, task allocation, and image detection, recognition, and classification. ML is often used when the task to be performed cannot be adequately described using more traditional functional or algorithmic descriptions. Instead, data is used to train the ML module as to what constitutes correct behavior with respect to the task.

Three of the major ML approaches are Supervised Learning, Unsupervised Learning, and Reinforcement Learning. Supervised Learning relies on training data in which each training instance is labeled with the answer that should be provided when this instance is the input to the module. For example, if the task is to use a set of features (e.g., weight and color) to classify objects, the correct object will be included as part of each training instance. Unsupervised Learning uses unlabeled training data and deduces patterns from that data. In Reinforcement Learning, each training instance is a sequence of inputs, and in addition to the training data, the learning process uses a representation of the task's environment and a reward function. The purpose of training is to produce a *policy* which describes the task behavior as a map from states (containing the module's knowledge of its environment) to actions (the module outputs). Reinforcement Learning attempts to maximize the cumulative reward function during training by executing each training instance in a representation of the operating environment.

In all three approaches, the ML development process contains four phases. During the first phase, data to be used in training the ML system is collected and preprocessed. This phase may include some data standardization and may add synthesized data to fill out the training dataset. For Supervised ML, this stage also includes the labeling of training data. The preprocessed data is then used as input to the second phase during which an ML algorithm is used to build a model representing the task described by the training data. Common model representations include decision rules, decision trees (i.e., a tree where the leaves are the decisions), and different variants of Neural Networks. A Neural Network (NN) is a set of nodes with weighted directed edges that represent the connections between the nodes. The weights on the edges are adjusted during the learning process. The nodes are often organized into layers; a Deep NN is an NN with many layers. In the third phase, the trained model is evaluated to determine how well it performs on data not previously seen. This is often done by testing the model using a new set of test data. In rare cases, formal verification is applied to evaluate the model. In the final phase, the ML module is deployed. While in currently operational systems, the learning process stops prior to system deployment, an ML module may continue to learn after deployment. In this case, the model will continue to be updated during system operation.

The use of ML introduces many assurance challenges. As described in two recent surveys [6, 77], software components developed using ML generally rely on structures such as Deep Neural Networks to make decisions about classifying new inputs or choosing the next action to perform. Such structures are too complex for human understanding, resulting in a lack of clarity about why an ML component makes the decisions it makes. This makes it difficult to trust an ML component, and the system that relies on it, to behave correctly. One challenge is assuring that the system satisfies critical properties such as safety.

Beyond assuring that an ML module satisfies critical properties, we must determine under what conditions the module can be trusted to produce "good" results once it is deployed. The quality of an ML module depends on the data used during training. This data could be incomplete, inconsistent, or inaccurate (e.g., incorrectly labeled). Further, the operating environment will likely differ from the training environment. It may violate assumptions about the world in which it will be deployed, and the data in the operating environment may not be distributed similarly to the distribution used during training. In addition, some ML modules continue to learn after deployment, meaning that the ML module and the system relying on it cannot be verified once to provide assurance—assurance must continue throughout system development and operation.

3.3.4 What Development Process to Use

A critical question in developing and assuring autonomous systems is which of the many available development methods to use. Another related critical question is whether, and if so how, to include formal methods in the development process. This section reviews the available methods, briefly describes how they evolved historically, and addresses the question of whether and how to include formal methods in the development process.

Until the late 1980s, the predominant approach to software development was the waterfall model, a sequential approach that required one development phase to be completed and documented before the next phase begins. Thus, for example, the requirements phase needed to be completed and documented before the design phase. Beginning in 1988, two other methods were proposed to augment the waterfall model, the V model, which adds verification and validation to each development phase [33], and Model Based Systems Engineering (MBSE) [30], which supplements the traditional waterfall method with increasingly detailed models. MBSE associates one or more *models* with each development phase. Claimed advantages of MBSE are significant productivity increases, improvements in system quality, and cost savings [16]. An additional and widely recognized advantage of MBSE is that models use abstraction to hide design and implementation detail, a practice critical in managing complex systems [65].

Also occurring in the late 1980s was growing criticism of the waterfall model. There was widespread agreement among many software practitioners about the model's major flaws—the failure to produce executable code until late in development, the inflexibility to change, and the heavy focus on documentation. To overcome these perceived flaws, more incremental development methods were proposed. In 1988, the first of these, the *spiral model*, integrated incremental development with risk management [13]. In 2000, more than a decade later, *agile* software development was introduced. In contrast to the waterfall model, agile methods, such as extreme programming [10], recommend small software releases, rapid development cycles with feedback loops involving developers and customers, and little to no documentation. Unlike the waterfall method, viewed as document-driven, and MBSE methods which are model-driven, agile methods (estimates are that there are more than 20 different agile methods) are test-driven; test cases are created before a piece of code is implemented and executed on the code as soon as it is implemented. The most recent development method, DevOps [7], integrates system development and the operational system into an overall agile framework.

3.4 Addressing the Challenges

This section addresses the first three challenges in assuring autonomous systems: how to specify, verify, validate, and elicit requirements, how to handle changing environments, and how to assure machine learning. Section 3.5 addresses the fourth challenge—what process to use in developing and operating autonomous systems.

3.4.1 Requirements Specification, Analysis, and Elicitation

This section answers three questions: how to formally specify, verify, and validate autonomous system requirements, what requirements framework should one use, and how to elicit the requirements of autonomous systems.

3.4.1.1 Specifying, Formally Verifying, and Validating System Requirements

One effective way to overcome the imprecision and ambiguity of natural language requirements is to specify the system requirements formally. Two complementary techniques are available to express requirements formally. First, the requirements may be specified as a set P of *system properties*, such as safety properties, each property in P expressed as a mathematical formula (see, e.g., [73]). Properties in P may be derived from natural language descriptions, perhaps a list of shall statements. A second approach to representing requirements formally is to create a *formal requirements model*, e.g., a state machine model. The essential difference between requirements expressed as a set of properties P and a requirements model is that the behavior specified by a requirements model may be checked for completeness, whereas the behavior defined by the set P, except in the case of very simple systems, will be inherently incomplete. By completeness, we mean that every function in the model (every function defined in P) is total. Ideally, one would like to have both a set P of system properties and a requirements model S so that one can verify that the system model S satisfies each property in P.

The major advantage of a formal requirements model is that one may use software tools to check, often automatically, that the model satisfies properties of interest. In 1995, more than 25 years ago, our research group assigned a state machine semantics to the tabular notation used in the A-7 requirements document [44, 47]—up until then, the meaning of the tables had only been described in natural language [54]. Based on our state machine semantics, we developed a suite of tools, called the SCR (Software Cost Reduction) toolset, for formally verifying and validating formal requirements models. One tool in the initial SCR toolset was a consistency checker. Applying consistency checking to the tables in the A-7 requirements document [44, 47] detected, to our surprise, 17 missing cases and 57 instances of non-determinism! That so many defects were uncovered was unexpected given that the document had been inspected manually for defects by two independent review teams.

A major benefit of a state machine semantics is that a state machine can be automatically translated into executable code that "simulates" the system behavior by symbolically executing the formal model. This technique was used to build the SCR simulator, a second tool in our initial toolset [46]. The simulator starts in the initial state and then updates the system state each time the user inserts a new input event. At each simulation step, one can check that selected system properties hold. As stated above, a customer can use such a simulator to validate the formal model.

After introducing the consistency checker and simulator in 1995 [46], we extended the SCR toolset from 1995 onward to include many other tools. Among these are two model checkers [48], an invariant generator [62], test generator [35], theorem prover [4], source code generator [69], and a tool that synthesizes a formal requirements model from scenarios [52]. To facilitate customer validation, we also developed a tool for designing a graphical interface for the simulator which looks like the interface of the system under development [49]. Because all SCR tools share the same semantic foundation and because the toolset includes heterogeneous formal methods, the SCR tools may be classified as "integrated" formal methods as defined by Gleirscher et al. [40]: One may use the tools together to analyze, correct, and verify a formal model. While our group was the first to develop tools that integrated different formal methods based on a common semantics, other researchers have subsequently constructed similar toolsets (see, e.g., [41, 45]).

As noted in Sect. 3.3.1, a formal requirements model is mathematically based—that is, it has an explicit formal semantics. Translating natural language requirements, such as user stories, or requirements in semi-formal languages, such as UML and SysML, which lack a formal semantics, into a formal representation is difficult due to the inherent ambiguity and lack of detail in these representations. Recently, our group tried to translate requirements in SysML for an autonomous ground vehicle into a formal model but failed due to missing details in the SysML documents. What ultimately proved useful was the vehicle's user manual.

3.4.1.2 Two Classical Requirements Frameworks

Reviewed in this section are two classical frameworks, introduced in the 1990s, for modeling and reasoning about system and software requirements: Parnas' Four-Variable Model [76] and the requirements model of Zave and Jackson [88]. This section describes the correspondence between these two classical models and the properties and requirements models described in Sect. 3.4.1.1.

Two major components of Parnas' Four-Variable Model (FVM) [76] are the system `SYS` and the part of the system environment, `NAT`, which `SYS` observes and changes. `NAT` contains *monitored quantities* and *controlled quantities*, environmental quantities that `SYS` monitors and controls. In the FVM, `REQ` specifies the system requirements as the required relation between the system's monitored and controlled quantities. REQ corresponds to what we call a requirements model. The FVM does not explicitly mention properties.

The requirements model formulated by Zave and Jackson [88] also contains two entities, the `Machine` and the `World`, which loosely correspond to `SYS` and `NAT`. What the Zave-Jackson model calls requirements corresponds to what we call properties. What they call specifications corresponds roughly to what we call the requirements model. These two classical models differ in one important respect. Whereas, in the FVM, `SYS` includes both the software *and* the system's hardware devices, e.g., the sensors and actuators that observe and control environmental quantities, Zave and Jackson's `Machine` contains only software; sensors and actuators are part of the `World`. We prefer the FVM because our goal is to define the *system* requirements, not the *software* requirements. We also want to prove *system properties*, not just *software properties* because the properties that we care about—safety, security, fault tolerance, and timing—are system properties.

3.4.1.3 How to Elicit the Requirements

One of the most difficult challenges in requirements engineering is eliciting the system requirements. Many techniques for requirements elicitation have been used in the past, including interviews with domain experts and user stories. As shown in the case of the A-7 requirements, understanding system requirements is much easier if the system already exists. One method proven to be effective for eliciting the requirements of new systems is the agile method, which uses many implement-test cycles and early customer feedback to better understand the system requirements early in development. In our experience, a second effective method for eliciting the requirements of new systems, including autonomous systems, is to interview experts and, based on interview data, to build finite state diagrams of the required system behavior. Finite state diagrams have two major advantages. First, engineers find them easy to understand and validate. Second, state diagrams can be translated easily into a formal representation that can serve as the basis for automatically checking requirements for consistency, completeness, and critical system properties.

3.4.2 Maintaining Safety in Dynamic, Rapidly Changing Environments

In current software practice, testing is the technique used most often to find and correct software defects. A major advantage of testing is scalability—testing can detect and correct defects in very large systems. A significant problem with testing, however, is that it is largely ad hoc. As a result, some critical aspects of the software behavior may remain untested. While formal verification is designed to address this problem, formal verification is not widely used in practice, due in part to the significant manual effort required to formulate a formal model.

One promising technique that addresses this and other problems with formal verification is *runtime verification*, a combination of testing and formal methods which monitors the runtime behavior of both the system and its environment, and, when it detects the potential for a safety violation, it either reports the safety hazard or takes action to mitigate or eliminate it. By combining testing and formal methods, runtime verification obtains the benefits of both—it avoids ad hoc testing *and* the problems of formal verification. Moreover, unlike testing and formal methods, runtime verification addresses a fundamental problem: how to detect and respond to safety hazards during runtime which could not be predicted during development.

Establishing a runtime monitor requires the following steps: (1) formulate, validate, and formally verify a formal model of the required behavior of the monitor, (2) automatically synthesize executable source code from the model, and (3) instrument the autonomous system to contain the synthesized monitor code [31]. Once the autonomous system is deployed, the monitor analyzes the system's incoming events to identify potential safety violations and, if it detects a potential violation, determines an action that can avoid or mitigate the violation.

As noted above, the action taken by a runtime monitor varies: The monitor could report the problem to the autonomous system, relying on the system to design a fix (*runtime monitoring*), or alternatively, the monitor could itself determine how to fix the problem (*runtime enforcement*). Thus the main idea is that the autonomous system performs the system's nominal operations, whereas a high assurance component, i.e., the monitor, guarantees the system's safety. Clearly, runtime monitoring will be useful not only during system runtime but during the system development as well. During testing, a runtime monitor can detect situations in which the system under development could violate a safety property.

3.4.3 Assuring Autonomy That Relies on Machine Learning

This section describes two methods useful in assuring autonomous systems that rely on machine learning (ML): formal verification and runtime verification.

3.4.3.1 Formal Verification

The formal verification of machine learning modules to check critical properties is one method of increasing assurance in systems using ML. This approach only applies to ML modules in which all learning occurs prior to training—the module does not continue to learn after system deployment. The focus of recent research has been on ML policies represented as Deep Neural Networks (DNNs). The approaches are all similar: An equivalent form or over approximation of the DNN is represented in a verification tool, and properties of interest are evaluated against the DNN representation. Many different analysis techniques have been applied, including SMT solvers [9, 56, 64, 67, 89], Linear Programming solvers plus SAT solvers [27], Mixed Integer Linear Programming solvers [83], abstract refinement [28], abstract interpretation [36], symbolic interval propagation [86], invariant inference [61], and combinations of reachability analysis algorithms [84]. In some approaches, e.g., [58, 89], the results of verification are used to guide additional training that leads to an improved policy.

Verisig [57], which focuses on ML for controllers (i.e., action selection tasks), learns a policy, represents it as a DNN, and then transforms it into a formal state machine model. It uses hybrid automata as the formal representation, combines the model of the policy with a hand-constructed hybrid automata model of the environment, and analyzes the combined model for desired properties. The results of analysis can be used to improve the policy by guiding additional training.

VERIFAI [24] is a toolkit for simulation-based verification of systems that use ML for perception and control. Three models are used, together with a simulator, to perform verification; the models of the perception and controller modules are represented as executable code, while the model of the environment is expressed in a probabilistic programming language. Properties to be evaluated against the model using simulation-based verification can be represented as metric temporal logic formulae, objective functions, or code. The results of analysis can be used to improve the learned system by, e.g., using counterexamples found during property checking, to construct data for use in retraining the learning module [25], or to help synthesize parameters for training algorithms.

For Reinforcement Learning (RL), formal analysis can be applied during the learning process to ensure that the learned policy behaves safely. A safety monitor (or shield) can be used during training to verify, at each training time step, that the action selected by the learning algorithm satisfies safety properties, and, if not, the policy is not updated [2, 34]. This method could also be used to provide continued assurance of RL modules that continue learning after system deployment.

3.4.3.2 Runtime Verification

As with environmental uncertainty, runtime monitoring and enforcement can be used to address the uncertainty that characterizes ML in an autonomous system. When used as part of an operational autonomous system, runtime verification is particularly well suited to ensuring that systems containing modules that continue to learn during system operation maintain safe behavior.

In one approach, an online shield is used during runtime to keep a learned policy inside a "safe area" of the state space [8, 39]. If the next step selected by the learned control policy can lead to the system eventually ending up in an unsafe area, then these approaches use a second mathematics-based policy to choose an action guaranteed to keep the system in the safe area; otherwise, the learned control policy performs its chosen action.

Another approach to handling situations in which a module is behaving unacceptably relies on Simplex [81], an architecture in which an "advanced" module is paired with a "baseline" module with similar, but simpler functionality than the advanced one but with guaranteed behavior. When the advanced module violates the specified requirements, control switches to the baseline module. In the original Simplex architecture, once the baseline module took control, it kept control. To obtain the full benefits of the advanced module, it is advantageous to return control to the advanced module once the violation that caused the switch to the baseline module has been addressed. This control logic needs to be designed carefully to ensure that the system does not thrash between the advanced and baseline modules.

Recent work on designing such control logic is described in [23, 63, 85]. A Simplex architecture for runtime verification of machine learning modules in autonomous systems is described in [78]; this approach includes mechanisms to return control to the ML module after a violation is handled. Finally, [20] describes an architecture for runtime verification of ML modules which includes multiple runtime monitors, a Monitor Selector that determines which monitor is used during each execution step, and a Contingency Manager that determines if a violation of the selected monitor has occurred, and if so, what action to perform to address the violation. The Monitor Selector and Contingency Manager were developed using formal specification, verification, and synthesis. This architecture was demonstrated on an avionics system that uses a DNN to estimate the distance of a taxiing aircraft from the centerline of the runway based on camera images taken onboard the aircraft. Runtime verification was used to detect and intervene when the ML module was no longer performing reliably or when system safety was threatened.

3.5 Assuring Autonomy with MBSE, Formal Methods, and Agile

While the number of autonomous systems that currently exist or are under development is steadily growing, there is still no agreed upon process for building these systems [40]. Because most autonomous systems are complex and safety-critical and because they perform important functions, the selected process must provide assurance that the autonomous systems produced behave safely and correctly. In our view and the view of others [32, 40], assuring autonomous systems is one of the toughest problems but a problem where formal methods could have huge benefits. To realize these benefits, however, the disadvantages of formal methods—that they are difficult to learn and apply, hard to integrate into existing development processes, and too costly—need to be overcome.

This section introduces a systematic, mathematically well-founded process for autonomous system development and operation which combines MBSE, formal methods, and the best agile and DevOps practices. Our recommended process was inspired by the software development process introduced by Ghezzi [37], who in 2018 proposed a "marriage" between formal methods and agile software development. Ghezzi's recommended process embeds formal methods within an overall agile and DevOps framework.

3.5.1 Overview of Our Recommended Process

Both the agile development process and DevOps, which have become increasingly popular in recent years, have introduced many effective practices into system and software development. These practices include incremental development, implement-test cycles, and early customer feedback. However, as noted by Meyer [72], agile processes have clear disadvantages, namely, their almost total elimination of requirements documents, specifications, and models, all critical in assuring safety-critical systems. Even their proponents acknowledge the limitations of agile and DevOps for building safety-critical systems. In May 2021, for example, N. Chaillan, who leads the U.S. Air Force's DevOps efforts, stated that software supporting nuclear vehicles should be developed separately from software built using DevOps, i.e., that DevOps is more suitable for less safety-critical systems [17].

Due to agile's serious deficiencies for developing safety-critical systems, our recommended process, unlike Ghezzi's, does not embed formal methods into an agile framework. Instead, we recommend embedding formal methods in an MBSE framework. The advantages of model-based methods—significant productivity increases, quality improvements, and cost savings—are well known, especially for systems performing similar functions [16]. Moreover, two recent surveys report that practitioners are increasingly using MSBE. A 2018 study found, for example, that 89 percent of embedded systems currently under development use model-based methods [1]. Further, a 2021 survey reports that the adoption of MBSE is growing in both industry and government, with more traditional document-based development processes being increasingly replaced by model-based processes [53]. Although practitioners use a wide range of software modeling approaches, from informal sketches to formal models, for the safety-critical components of autonomous systems, we recommend formal models give the added assurance one obtains by formally verifying and validating these models. In our view, the complexity and safety issues characteristics of most autonomous systems require this. In addition to an MBSE process, which includes formal methods, we recommend the adoption of the most effective agile practices. Using formal methods helps assure that the delivered autonomous system satisfies critical system properties such as safety; adopting agile practices helps shorten development time.

Our recommended development process differs in other important respects from Ghezzi's. First, unlike Ghezzi's process, our process includes validation of the formal model. Moreover, because powerful algorithms are critically important in autonomous systems, e.g., for navigation and collision avoidance, we anticipate the use of formal methods in proving the correctness of critical algorithms. Finally, in our proposed process, runtime verification plays an important role, not only for testing the system implementation but in addition for runtime monitoring and enforcement during system operation.

3.5.2 Phases of Our Recommended Process

Figure 3.1 shows the phases of our recommended process. The process begins by (1) iteratively constructing small models of the required safety-critical system behavior, formally verifying and obtaining customer validation of the formal models, and

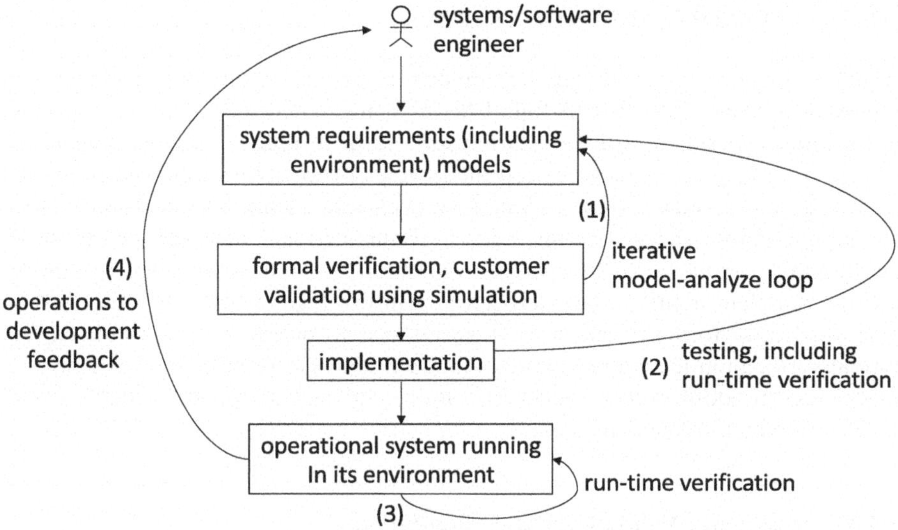

Fig. 3.1 Using MBSE, formal methods, and agile practices to build assured autonomous systems

using other methods to construct and assure the informal models. During this iterative process, smaller models are composed to form larger models, and, to the extent feasible, these larger models are formally verified and validated. Once high confidence is established that the composite model is correct, the system code is generated; some code, such as the control logic, may be automatically synthesized from a formal model. This system code is evaluated using standard testing and runtime verification as shown in (2), and any defects detected are used to correct the models or to correct the system code. Once testing is complete, the system implementation is deployed in the target environment. During system operation, we increase assurance by using runtime verification, as shown in (3), to handle environmental behavior unknown at development time. Similar to Ghezzi, in our approach, the operational system monitors the behavior of the environment and collects data that can be used to update the original environmental assumptions. These updated assumptions are then used to update the current system implementation (4).

3.5.3 Aspects of Our Process Inspired by Agile Practices

This section describes three aspects of our process inspired by agile practices, namely, repeated model-analyze cycles, customer validation using simulation, and incremental development.

3.5.3.1 Repeated Model-Analyze Cycles

A major aspect of agile software development is repeated cycles of implementation followed by testing. As Rear Admiral James Kilby stated recently, "Build a little, test a little, experiment a little—learn a lot" [26]. In place of agile's *implement-test cycles*, our process includes repeated cycles of modeling followed by model analysis, i.e., a sequence of *model-analysis cycles*. In each model-analysis cycle, a formal model of some system operation is formulated and analyzed, i.e., formally verified and validated. To paraphrase Admiral Kilby, our approach is to "model a little, simulate a little, formally verify a little—learn a lot." As described in Sect. 3.2.1.2, in formal verification, a set of system properties is formulated and formal verification techniques applied to check that the formal model satisfies the properties. In addition to formal verification, our analysis phase includes *model validation* which we describe next.

3.5.3.2 Customer Validation Using Simulation

Two important agile practices are early, continuous delivery of working software and frequent interaction between developers and customers. The role of customers is to provide feedback to developers about the latest implementation. One important question to be answered by customers is does the software behavior capture the intended system behavior? If the formal model is executable, rather than ad hoc code, we can have customers validate simulated system behavior, automatically generated from the system model. As shown in [50], when the formal model is executable, a simulation can be trivially constructed from the formal model. To facilitate customer interaction with the simulator, we can design a user-friendly simulator interface. Such an interface enables the customer to evaluate the running simulator code using an interface that looks like the interface of the system under development. For an example of a customized simulator front end, we used to validate a formal model of an aircraft's flight program, see [49].

3.5.3.3 Incremental Development

The advantage of incremental development in agile and other software practices is that one need not wait until the development process ends for a working system. Thus, incremental development enables early feedback to developers about the required system behavior. It is well known that the earlier the developers learn what needs to change, the cheaper it is to implement the change. As in agile, our process relies on incremental development, but instead of producing code, our process produces small models. To initiate development, we begin by formally modeling a single safety-critical operation of an autonomous system, for example, in an unmanned automotive vehicle, cruise control. Once the formal model of that operation has been formally verified and validated, a new formal model can be formulated of a second operation, for example, automatic parking, and that model formally verified. A major benefit of incremental development is that formal models of different operations can be composed and the composite model checked for property violations. In most cases, the interactions between operations, often called *feature interactions* or, alternatively, *emergent behavior*, will be benign, but, in general, these interactions can lead to violations of critical system properties. This approach can be highly effective in uncovering both design and implementation defects: for example, a 2009 analysis of the interaction between different operations in a semiautonomous automotive system uncovered eleven critical defects in the system design [12].

3.5.4 Three Additional Aspects of Our Process

This section describes three additional aspects of our process, namely, runtime verification, a system (or software) requirements document, and product lines.

3.5.4.1 Runtime Verification

The formal verification techniques used in the model-analyze cycles described above are applied to a formal model. As described in Sect. 3.4.2, *runtime verification* techniques are available which may be applied to running code. In our process, runtime verification may be applied during system testing, as shown in (2) in Fig. 3.1. Alternatively, runtime verification may be applied during system operation, as shown in (3) in Fig. 3.1. During system operation, the system may perform either runtime monitoring or runtime enforcement.

3.5.4.2 System Requirements Document

The formal model developed provides a solid foundation for a system (or software) requirements document. Such a document is a reference document that provides a precise, unambiguous statement, rather than an informal, narrative description, of the required behavior. Such a requirements document will contain all of the information in the formal requirements model, but, unlike the requirements model, it will be organized so that developers can easily find precise and unambiguous descriptions of the required behavior. The requirements document is a working document that must be kept up to date as understanding of the required behavior improves. It should eventually be approved by the system's customers. For an early example of such a document, see [3, 54]. Components of this requirements document include specifications of the monitored and controlled variables, user-defined data types, the required system operations specified as functions, a list of system properties, and environmental assumptions. If a natural language list of requirements exists, to the extent feasible, each natural language requirement should be linked to the section of the requirements document that specifies that requirement. The existence of such a requirements document addresses a major problem with agile practices—although the latest implementation reflects the developers' most recent understanding of the required system behavior, those aspects of the system behavior that are required are not separated from those aspects needed to produce working software. For example, although a nice user interface is useful for demonstrating and evaluating the required system behavior, i.e., the most recent version of the running software, in agile practices, one cannot determine whether the particular user interface used is required or if other interfaces are possible. In most cases, our guess is that the details of the interface design are preliminary and subject to change.

3.5.4.3 Product Lines

The development process in Fig. 3.1 should prove especially effective in developing *software product lines* [19]. In a software product line, the notion of reuse is used to increase the quality and to decrease the cost and time to market of related software systems. To achieve this, the basic approach is to exploit the similarities between systems. In a software product line, individual software systems are developed in two stages. In the first stage, commonalities and differences among product line members are identified, and a set of shared software artifacts is constructed. During the second phase, individual systems are built using a subset of the shared software artifacts. The assumption is that the investments required to develop the reusable software artifacts are outweighed by the benefits obtained in deriving the individual systems. Using formal methods to develop product lines has the advantage that, from the second product onward, the benefits of having a validated, formally verified initial system model significantly outweigh the efforts required to modify and extend the formal model for each additional product line member [60].

3.5.5 Barriers in Applying Formal Methods and How to Overcome Them

A wide range of formal methods have already been applied successfully to practical systems; researchers leading these efforts include Gerard Berry of Esterel; Jean-Raymond Abrial, the creator of the B-Method; Nicolas Halbwachs who invented the Lustre formal method and tools; and our own group, just to name a few. However, for the development process described in Sects. 3.5.1–3.5.2 to assure the safety and correctness of autonomous systems, several existing barriers to applying formal methods in industry must be overcome. The process described above overcomes one barrier, i.e., it describes how to integrate formal methods into software development in industry. However, progress must be made in several other areas for the above process to become feasible. First, the quality, usability, and robustness of the available tools must be improved. There is widespread agreement that currently the tools supporting formal methods have limitations. In addition, means are needed for professional maintenance of these tools. Second, although practitioners of MBSE are already formulating models, work is needed to put these models on a more solid mathematical foundation. Third, better education and training of software developers is urgently needed—currently, most software developers lack proper training in formal methods. Addressing the preceding problems should help overcome another major barrier to adoption of formal methods in industry, i.e., the current high cost of applying formal methods.

3.6 Summary and Conclusions

This chapter identified four major challenges in assuring autonomous systems—how to acquire and represent the system requirements, how to handle environments that change during runtime, how to assure machine learning components, and what process to use in developing and operating autonomous systems. Clearly, other major issues must also be addressed in assuring autonomous systems. Two additional challenges are how to design effective human-autonomy interaction and how to guarantee that autonomous systems behave securely. Recently, formal methods have been used by many researchers to check the human-computer interaction (HCI) in systems, such as safety-critical medical devices, for defects. The consensus is that formal methods can significantly reduce the defects in HCI designs (see, e.g., [42]). Regarding security, formal methods have been used for decades to verify that a critical system behaves securely, e.g., that it prevents unauthorized disclosure and modification of sensitive information (see, e.g., [51]), interference by external intruders, etc. Verifying the safety and security of autonomous systems, including their interactions with humans, is extremely difficult and will require new research.

Fortunately, much progress has been made recently both by software practitioners and by formal methods, software engineering, and machine learning researchers that addresses some of these challenges. Software practitioners have introduced agile methods and DevOps to overcome well-known deficiencies in the traditional waterfall model and recommended useful practices—incremental development, early and continuous delivery of working software, short development cycles, and frequent interaction between customers and developers to accomplish this. Practitioners have also introduced model-based methods and the abstraction inherent in models to manage the complexity of large, safety-critical systems. Concurrently, researchers have made significant advances in formal verification, such as SMT solvers and runtime verification. Moreover, machine learning researchers have invented powerful new techniques that can improve the performance of future autonomous systems.

To address three of the challenges, we identified state-of-the-art techniques for assuring that autonomous systems behave safely. We addressed the fourth challenge by proposing a development process which combines MBSE, including formal methods, with effective practices of standard agile processes and DevOps. The details of the process described in Sect. 3.5 still need to be worked out as do the details of how to apply the suggested process to produce a product line. Further, more work is needed to make existing formal methods and their support tools usable by practitioners. Although some progress has been made in designing usable formal methods tools, see, e.g., [50], significant challenges remain in improving the usability of tools such as model checkers, interactive theorem provers, and SMT solvers.

However, we believe strongly that the development process introduced in Sect. 3.5 is feasible and highly promising. In this book chapter, we reviewed both past research and some of the current research moving in this direction, but more remains to be done. In his 2018 article, Ghezzi called for a collective effort that includes both researchers and practitioners in developing and assuring critical systems. Similarly, we call for a collective effort among researchers and practitioners to extend and apply MBSE, formal methods, and the best agile and DevOps practices to develop, operate, and assure future autonomous systems.

Acknowledgments We are grateful to our colleagues, Signe Redfield, Mark Cavolowsky, Laura Hiatt, and Mak Roberts, for useful discussions on many of these topics. This work was funded by the Office of Naval Research.

References

1. Akdur, D., Garousi, V.: A survey on modeling and model-driven engineering practices in the embedded software industry. J. Syst. Archit. **91**, 62–82 (2018)
2. Alshiekh, M., Bloem, R., Ehlers, R., Kónighofer, B., Niekum, S., Topcu, U.: Safe reinforcement learning via shielding. In: Proc., 32nd AAAI Conf. on Artificial Intelligence, vol. 32 (2018)
3. Alspaugh, T., Faulk, S., Britton, K., Parker, R., Parnas, D., Shore, J.: Software requirements for the A7-E aircraft. Tech. Rep. NRL-9194, Naval Research Lab., Wash., DC (1992)
4. Archer, M.: TAME: Using PVS strategies for special-purpose theorem proving. Annals of Mathematics and Artificial Intelligence **29**(1–4), 139–181 (2000)
5. Avizienis, A., Laprie, J., Randell, B., Landwehr, C: Basic Concepts and Taxonomy of Dependable and Secure Computing. IEEE Trans. Depend. Secur. Comput. **1**(1), 11–33 (2004)
6. Ashmore, R., Calinescu, R., Paterson, C.: Assuring the machine learning lifecycle: Desiderata, methods, and challenges. CoRR **abs/1905.04223** (2019)
7. Bass, L., Weber, I., Zhu, L.: DevOps: A Software Architect's Perspective. SEI Series in Software Engineering. Addison-Wesley, New York (2015)
8. Bastani, O.: Safe reinforcement learning via online shielding. CoRR **abs/1905.10691** (2019)
9. Bastani, O., Pu, Y., Solar-Lezama, A.: Verifiable reinforcement learning via policy extraction. In: Proceedings of the 32nd International Conference on Neural Information Processing Systems, NIPS'18. Curran Associates Inc., Red Hook, NY, USA (2018)
10. Beck, K.: Extreme Programming Explained. Addison-Wesley, Boston (2000)
11. Bieber, C.: Panelist comments given at Should you rely on that AI? On-line workshop to explore the multi-domain challenge of AI operational testing. Univ. of MD Applied Research Laboratory for Intelligence and Security, January 28, 2021 (2021)

12. Black, J.A.: System safety as an emergent property in composing systems. Ph.D. thesis, Carnegie Mellon Univ. (2009)
13. Boehm, B.W.: A spiral model of software development and enhancement. Computer **21**(5), 61–72 (1988)
14. Boulet, M.T.: The autonomous systems tidal wave. Lincoln-Lab Journal **22**(2) (2017)
15. Brooks Jr, F.: No silver bullet essence and accidents of software engineering. IEEE Computer **20**, 10–19 (1987)
16. Broy, M., et al.: What is the benefit of a model-based design of embedded software systems in the car industry? Emerging technologies for the evolution and maintenance of software models In: Rech, J., Bunse, C. (eds.) IGI Global, pp. 343–369 (2012)
17. Chaillan, N.: Navy large unmanned surface and undersea vehicles: Background and issues for Congress. Keynote, High Confidence Software and Systems Conf., May 6–8, 2021 (2021)
18. Clarke Jr., E.M., Grumberg, O., Kroening, D., Peled, D., Veith, H.: Model Checking, Second Edition. MIT Press (2001)
19. Clements, P., Northrop, L.: Software Product Lines: Practices and Patterns. Addison-Wesley, Boston (2001)
20. Cofer, D., Amundson, I., Sattigeri, R., Passi, A., Boggs, C., Smith, E., Gilham, L., Byun, T., Rayadurgam, S.: Run-time assurance for learning-enabled systems. In: R. Lee, S. Jha, A. Mavridou, D. Giannakopoulou (eds.) NASA Formal Methods, pp. 361–368. Springer International Publishing, Cham (2020)
21. Computing Community Consortium: Assured autonomy: Path toward living with autonomous systems we can trust. Report by the U.S. Networking and Information Technology Research and Development Program (2020)
22. Defense Science Board: The role of autonomy in DoD systems. Report of the Defense Science Board Task Force. Office of the Under Secretary of Defense for Acquisition, Technology and Logistics, Wash., DC (2012)
23. Desai, A., Ghosh, S., Seshia, S.A., Shankar, N., Tiwari, A.: SOTER: A runtime assurance framework for programming safe robotics systems. In: IEEE/IFIP International Conference on Dependable Systems and Networks (DSN) (2019)
24. Dreossi, T., Fremont, D.J., Ghosh, S., Kim, E., Ravanbakhsh, H., Vazquez-Chanlatte, M., Seshia, S.A.: VerifAI: A toolkit for the formal design and analysis of artificial intelligence-based systems. In: I. Dillig, S. Tasiran (eds.) Computer Aided Verification, pp. 432–442. Springer International Publishing, Cham (2019)
25. Dreossi, T., Ghosh, S., Yue, X., Keutzer, K., Sangiovanni-Vincentelli, A., Seshia, S.A.: Counterexample-guided data augmentation. In: Proceedings of the Twenty-Seventh International Joint Conference on Artificial Intelligence, IJCAI-18, pp. 2071–2078. International Joint Conferences on Artificial Intelligence Organization (2018)
26. Eckstein, M.: Lawmakers question who will shepherd navy's future unmanned fleet. USNI News, March 18, 2021 (2021)
27. Ehlers, R.: Formal verification of piece-wise linear feed-forward neural networks. In: Proceedings Fifteenth International Symposium on Automated Technology for Verification and Analysis (ATVA) (2017)
28. Elboher, Y.Y., Gottschlich, J., Katz, G.: An abstraction-based framework for neural network verification. In: S.K. Lahiri, C. Wang (eds.) Computer Aided Verification, pp. 43–65. Springer International Publishing (2020)
29. Erwin, S.: DARPA picks Northrop Grumman as its commercial partner for satellite servicing program. Space News, March 4, 2020 (2020)
30. Estefan, J.A.: Survey of model-based systems engineering (MBSE) methodologies. INCOSE MBSE Focus Group 25 (2007)
31. Falcone, Y., Havelund, K., Reger, G.: A tutorial on runtime verification. Engineering Dependable Software Systems **34**, 141–175 (2013)
32. Farrell, M., Luckcuck, M., Fisher, M: Robotics and integrated formal methods: Necessity meets opportunity. In: Proc., 14th Internat. Conf. on Integrated Formal Methods (iFM), Lecture Notes in Computer Science (LNCS), **11023**, 161–171 (2018).

33. Forsberg, K., Mooz, H.: Application of the 'Vee' to incremental and evolutionary development. INCOSE International Symposium **5**(1), 848–855 (1995)
34. Fulton, N., Platzer, A.: Safe reinforcement learning via formal methods: Toward safe control through proof and learning. In: Third Conference on Artificial Intelligence and Theorem Proving (AITP '18) (2018)
35. Gargantini, A., Heitmeyer C.: Using model checking to generate tests from requirements specifications. In: ACM SIGSOFT Software Engineering Notes **24**(6), 146–162 (1999)
36. Gehr, T., Mirman, M., Drachsler-Cohen, D., Tsankov, P., Chaudhuri, S., Vechev, M.: Ai 2: Safety and robustness certification of neural networks with abstract interpretation. In: Security and Privacy (SP), 2018 IEEE Symposium on (2018)
37. Ghezzi, C.: Formal methods and agile development: Towards a happy marriage. In: The Essence of Software Engineering, pp. 25–36. Springer International Publishing, Cham (2018)
38. Gilday, M.: DoD enterprise DevSecOps initiative and Platform One. Congressional Research Service report 45757 (2021)
39. Gillula, J.H., Tomlin, C.J.: Guaranteed safe online learning via reachability: tracking a ground target using a quadrotor. In: 2012 IEEE International Conference on Robotics and Automation, pp. 2723–2730 (2012)
40. Gleirscher, M., Foster, S., Woodcock, J.: New opportunities for integrated formal methods. ACM Comput. Surveys **52**(6) (2019)
41. Halbwachs, N.: A synchronous language at work: the story of Lustre. Third ACM and IEEE Internat. Conf. on Formal Methods and Models for Co-Design, 3–11 (2005)
42. Harrison, M. D., L. Freitas, M. Drinnan, J. C. Campos, P. Masci, C. Maria, M. Whitaker: Formal techniques in the safety analysis of software components of a new dialysis machine. Science of Computer Programming **175** 17–34 (2019)
43. Havelund, K., Rosu, G.: Monitoring Java programs with Java PathExplorer. Tech. Rep. 01.19, RIACS (2001)
44. Heitmeyer, C., Labaw, B, Kiskis, D.: Consistency checking of SCR-style requirements specifications. In: Internat. Symp. on Requirements Eng. (1995)
45. Heimdahl, M., Whalen, M., Thompson, J: Nimbus: A Tool for Specification Centered Development. In: Proc., 11th IEEE Internat. Requirements Eng. Conf. (2003)
46. Heitmeyer, C., Bull, A, Gasarch, C., Labaw, B.: SCR: A toolset for specifying and analyzing requirements. In: Proc., 10th Annual Conf. on Computer Assurance (COMPASS) (1995)
47. Heitmeyer, C.L., Jeffords, R.D., Labaw, B.G.: Automated consistency checking of requirements specifications. ACM Trans. Software Eng. and Methodology (1996)
48. Heitmeyer, C., Kirby, J., Labaw, B., Archer, M., Bharadwaj, R.: Using abstraction and model checking to detect safety violations in requirements specifications. IEEE Trans. on Softw. Eng. **24**(11), 927–948 (1998)
49. Heitmeyer C., Kirby J., Labaw B., Bharadwaj R.: SCR: A toolset for specifying and analyzing software requirements. In: Hu A.J., Vardi M.Y. (eds.) Computer Aided Verification, *Lecture Notes in Computer Science*, vol. 1427, 57–74. Springer (1998)
50. Heitmeyer, C., Archer, M., Bharadwaj, R., Jeffords, R.: Tools for constructing requirements specifications: The SCR toolset at the age of ten. International Journal on Computer System Science and Engineering **20**(1), 19–35 (2005)
51. Heitmeyer, C., Archer, M., Leonard, E., McLean, J.: Applying formal methods to a certifiably secure software system. IEEE Trans. on Softw. Eng. **34**(1) (2008)
52. Heitmeyer, C., Pickett, M., Leonard, E., Archer, M., et al.: Building high assurance human-centric decision systems. Autom. Softw. Eng. **22**(2) 159–197 (2015)
53. Henderson, K., Salado, A.: Value and benefits of model-based systems engineering (MBSE): Evidence from the literature. Syst. Eng. **24**(1), 51–66 (2021)
54. Heninger, K.L.: Specifying software requirements for complex systems: New techniques and their application. IEEE Trans. Softw. Eng. **SE-6**(1), 2–13 (1980)
55. Holley, P.: Amazon's autonomous robots have started delivering packages in a new location: Southern California. Washington Post, August 12, 2019 (2019)

56. Huang, X., Kwiatkowska, M., Wang, S., Wu, M.: Safety verification of deep neural networks. In: R. Majumdar, V. Kunčak (eds.) Computer Aided Verification, pp. 3–29. Springer International Publishing, Cham (2017)
57. Ivanov, R., Weimer, J., Alur, R., Pappas, G.J., Lee, I.: Verisig: verifying safety properties of hybrid systems with neural network controllers. CoRR **abs/1811.01828** (2018)
58. Ivanov, R., Weimer, J., Alur, R., Pappas, G.J., Lee, I.: Verisig: Verifying safety properties of hybrid systems with neural network controllers. In: Proceedings of the 22nd ACM International Conference on Hybrid Systems: Computation and Control, HSCC '19, p. 169–178. Association for Computing Machinery, New York, NY, USA (2019)
59. Ivanova, K., Gallasch, G.E., Jordans, J.: Automated and autonomous systems for combat service support: Scoping study and technology prioritisation. Report DST-Group-TN-1573, October 2016 (2016)
60. Jackson, D., Millett, L., Thomas, M.: Software for Dependable Systems: Sufficient Evidence? National Academies Press (2007)
61. Jacoby, Y., Barrett, C.W., Katz, G.: Verifying recurrent neural networks using invariant inference. In: D.V. Hung, O. Sokolsky (eds.) Automated Technology for Verification and Analysis—18th International Symposium, ATVA 2020, Hanoi, Vietnam, October 19–23, 2020, Proc., *Lecture Notes in Computer Science*, vol. 12302, pp. 57–74. Springer (2020)
62. Jeffords, J., Heitmeyer, C. Proc., ACM SIGSOFT Internat. Symp. on Foundations of Software Eng. 56–69 (1998)
63. Johnson, T.T., Bak, S., Caccamo, M., Sha, L.: Real-time reachability for verified simplex design. ACM Trans. Embed. Comput. Syst. **15**(2) (2016)
64. Katz, G., Huang, D.A., Ibeling, D., Julian, K., Lazarus, C., Lim, R., Shah, P., Thakoor, S., Wu, H., Zeljić, A., Dill, D.L., Kochenderfer, M.J., Barrett, C.: The Marabou framework for verification and analysis of deep neural networks. In: S. Dillig Isiland Tasiran (ed.) Computer Aided Verification, pp. 443–452. Springer International Publishing (2019)
65. Kramer, J.: Is abstraction the key to computing? Commun. ACM **50**(4), 36–42 (2007)
66. Krishner, T.: 3 crashs, 3 deaths raise questions about Tesla's Autopilot. AP, (2020)
67. Kuper, L., Katz, G., Gottschlich, J., Julian, K.D., Barrett, C., Kochenderfer, M.J.: Toward scalable verification for safety-critical deep networks. **https://arxiv.org/abs/1801.05950** (2018)
68. Landwehr, C., Ludewig, J., Meersman, R., Parnas, D.L., Shoval, P., Wand, Y., Weiss, D., Weyuker, E.: Software systems engineering programmes a capability approach. Journal of Systems and Software **125**, 354–364 (2017)
69. Leonard, E.I., Heitmeyer, C.L.: Program synthesis from formal requirements specifications using APTS. Higher-Order and Symbolic Computation 16, 2–13 (2003)
70. Martin, B., Tarraf, D.C., Whitmore, T.C., DeWeese, J., Kenney, C., Schmid, J., DeLuca, P.: Advancing autonomous systems: An analysis of current and future technology for unmanned maritime vehicles. Tech. Rep. RR2751, RAND (2019)
71. McDermid, J.: Autonomous systems. In: Human-Machine Trust: Risk-Based Assurance and Licensing of Autonomous Systems, SCI-313 Specialist Meeting Report, NATO STO-MP-SCI-313, 3–5 December 2018 (2018)
72. Meyer, B.: Agile!: The Good, the Hype, and the Ugly. Springer (2014)
73. Miller, S.P., Tribble, A.C., Whalen, M.W., Heimdahl, M.P.E.: Providing the shalls: Early validation of requirements through formal methods. International Journal on Software Tools for Technology Transfer **8**(4/5), 303–319 (2006)
74. de Moura, L., Bjørner, N.: Z3: An efficient SMT solver. In: C. Ramakrishnan, J. Rehof (eds.) Tools and Algorithms for the Construction and Analysis of Systems, 337–340. Springer (2008)
75. Parnas, D.L., Clements, P.C.: A rational design process: How and why to fake it. IEEE Trans. Softw. Eng. **SE-12**(2), 251–257 (1986)
76. Parnas, D.L., Madey, J.: Functional documentation for computer systems. Science of Computer Programming **25**(1), 41–61 (1995)
77. Pereira, A., Thomas, C.: Challenges of machine learning applied to safety-critical cyber-physical systems. Machine Learning and Knowledge Extraction **2** (2020)

78. Phan, D.T., Grosu, R., Jansen, N., Paoletti, N., Smolka, S.A., Stoller, S.D.: Neural simplex architecture. In: R. Lee, S. Jha, A. Mavridou, D. Giannakopoulou (eds.) NASA Formal Methods, pp. 97–114. Springer International Publishing, Cham (2020)
79. Pinelis, J.: Should you rely on that AI? On-line workshop to explore the multi-domain challenge of AI operational testing. U of MD ARL for Intellig. and Security, Jan 28, 2021 (2021)
80. Rozier, K.Y.: Specification: The biggest bottleneck in formal methods and autonomy. Aerospace Engineering Conference Papers, Presentations and Posters. 8th International Conference, VSSTE 2016, Toronto, ON, Canada, July 17–18,2016. (2016)
81. Sha, L.: Using simplicity to control complexity. IEEE Softw. **18**(4), 20–28 (2001)
82. Stanley, B.M., Froehlich, E.M., Koechling, J.C., Cornick, M.T.: Automation of armored four-wheel counter-steel vehicles. Lincoln-Lab Journal **22**(2) (2017)
83. Tjeng, V., Xiao, K.Y., Tedrake, R.: Evaluating robustness of neural networks with mixed integer programming. In: 7th International Conference on Learning Representations, ICLR 2019, New Orleans, LA, USA, May 6–9, 2019. OpenReview.net (2019)
84. Tran, H.D., Yang, X., Manzanas Lopez, D., Musau, P., Nguyen, L.V., Xiang, W., Bak, S., Johnson, T.T.: NNV: The Neural Network Verification tool for deep neural networks and learning-enabled cyber-physical systems. In: S.K. Lahiri, C. Wang (eds.) Computer Aided Verification, pp. 3–17. Springer International Publishing, Cham (2020)
85. Vivekanandan, P., Garcia, G., Yun, H., Keshmiri, S.: A simplex architecture for intelligent and safe unmanned aerial vehicles. In: 2016 IEEE 22nd International Conference on Embedded and Real-Time Computing Systems and Applications (RTCSA), pp. 69–75 (2016)
86. Wang, S., Pei, K., Whitehouse, J., Yang, J., Jana, S.: Formal security analysis of neural networks using symbolic intervals. In: Proc., 27th USENIX Conf. on Security Symp., SEC'18, 1599–1614. USENIX Association, USA (2018)
87. Woodcock, J., Larsen, P.G., Bicarregui, J., Fitzgerald, J.: Formal methods: Practice and experience. ACM Comput. Surv. **41**(4) (2009)
88. Zave, P., Jackson, M.: Four dark corners of requirements engineering. ACM Transactions on Software Engineering and Methodology **6**, 1–30 (1997)
89. Zhu, H., Magill, S.: Applying formal methods to reinforcement learning. Tech. rep., Galois, Inc. (2018)

Chapter 4
Specification of Autonomous Behavior

Signe Redfield and Jamie Lennon

Chapter 4 Editorial Note

This chapter focuses on the development of specifications for functional autonomous behavior, emphasizing aspects related to how useful a behavior is and understanding of what to verify and how to verify it. Within that context, it answers stakeholder questions related to defining the product's usefulness, what should be verified, and how it should be verified.

Technical challenge 7 (the dividing line between specification and design) is addressed through the presentation of two tools that together enable users to define the level at which the transition from specifying requirements to designing the system occurs and support the development of principled requirements that do not force a specific system design. One tool discussed in this chapter, the Capability Analysis Tables, addresses both challenge 22 (verifying the mapping between a specific system and behaviors, the task it is attempting to accomplish, and the environment within which it is expected to operate) and challenge 24 (development of a structured process connecting the physical system and formal verification tools). This tool enables representation of the physical/ground truth layer via the sensor input and control output elements and represents behavior integration via logical elements within the cells. These tables enable explicit specification of what is meant by "fitness" in the context of a particular task and enable connections to formal methods verification tools via conditions on their use. Because the tools presented in this chapter may be defined at arbitrary levels of abstraction, this chapter also partially addresses challenge 23 (tools that balance broad and specific representations), enabling analysis of systems across varied levels of abstraction. Stakeholder challenges are presented in the Chap. 1, Introduction. Technical challenges are introduced in Chap. 2, Verification Challenges.

S. Redfield (✉) · J. Lennon
Naval Research Laboratory, Washington, DC, USA
e-mail: signe@ieee.org; jamie.a.lennon2.civ@us.navy.mil

S. Redfield et al. (eds.), *Verification of Autonomous Systems*,
https://doi.org/10.1007/978-3-031-88546-4_4

Abstract Verification is impossible without a specification of a desired function. You must define what a task and its properties are before you can verify whether something is capable of performing that task while providing those properties. This chapter provides a Capability Representation for defining the precise parameters of the task, explicitly including (1) areas of uncertainty, where the autonomy is expected to decide on the correct course of action, (2) areas of abstraction, where the task specifier is aware of contextual uncertainty, and (3) implicit assumptions about desired system behavior and how the system should be evaluated. This capability representation is then used to support a documentation form, the Capability Analysis Table, that defines relationships between capabilities and enables translation from an arbitrary behavior design to simple logical assessment of basic structural properties of an autonomous behavior. Together, these elements enable specification of autonomous functionality.

4.1 Introduction

The majority of verification efforts have understandably been focused on assuring the safety and security of the systems under test, and they typically assume that by the time a system has reached this stage, it provides the correct function. In the autonomous systems domain, however, functionality cannot be taken as given. (For this document, we will use the term "autonomous system" as shorthand for "system utilizing one or more autonomous behaviors in the accomplishment of a task.")

By definition, autonomous systems are responsible for decision-making in situations when no explicit, lookup level response is provided. Autonomous behaviors are predicated on the assumption that actions will need to be selected at run-time on the basis of locally gathered sensor inputs that define whatever the current situation is. The number of potential states of the system is a function of the complexity of its sensor and communications input, the complexity of the environment in which it is operating, and the complexity of any internal representations of that environment it is capable of generating.

A system that cannot represent its environment and operates in a purely reactive way can still exhibit intractably complex behavior because the effect of its actions on both itself and the environment can be unpredictable (icy environments, for instance, have very unpredictable relationships between action and result).

A system that can represent an environment in great detail and perform complex computations about its actions has a very large internal state space representation of the world, is capable of very complex decision-making, and can likewise produce extremely complex behavior even in the context of a very predictable relationship between its actions and the results of those actions.

In both of these cases, specification (and subsequent verification) of system behavior is necessarily complex, as it does not reduce to a small, tractable number of potential system/environment states.

Our goal is to address the problem of specification development as a necessary precondition to autonomy verification. This is particularly important for tasks where the situation is sufficiently important that the autonomy cannot be allowed to introduce excess risk, such as operations in space.

4.2 Background

4.2.1 Why Specification?

Specification should not be a verification problem—but in practice, it is. Ideally, the engineering process would proceed through three phases: specification of requirements, design, and verification. Specification would be the first step, and verification would be performed against the specified requirements. However, developing good requirements is a well-known challenge. Formal methods may be employed to check sets of specifications for consistency or coverage, but they do not address another major difficulty: The specification must be specific enough to be verifiable, yet it should not be so detailed as to require a particular design. As we start to attempt to verify autonomous systems, we quickly find that these inadequate specifications become the most pressing problem we have.

It is extremely difficult to specify the *correct* behavior of an autonomous system. Even a simple-sounding behavior like "avoid obstacles" quickly becomes intractable when you get into the details of attempting to specify the desired behavior.

What counts as an obstacle? If your system also needs to deliberately contact some items, this is a nontrivial question. The metric used to define "obstacle" is critical. If the higher level goal is to make sure the robot does not damage itself, then an obstacle is "anything that could damage the robot," and the robot might drive right through a drywall panel but avoid raindrops. If the higher level goal is to make sure the robot does not damage its environment, then an obstacle is "anything that the robot could damage," which might mean that the robot avoids drywall panels but not raindrops. The specification must incorporate the aspects of these higher level goals that will affect the behavior design and implementation.

What additional constraints must be observed? If the system not only never hits anything but also never accomplishes its other goals because it does not support the ability to incorporate other externally provided goals as a method to select desired avoidance actions, is this behavior truly functionally successful? If the avoidance behavior is used in a subsumption architecture [1], then the behavior performs the best possible avoidance regardless of higher level goals, where "best" is defined in terms relevant only to avoidance and not with respect to any higher level goals. If the avoidance behavior is designed in the context of higher level goals and behaviors; on the other hand, it might incorporate a desired heading or a desired location in a specified coordinate frame. It might even incorporate simple handedness, where if it is given a choice, it will always turn in the same direction. Which of these is acceptable is driven by the specification. Without an in-depth understanding of how autonomous systems are designed and how their physical platform and operation in a particular environment affect them, requirements for autonomous system behavior often prove close to impossible to develop.

Because systems engineers often lack autonomous systems experience, these functional requirements are often rejected for being too complicated or untestable. The interactions between the individual behaviors and their expression in a real environment are ignored, and the resulting system is tested component by component, at the integrated system level, and at the operator/mission level rather than at the behavioral level through the entire hierarchy of behavioral abstraction layers.

Where functional autonomous behavior requirements are developed, there is a strong bias toward performing the behavior design as part of the requirement generation process. Ideally, the systems engineers developing the requirements would accept the uncertainty inherent in autonomous systems by defining particular regions of operation as the responsibility of the autonomous system and verifying the system performance at the level above this. In practice, however, in order to obtain testable specifications and linkages to lower level components (physical and software), engineers instead spend weeks, months, or years defining the precise sequence of actions they want the autonomous system to take as part of the specification development process.

The purpose of the tools presented here is to provide an alternative approach that generates a testable specification at the desired level of abstraction without requiring the people generating the specifications to design the autonomous behaviors.

4.2.2 Terminology and Definitions

The following terms are used to describe relevant aspects of the specification and verification problems presented here:

- *System*—the component, agent, or aggregation of components and/or agents that is the subject of the verification effort
- *Specification*—we are primarily focused on the business meaning of *product specification* defined as the required properties, components, and abilities of a system. This is distinct from the *formal specification* meaning used in the computer science community, which defines a specification in the context of mathematical languages supporting the description, analysis, and verification of key properties of software.
- *Capability* covers both *provided* capability, which is the latent ability to accomplish a task enabled by a robot's physical composition and complement of behaviors, and *desired* capability, which specifies what complement of behaviors will be adequate to accomplish the task. The Merriam-Webster dictionary defines capability as "the power or ability to do something." A provided capability

defines the *power or ability* that a robot provides; a desired capability defines the *something* that the power or ability is required to accomplish.
- *Verification*—the stage in the assurance process in which the designed system is evaluated against the provided specification
- *Autonomous*—mode of operation that relies on locally available sensor information rather than on commands from an external agent for run-time action selection decisions

4.2.2.1 What Is a Specification?

The purpose of a specification of a system is to ensure that the various parties concerned in the system's development and approval for use:

- Have defined the customer or larger system needs that the subject system or component is supposed to meet (which includes the necessary functions and features of the system from the perspective of the end user)
- Understand the role and function of the subject system or component in its operational context (includes the necessary functions and features of the system from the perspective of the designer)
- Can distinguish between necessary properties and optional or nice-to-have properties (affects cost and schedule/budget during development)
- Know the metrics that will be used to evaluate whether the system is acceptable prior to integration and user testing (supports assurance)

The specification is the documentation of the customer or larger systems' needs. Verification is the process by which the system or component is compared to the specification to determine whether it meets those documented needs. Validation is the process by which the completed system is tested to ensure it meets the customer's actual needs—validation bridges the gap between the customer's expressed needs as captured in the specification and the actual customer goals as expressed in a valid test environment.

In order to verify a system, we need tools and processes that support the generation of specifications. These tools and processes should enable the customer to express their needs in a way that supports easy comparison with the autonomy design. The tools should not require the customer to learn autonomy design. The tools and processes should ensure that a clear boundary between the specification and the design can be defined. The specifications they generate should not require the autonomy designer to have specialized knowledge in assurance or verification or to use specific design paradigms. Improvement of the mapping between the specification and the customer's actual goals is also desirable. The capability representations presented in this chapter and their connection to Capability Analysis Tables meet these goals.

4.2.3 Core Challenges

In the Systems Engineering Community, specifications are typically defined from the highest level system description to the lowest level component in the system. The role of the specification is to establish what is to be achieved, not to determine the design of a component or a system. Often, however, the development of component-level specifications can only be accomplished by defining a preliminary design for the system. While all systems experience problems due to the interconnection between physical structure and environmental properties, this is a particular problem for autonomous systems because we lack (1) a clear definition of the desired task performance, (2) a clear understanding of how environmental factors and physical system design affect system performance at that task when the system's components are integrated, even when the components are well understood, and (3) principled evaluation mechanisms for system performance.

If the system under test is a mechanism, we have theoretical principles that guide our understanding of how to evaluate it and what aspects of its performance we need to specify in order to ensure it will do the correct job adequately. If the system under test is a piece of software, its interface specification completely defines its environment, and while the correct job may be difficult or complex to define, its desired function may be specified in a testable way. If the system under test is an autonomous behavior, we move even further into uncertainty. The environment in which an autonomous system operates is by definition ill-constrained and nondeterministic; the system's performance measures are likely to be unclear and often fail to capture key aspects of its intended function; and the circumstances under which it is expected to work and the degree to which its performance degrades under environmental changes may or may not be well defined and is almost certainly incompletely known. We lack theoretical tools to predict its performance and principled design approaches. As a result it is not often obvious when high-level design decisions made by systems engineers based on system-level goals as part of the specifications development process will introduce internal inconsistencies that prevent the development of the required autonomous behavior.

Because of the uncertainty accompanying autonomous behavior design, the team developing the specifications is continually tempted to define the design of the autonomous behavior rather than just its goals. Systems engineers rarely have training in autonomous behavior design. That lack of training has outsized implications because even experts often lack intuition about how a given change in environmental assumptions, platform design, or task context will affect the autonomy design. Changing a given measurement by half an inch has predictable and intuitive implications for mechanical and structural aspects of the design. Changing that measurement by half an inch has unpredictable and counterintuitive implications for the performance of the autonomous behavior.

The design of autonomous behavior is currently as much an art as a science. With experience, designers begin to recognize how aspects of the task, the platform, and the environment play into changes in the behavior. Because the systems engineers lack understanding of the impact of these changes, these teams often do not realize when the specification of desired functionality has transitioned into specifying the design or when they are developing specifications based on an unworkable point design. Determining where the specification ends and the design begins is a more difficult challenge for the specification of autonomous behavior than for other aspects of the system partly because the desired functionality of the system is poorly understood and poorly expressed, partly because the design process is poorly defined and immature, and partly because the systems engineering team's lack of expertise and intuition drives their inability to recognize where the specification of functionality ends and the design process begins.

The *capability representation* (CR) defined in Sect. 4.3.4 addresses this challenge. But once the task, the desired capability, is specified, we still need to verify that the design meets that specification.

Autonomous behaviors are desirable because they adapt their actions to their local sensor data, to their local representation of the environment. It is this very flexibility that generates the core challenges with verification—the definition of autonomous behavior implies an inadequate specification of that behavior. If it were possible to completely specify the *correct* action for every situation in which the robot might find itself, we would not need autonomy. Instead we would develop intricate hierarchies of perception and provide a simple lookup table to connect the perceived state to the desired action.

In practice, however, it is impossible to specify every possible environmental configuration, from temperature to visibility to earthquakes to torrential downpours to bystanders and objects (both in terms of semantic certainty and in terms of location certainty). It is impossible to specify every possible system configuration, from battery level to tire tread to occlusions and wear on sensors. It is therefore likewise impossible to define a priori and specify a desired action for every relevant state in which the autonomous system might find itself.

As a result, there are significant differences between the software used to drive autonomous systems (the autonomous behaviors) and more traditional software. We can state *possible* inputs and outputs a priori. For example, the autonomous system must be able to move in these ways, or it must be able to sense these aspects of its environment. What we cannot do is state a priori what output should be selected in the face of what inputs. Specification development for autonomous behaviors is particularly challenging because the definition of the desired output action is delegated to the system and is therefore effectively unknown to the specification developer. Requirements for autonomous systems have to be goal-oriented in order to avoid designing the autonomy as part of the requirements development process. Defining a high-level specification of goals or constraints in the form of a rule or utility function would seem to be an intuitive approach to specifying autonomous behavior. But the behaviors themselves, the decision mechanisms that define the actions to be taken, are often expressed via rules and utility functions which typically produce fragile, rather than optimal, behavior.

For a traditional software application, like a word processor, the input interface consists of the underlying operating system state and the actions the user can take with a mouse within the context of the pixels on the screen or with a keyboard. The environmental state consists of known or knowable system properties, with a relatively small number of well-defined failure modes. The outputs of the software include updated state (stored temporarily or permanently, including both application state and document state) and updated visual information to the monitor. The most important aspect of the problem is that it is essential that the software be predictable—the result of a given input must always map to the same output. If someone types "k" into a document, the internal document representation and the monitor display must both register a "k," not an "m." Any discrepancy between the document and the display or between the input and the document/display is a critical problem.

For a more complex software application, the interfaces can be more difficult to manage. The software may have to balance the needs of multiple inputs from different interfaces simultaneously, and the goals of the software may include more complex actions. But again, the goal is for those actions to be predictable for a given combination of internal state and inputs.

Those predictable input/output relationships are used to make the verification process tractable. We *know* what the goal is, what the desired action is, and what the combinations of system state and output ought to be, to a degree that is not possible to attain with autonomous systems.

For an autonomous system, the effectiveness of any given behavior is fundamentally dependent on the environmental conditions and the physical system state, as well as the details of which action is chosen at which time within the behavior, while the performance of any given behavior is fundamentally dependent on the effectiveness of the behavior in the context of the current state of the task it is attempting to accomplish. The software itself is dependent on the hardware interface in the same way that traditional software is dependent on its hardware interface, but the actions we ask that software to take, the complexity of its decision-making, and the degree to which we know a priori what action it *should* take are significantly different.

As compared to normal software applications, in autonomous systems the fundamental link between context and desired action is complicated by the need to select different actions under circumstances that appear, on the surface, to be identical and, in some cases, to select different actions under circumstances that are actually identical. These systems are unpredictable *by design*. Identifying what information should be collected or generated and how it should be used to drive that action selection process is critical to the design of autonomous behaviors. Autonomous behavior cannot be effectively evaluated out of its operational context.

Because operational context (the effect of the system's actions on its environment) is essential to the evaluation of these systems, functional specifications for the system's desired behavior must be established.

The desired outcomes of autonomous behaviors are often described in the negative: The system *should not* hit things. But we fail to specify what the system *should* do, because what the system should do is fundamentally unknowable at design time. We know that the system should stop, or turn left, or turn right, or back up, depending on the specific combination of goals and actions and physical system state and environmental state occurring at the time, but until we know the precise situation, we cannot define which of those options the system should use.

This drives an inherent level of uncertainty and abstraction in the autonomous behavior specification. We need to be able to explicitly define and verify desired behavior without needing to dive into details that are only knowable at run-time.

4.2.4 Current Approaches to Specification

4.2.4.1 Formal Specification

In formal approaches to specification, the user input is used to develop a formal model of the desired system. This drives the necessary abstraction and forces verification at the level to which the model is developed. There are several types of models that can be developed.

Luckcuck et al. [2] have developed an excellent review of the state of the art in formal specification of autonomous systems. Set-based models like FORMS [3] (focusing on architectural characteristics of self-adaptive systems) and Event-B [4, 5] use a set-theoretic representation to capture system properties but typically lack mechanisms to capture behavior.

Logical approaches, especially those using temporal logic [6], are often focused on task specification and assessment of whether a given behavioral choice does or does not violate task rules.

These approaches are necessarily conceptual in nature and can model high-level aspects of the problem like "system must be able to stop," "system must be able to turn right," "system must be able to turn left," "system must be able to back up," and, for cases where there is no obstacle present, "system must be able to move forward." The autonomous decision-making can then be encapsulated into "system must be able to choose which motion to make based on the presence or absence of an obstacle" with more detailed specifications related to when to choose which motion option.

But this does not clarify the line between abstraction and behavior design; in fact, it blurs it further. Instead of giving us a simple metric for "here be specification" and "here be design," it says "specification captures both the goal and the design."

This mathematical and logical model forms a specification of the system, but it also defines a design—only designs compatible with this specification are acceptable, and the process of developing the model typically constrains the system to a specific design paradigm. The resulting logical specification only holds if the design can be proven to match it, which means that only designs that can be evaluated against the model are acceptable.

SMT solvers and formal approaches to system modeling can be used to evaluate properties of autonomous systems [7], providing formal specifications in the software sense to help identify problems with the design and generate missing requirements. The process of modeling the system often brings underlying inconsistencies and incongruities to light, enabling the generation of a consistent and complete requirement set. If the system meets this formal specification, then we have proof of coverage and correctness and consistency and lack of loops in the design. The properties that these approaches address are typically meta-requirements of the system. These include safety properties and specific requirements that define the ways in which the system is expected to address the safety of itself or its environment. If the SMT solvers or other tools identify no counterexamples, the system meets the requirements and is therefore safe.

But in many cases, the autonomous behavior may assume safety provided by lower level components (a path-planning behavior that assumes a lower level obstacle avoidance behavior, for example). In other cases, safety requirements are not explicitly stated but are built into the autonomous behavior through a design spiral that involves identifying the safety issues the system faces as it operates in the world and addressing those issues as they arise.

One significant challenge with these approaches is that it is impossible to determine whether the correct tests have been run. Some autonomous systems will have design flaws related to their interaction with the environment that cannot be caught by formalized or logical models, since the problem is in the underlying properties of the environment rather than in the assumptions about the system's responses to the environment. The best these models can do is flag the situations where the environment is not conforming to assumptions, and even that relies on sensor data that is capable of identifying those situations. By definition, that information is not available to the system (since if it was, the system would be using it to avoid those situations). This produces gaps or blind spots in the formal analysis process.

4.2.4.2 Behavior Specification

Behavior specification is typically accomplished via goal-based specification development. Requirements engineering (RE) has been used to define the goals of complex software systems and assumes that by defining the goals of a system, "eliciting, elaborating, structuring, specifying, analyzing, negotiating, documenting, and modifying requirements" can be accomplished [8].

In the dynamically adaptive systems community goal-based modeling has been used to perform requirements engineering [9]. This approach relies on four levels of specification development effort, each representing a different level of abstraction in the task decomposition process. Level 1 represents the goals, level 2 establishes how level 1 artifacts are to be composed, level 3 defines the adaptation mechanisms used by level 2, and level 4 defines the adaptation mechanisms available. It defines both "goals" (the functional goals, what the system needs to do) and "soft goals" (the nonfunctional goals, like performance and reliability). However, because the composition of the adaptation mechanisms (behaviors) is defined by the requirements, it is effectively designing the autonomous behavior rather than simply specifying the constraints within which it should be designed.

These goal-based approaches to autonomy specification can also be policy-based. In the Autonomy Requirements Engineering (ARE) approach [10], objectives are defined for the self through policies associated with goals, environmental and task context, evaluation methods, and behaviors using the KnowLang language [11]. These policies define what the system should do in particular situations, using probability distributions to connect individual policies with specific situations. These probability distributions partially address the state space issue arising from the interactions between autonomous systems and their environments, but the problem of specifying the autonomy design as part of the requirements generation process remains.

If one starts with the Remote Agent Experiment, often called the first autonomous spacecraft, verification and validation of autonomous systems in the space community initially focused on scenario-based testing and model-based validation [12]. More recent work in the context of Martian rovers [13] focused on verification of systems using a particular model-based approach to behavior design and selection and did not generalize to autonomous systems in general. High-level decision-making and low-level control are assumed to be separately specified, separately developed, and separately evaluated. Their functional verification approach for the low-level control elements includes static analysis of the code and model checking for assessment of event-based function, with field testing as an indispensable component of an effective strategy due to the inability to effectively test environmental assumptions in any other way. The high-level decision-making is assumed to generate plans that can be independently evaluated—not an assumption that our approach requires.

4.3 Capability-Based Specification and Design

The purpose of this approach is to enable the evaluation and assessment of autonomous behaviors by constructing a representation that allows a direct comparison between desired and provided capabilities. It allows designers to minimize the gap between the desired system and the provided system by expressing both sets of needs in the same language and by providing a simple documentation and specification mechanism that elicits those requirements in a consistent, clear form that can easily be exported for formal logical checking.

This approach to behavior specification is independent of the autonomy design approach guiding the designer and independent of the platform, environmental domain, or mission area in which the behavior is designed to be used.

4.3.1 Task/Behavior Duality Principle

The underlying assumption behind the work presented here is that both the tasks that need to be accomplished and the behaviors that accomplish them require the same information in order to be specified. The contextual goals that are part of a task definition must be known in order to apply a specific behavior to the accomplishment of that task. Not only the interfaces but also the details of decision-making within a behavior are essential to determine whether or not that behavior can be appropriately applied to that specific task.

We define the concept at the core of this duality as the *capability*.

Dictionaries define capability as "the power or ability to do something." A *provided* capability defines the "power or ability" that is provided; a *desired* capability defines the "something" that the power or ability is required to accomplish.

Behaviors and tasks are often framed as functions of defined inputs and outputs. But a capability is not just a function connecting the inputs of the behavior to the outputs of the behavior or even a function connecting an input scenario and physical context to desired or probable output scenarios. A capability must inherently be defined in terms of the user's expectations and assumptions.

In that context, when we think of a behavior as a provided capability, we understand that it is available to be run and can cause actions to be taken in the world that result in the accomplishment of a job *defined by a user*. Similarly, when we think of a task as a desired capability, it defines what power or ability is *required by the user* from the robot by defining what changes in the world need to happen. In both cases, the definition of *successful* is determined by context. Depending on the context in which the behavior is used, it may or may not successfully achieve either its internally defined goals or the task's goals.

The key to this duality principle is to recognize that the information required to specify a provided capability and the information required to specify a desired capability are the same. They need to be defined at the same level of abstraction, the same level of uncertainty, with the same constraints and the same metrics. We match tasks to behaviors by matching desired capabilities to provided capabilities. In this way, the capability can be thought of as the set of information that is common to both the specification of the autonomous behavior and the specification of the task.

This is a strict duality—every capability represents both a desired capability and a provided capability. The desired capability is the specific task represented by this capability. The provided capability is the specific behavior represented by this capability. Any change in either the task or the behavior defines a new, distinct capability, since the new task or behavior is no longer the dual of the previous behavior or task.

Once we recognize this relationship and develop this strict definition of "capability" as the instantiation of this duality, we can build on it to capture similarities and differences between different capabilities and the resulting implications for tasks and behaviors. Exploring these implications will be future work.

4.3.1.1 Duality Principle Example: *grasp cup*

The task associated with the *grasp cup* capability is defined as a job where the robot starts with its gripper not in contact with a cup and ends with the gripper completely constraining that cup from moving along or rotating around any axis in three-dimensional space.

The behavior associated with this capability is defined as a sequence of actions (or rules that generate actions) that move the gripper from a state where it is not in contact with a cup to a state where it is completely constraining the motion of that cup.

If we change the definition of this behavior in any way, for instance, to specify cups with handles rather than any cup, we have fundamentally changed the provided capability. Instead of providing the capability *grasp cup*, we are now providing the capability *grasp cup-with-handle*. This new provided capability is no longer the dual of the desired capability *grasp cup*. Making this change breaks the duality between the provided and desired capabilities and defines a new capability with a new dual task of *grasp cup-with-handle*.

We can now begin to define relationships between different capabilities. The goal of behavior design is to minimize the distance between what the robot does and what the robot is supposed to do. In capability space, this is the distance between a desired capability such as *grasp cup* and a provided capability such as *grasp cup-with-handle*. From a verification standpoint, we need to assess whether a provided capability is both functionally correct (in that it will cause the system to take correct actions) and adequate (in that it is close enough to the desired capability to meet the desired capability's performance requirements). In this chapter, we will focus on the first problem—assessing functional correctness—and will leave adequacy for a future discussion.

In order to define a *capability* in terms where the distance between two capabilities can be assessed, we must first identify the information necessary to define a task and the information necessary to define a behavior. The following ontology standard development efforts provide starting definitions of the minimum information necessary to define a capability.

4.3.2 Defining Tasks: The P1872.1 Robot Task Representation Ontology

The IEEE P1872.1 Standard Development Working Group has developed a robot task representation ontology. They are basing their work around a framework capturing five *frames* or categories of key concepts necessary to specify a robot's job [14]:

- *Command*—this frame captures information needed a priori to define the initial state, preconditions, and input information.

- *Action*—this frame captures information about *how* the job should be done; for autonomous operations, this may consist only of fault responses, a primary behavior, and exit conditions.
- *Result*—this frame captures information about the desired state of the system and the environment when the task is over, what information should be provided by the robot(s) attempting to accomplish the task, and the metrics and evaluation tools used to determine whether or not the robot(s) have successfully accomplished the task.
- *Constraints*—this frame captures the types of constraints that may be needed to define a task.
- *Resources*—this frame captures information about the environment, the platform, the capabilities provided by whatever system is used to accomplish the task, objects in the environment, and any exhaustible resources that may need to be used or define trigger points in the task itself.

The task ontology specifies the information necessary and sufficient to define a desired capability.

4.3.3 Defining Behavior: The P1872.2 Autonomous Behavior Ontology

The information necessary to capture an autonomous behavior [15] includes:

- The initial state of the environment including the system itself and the inputs to the system
- Any information related to how the behavior functions, including:
  - What actions can be taken
  - How the decision-making action-selection mechanism is structured
  - What functions and information are used to drive decision-making
  - What perception algorithms are used to gather information about the environment including robot state
  - How that information is or is not integrated into a larger model of the current state of the entire ecosystem
- The desired end state of the environment including the system itself and the outputs from the system

4.3.4 Capability Representation (CR)

Building from the information identified as essential in the task ontology development work and in the autonomous system ontology development work, we identify four major categories of information that must be present to create a complete capability representation.

These four categories are:

- *Context* (T)—this category holds the information related to the task as understood by the user. It includes goal definitions and desired outcomes for nominal and faulted conditions, metrics that are appropriate for evaluating behavior performance, and any contextual constraints on when or how it can be used.
- *Action* (A)—this category captures the information related to how the behavior accomplishes the corresponding task. For autonomous operations, this category determines the boundaries of autonomous operation and identifies any aspects of the task that must be completed in particular or constrained ways.
- *Platform* (P)—this category defines the relevant information about the physical system providing the capability. In the case of an autonomous behavior, it defines either the platform(s) in which the behavior is implemented or the specification for the platform(s). Some behaviors, for example, will require color imagery, while other will require grippers. This category captures that information.
- *Environment* (E)—this category defines the relevant information about the world in which the capability operates. If the capability needs to work in a simulation, the relevant parameters of that simulation are defined here. If the capability is limited to flat two-dimensional representations or must operate in a three-dimensional fluid, that information is captured here. This category also includes resources and elements that may be used up over the course of the task, objects that the robot must or must not interact with, and any other salient information about the environment in which it works. If a given capability requires interaction with some entity that is not captured in the Platform category, the specification for that entity is captured in the Environment category.

These four categories represent a capability x as Capability Representation (CR) C_x:

$$C_x = \begin{cases} T_x \\ A_x \\ P_x \\ E_x \end{cases} \tag{4.1}$$

These four categories can be considered facets in a faceted classification system or taxonomy. In such a system, each facet supports a branching, hierarchical taxonomy of options [16]. Terminating a facet before the tip of the taxonomy is reached encodes a range of options, e.g., the Action may be a *grasp* (a gripper-centric action as opposed to a body-centric action or an arm-centric action) without specifying a particular type of grasp. Terminating the facet at the root effectively specifies "any" of the potential options.

4.3.4.1 Mapping Between Capability Representation and Task Representation

Task representation requires defining a *desired* capability. Instead of specifying a capability that already exists, the role of task representation is to specify the requirements on a system that could provide that capability and the criteria by which its performance will be judged and the activities and state changes that must occur as part of accomplishing the task.

The *Constraints* frame of the Task Representation ontology is the most complex to represent in this framework. Effectively, each constraint instantiated in a specific capability instance is represented by the structure of the classification and by any constraints applied as success or termination criteria.

Uncertainty and abstraction are explicitly captured by prematurely truncating a branch of the classification tree with a terminator.

The *Resources* frame is, on the other hand, simple to represent. The Environment category captures all the physical information about the robot's surroundings, while the Platform category captures all the physical and nonphysical constraints on the robot or robot team providing the capability, as well as any sub-capabilities it must provide.

The *Command* and *Results* frames are captured in the Context category, while the information specified by the *Action* frame is captured in the Action category.

4.3.4.2 Mapping Between Capability Representation and Autonomous Behavior

Autonomous behavior representation requires defining a *provided* capability. Instead of specifying constraints on the capability to define a potential, the role of autonomous behavior representation is to capture that which has already been designed and discover what tasks it could be used for. The purpose of autonomous behavior representation is to specify the system itself, the environment in which it operates, the criteria by which it makes decisions, and the mechanisms that translate information from the environment to the system and actions from the system to the environment.

The Platform category (P) captures the details of the system on which the behavior is instantiated. In cases where the behavior has been designed but not yet instantiated, it captures the constraints on that system, in exactly the same way those constraints are captured in the task representation case. In cases where the behavior has been instantiated in a physical system, it captures the details of that system.

The Environment category (E) captures any constraints associated with environments in which the behavior will or will not work. For example, an avoidance behavior that assumes the robot is free to avoid in three spatial dimensions will not work in a situation where the robot is constrained to maneuvering on the surface of a plane.

The Action (A) category captures the mechanisms by which the autonomous behavior is defined—the relationships between perception and action, the decision-making approach, and the optimization functions used to select actions.

The Context (T) category captures any information about the behavior that is defined externally to the system. It includes constraints on behavior initiation and behavior exit criteria, the initial state and the inputs, and the desired end state and the outputs.

Abstraction
Abstraction refers to the level of detail *necessary* in order to specify the capability. If a given capability has an Environment category that specifies a flat operating surface perpendicular to the force of gravity, then rugged outdoor terrain is excluded, but concrete parking garages, carpeted offices, and tiled kitchens are all included. Fundamentally, *abstraction* defines a region where all options must be supported. If the specification is further abstracted to remove the "perpendicular to the force of gravity" constraint, then a system instantiating the new capability would need to be able to operate on flat walls and floors.

Uncertainty
Uncertainty refers to the level of detail *available* to specify the capability. For autonomous systems, regions of uncertainty in the Action category define the scope within which autonomous operation is allowed. Fundamentally, *uncertainty* defines a region where any option is allowed. If the capability indicates that the selection of concrete parking garage, carpeted office, or tiled kitchen floor is uncertain rather than abstract, then it is not known a priori which environment is intended.

4.3.4.3 Simple Example: *move-motor*

We start with an extremely simple example of an autonomous behavior: *move-motor*. Even at this level, we can distinguish between autonomous operation and nonautonomous operation. We are exclusively concerned with autonomous operation, so we will only consider capabilities that involve some form of decision-making.

In the space domain, *move-motor* may be used in very different contexts. Assume a robot arm, mounted on a spacecraft. In order to use the arm, each motor must work at the desired accuracy. As part of the operation that moves the arm out of the locks that keep it safe during launch, a basic system checkout is performed that ensures that all the motors are receiving power and commands. During this system checkout, *move-motor* could be used to ensure that a particular motor is functional.

Once the arm is in its operating pose, a calibration routine is used to ensure that the arm's accuracy is still good enough to support the desired operation. As part of an automated calibration routine, *move-motor* could be used to ensure that a given motor still provides the required accuracy. In this case, it would be called as part of an automated script.

Any autonomous operation performed by this arm needs to consider the possibility of damage to the spacecraft it is mounted on. In this case, *move-motor* could be used in the context of an avoidance algorithm to enable the arm to reach a goal while avoiding damage to the spacecraft. Instead of calling the command and waiting for the behavior to finish, the avoidance algorithm would repeatedly call *move-motor* every time the avoidance algorithm's goal or context changed.

The core *move-motor* capability has a Capability Representation; additionally, each of the tasks it can be used to perform has their own Capability Representations.

Core *move-motor*

The core functionality of the *move-motor* behavior is to move a motor. Therefore, even before we consider the different tasks or applications that it may be used in, we can define certain fundamentals of the Capability Representation.

At the most basic level, any platform element capable of *move-motor* requires a motor. Most simple capabilities that address basic function required to support a wide range of tasks will also require some mechanism to focus them on the correct output. In this case, the generic *move-motor* capability requires the ability to identify which motor is being moved.

However, the existence of a *move-motor* capability does not imply specific sensing or additional platform elements unless autonomous operation is necessary. Because we are assuming the need for autonomy, the Platform category of the CR must include at least one sensor. Because we are not specifying what type of motor is to be used, we truncate the representation at *sensor* rather than attempting to specify what sensor data it provides (e.g., whether it is based on time, current, or angular distance moved). Coordinate frames relevant to the Platform category are also defined here.

We have already begun adding *Constraints* to the task by defining the structure of the capability instance.

The Environment category is defined as meeting the requirements for motor operation, so that category is defined with an acceptable temperature range. Specification of a range of temperatures indicates that the behavior is not expected to function properly when the environmental temperature is outside those limits. The use of S-tag *all* indicates that the behavior needs to work properly at every temperature within that range. Because evidence exists (in the form of certification from the manufacturer) that the motor is expected to operate properly in that range, the assurance case can point to that certification instead of testing the motor at every temperature.

The Action category defines that we are interested in *self-motion relative to self-frame* and defines the *self* in *self-motion* and the *self-frame* in terms of the motor and associated coordinate frames defined in the Platform facet.

Until we know which task or application is required, we cannot further define the CR for *move-motor*.

This gives us a partial capability representation for the *move-motor* capability CR_{mm}:

$$CR_{mm} = \begin{cases} T_{mm}: & \text{(:)} \\ A_{mm}: & \text{motion (self in self-motion frame, rel. to self-fixed frame)} \\ P_{mm}: & \text{motor (def. self-motion frame, self-fixed frame, motor-id),} \\ & \text{sensor (:)} \\ E_{mm}: & \text{temperature } (T_L < \text{temperature} < T_H,\ \text{S-tag = all}) \end{cases} \tag{4.2}$$

where (:) denotes uncertainty (any option supported) rather than abstraction (all options supported). The motor specification defines the self-motion and self-fixed coordinate frames relative to the motor, and the motion specification defines that the motion is of the self in the self-motion frame relative to the self-fixed frame.

move-motor for System Checkout

The System Checkout operation is defined as ensuring that a motor is capable of moving. In this context, evaluation is based on whether or not the motor moves at all, and the specific details of starting and ending position are irrelevant as long as the difference between the start and end points is sufficiently large that the sensor can measure it. This is represented in the Context category as

$$\begin{aligned} T_{mmSC}\ :&\text{goal (waypoint (start-value = any,} \\ &\text{end-value = [(any - start-value)>threshold]),} \\ &\text{S-tag = any)} \end{aligned} \tag{4.3}$$

where the *threshold* is defined as a motor parameter. In this case, the start-value and end-value variables may be defined in terms of the self-motion frame or the self-fixed frame as long as they are both defined in the same frame.

From a verification standpoint, the inclusion of the *success tag* "S-tag = any" means that verification of this behavior will require showing only that there exists at least one combination of start value and end value that satisfies the constraints and causes the motor to move. Identification of a valid case that exists is sufficient for verification, and validation can be accomplished through a simple demonstration of that case rather than a full suite of tests or an extensive analytic regime.

move-motor for Calibration

In the Calibration operation, however, the goal is to ensure the motor is capable of moving to particular positions with the required accuracy. This is represented in the

Context category as

T_{mmCC} : goal (waypoint (start-value = specific angular point, end-value =
specific angular point), accuracy (angular distance
[actual end-value - desired end-value < threshold]),
S-tag = accuracy)

(4.4)

where the *threshold* is defined as the accuracy with which the motor must attain the desired position and the start-value and end-value variables are defined in terms of the platform's self-fixed frame.

From a verification standpoint, the inclusion of the success tag "S-tag = accuracy" means that to verify the behavior the motor must reach a specific desired position with the desired accuracy. This may require multiple tests in order to develop a statistically significant sample, or it may be sufficient to prove via demonstration that it is capable of the required accuracy.

***move-motor* for Autonomous Operation**

In the Autonomous Operation case, when the motor moves or stops is defined by external factors that change as a function of environmental situations and mission goals at a higher level. For the case of an autonomous operation where the motor is moving because the arm is avoiding damage to its environment, the goal of the action is to satisfy these larger goals. The Context category would therefore be defined in terms of the interface to the motor:

T_{mmAO} : goal (action (speed = input-value-sp, direction = input-value-dir),
metric (| input-value-sp - actual-value-sp |< threshold AND
input-value-dir == actual-value-dir),
S-tag = range)

(4.5)

From a verification standpoint, the inclusion of the success tag "S-tag = range" means that to verify the behavior the motor must be capable of responding to the full range of possible input parameter values. This will require multiple tests in order to ensure that sufficient evidence is captured to support the statement that the motor responds correctly across the full range of potential inputs.

4.3.5 Comparison of Capabilities

Because we have now specified different variations on the *move-motor* capability, we can start to look at the differences between them. How similar is C_{mmCC} to C_{mmSC}? From our definitions in Sect. 4.3.4.3, we can see that the only difference between the two is the Context category defining their goals.

T_{mmSC} defines a waypoint goal with any start value, an end value that is a function of any value, a simple threshold constraint, and a success tag that states that any value that meets these criteria is acceptable, while T_{mmCC} defines a waypoint with a specific start value, a specific end value, a specific accuracy, and a success tag that depends on accuracy.

Because T_{mmCC} specifies details that are left vague in T_{mmSC}, the provided capability C_{mmCC} can be used to satisfy the desired capability C_{mmSC}, while the reverse is not true. The desired capability C_{mmCC} cannot be satisfied by the desired capability C_{mmSC}.

T_{mmAO} defines an action goal rather than a waypoint goal and is therefore qualitatively different from both T_{mmSC} and T_{mmCC}. From this, we can conclude that the provided capability C_{mmAO} cannot be used to satisfy either desired capability C_{mmSC} or C_{mmCC}.

4.3.6 Capability Analysis Tables

The Capability Representation described in Sect. 4.3 captures the known and unknown information about the desired or provided capability. It provides a mechanism to define the platform requirements and a way to capture hierarchical relationships between capabilities in either design or task specification (via the Action category through tags representing required supporting capabilities), but these representations are not easily readable or human-friendly.

The CR definition determines where the specification ends and the design begins by defining the point where the specification explicitly states that further detail is not provided and must be developed by the designer. It provides the evaluation metric to be used in the verification process but does not define the specific autonomy design or the performance of the designed system. Because it does not define the autonomy design, it cannot define the supporting requirements hierarchy. It is also not sufficient to address the core problem of defining a behavior in a way that supports formal verification. A mechanism is needed that forces specification developers and behavior designers to explicitly define the boundary between them in a human-readable way and to document the details of the designed behavior.

For this, we need Capability Analysis Tables (CATs). These tables capture the design created to meet the specification and define sufficient context to support specification of the lower level behaviors used to accomplish it. The structure of these tables supports their use as a pass-through stage in the verification process, enabling both simple identification of obvious gaps in the design and future export to formal analysis tools.

4.3.6.1 CAT Structure

A single Capability Analysis Table (CAT) captures the design of a single autonomous behavior *at a specific level of abstraction* and defines the specification for each sub-behavior within that design [17]. The CAT sits at the intersection of the platform, the high-level behavior design, and the specifications for component behaviors within the design. It provides a human-readable representation of the relationship between these components. The redundancy inherent in the CAT enables human comprehension and review of these complex systems by ensuring that relevant information is visible even for complicated relationships.

Because CATs can be defined at whatever level of abstraction is appropriate, an individual CAT is typically small, representing only the relevant inputs and outputs for the behavior defined at that level of abstraction. Each sub-behavior can be specified using the Capability Representation in Sect. 4.3 and documented with its own CAT. Complex systems may be represented by a small number of complex CATs but are more effectively captured via a hierarchical structure of simpler CATs.

A simple CAT for a Braitenberg vehicle 2B [18] is shown in Fig. 4.1 as an introductory example. The high-level behavior captured in the CAT is identified in the upper left corner of the table—in this CAT we are documenting a behavior called *Braitenberg Vehicle 2B.*

In a CAT, the rows represent the potential outputs of an autonomous behavior—they are the things the behavior can affect. They include hardware control commands, direct outputs like voltage levels affecting any element of the system, outgoing communications, and information generated or stored by the behavior. For a Braitenberg vehicle 2B, as shown in Fig. 4.1, each row represents a control output to either the left or right motors.

The columns of the CAT represent the potential inputs to an autonomous behavior—they are the things that affect the behavior or that are represented as inputs to a function call. Variable inputs are derived from sensors, incoming communications, and information generated on-board the system. They include

Braitenberg Vehicle 2B / *Input*	A Priori	Left Illumination Sensor (LS)		Right Illumination Sensor (RS)		Notes
Output		Cond	Beh ID	Cond	Beh ID	
Right Motor (RM)	O: a, b	O: avail.	BR: RM = a*LS + b			Right motor turns faster when left sensor value is higher
Left Motor (LM)	O: a, b			O: avail.	BL: LM = a*RS + b	Left motor turns faster when right sensor value is higher

Fig. 4.1 Capability analysis table for Braitenberg vehicle 2B

sensor data, incoming communications, software instructions, state information, tasking changes, and information generated by the behavior in a previous iteration. The A Priori column captures fixed information that is provided to a behavior, like the structure of a map or a constant parameter value. For the Braitenberg vehicle, there are two variable inputs (the left and right light intensity sensors) and two a priori inputs (the parameters a and b used by the behaviors to convert the range of incoming sensor values to the range of outgoing motor values).

A human-readable *Notes* column is added to support additional comments and clarity for the human users of the table. This CAT uses the Notes column to ensure that human users know the relationship between the sensor inputs and the motor outputs.

In the body of the table, there are two cells at the intersection of each column and row. One cell captures any conditions related to the inputs in that column (the *Cond* sub-column), and the other defines which behavior should be active when those conditions are satisfied (the *Beh* sub-column). The behavior at that intersection requires that column as input and produces that row as output.

In our simple Braitenberg vehicle example, the left illumination sensor determines the magnitude of the control output to the right motor, and the right illumination sensor determines the magnitude of the control output to the left motor. Two functions are defined to instantiate its behavior (*BR* and *BL*), each one on its own row.

Information in the Conditions sub-column falls into three categories:

- Trigger Condition *T* identifies preconditions that must be met as a condition of entry but where that information is not passed into the behavior.
- Precondition *P* identifies preconditions and input parameters that are required to initiate the behavior and are passed to the behavior but do not need to be updated while the behavior is active. "P:avail." means that the information must be available when the behavior is initiated.
- Ongoing Conditions *O* identifies input information that must be updated (either continuously or intermittently) and is used by the behavior. A condition of "O:avail." means that the behavior requires the information to be passed to the behavior whenever it is updated.

Any thresholds or constraints in a given column apply to the information specified in that column. In this simple case, whenever a sensor value is available, the relevant motor command is updated. All the conditions in a row must be true in order for the behavior to produce that row's output.

Figure 4.2 shows a more complex table illustrating the basic structure and relationships that can be captured in a CAT. In order to capture behaviors that do not rely on incoming information, a *None Req.* (no input required) column is included. The *Old State* and *New State* columns track exit and entry from the behavior and internal state transition information. Because state transition information must be tracked for all behaviors, every row has an entry for that behavior in the New State column. The associated Old State column indicates the acceptable previous states. It is important to capture this information even when the acceptable previous states

Behavior ID	Input	None Req.	A Priori		Internal Variable 1		Internal Variable 2		External Variable 3		Sensor Input 1		Sensor Input 2		Comms In: Env.		Comms In: Task		Old State	New State	Notes
Output			Cond	Beh ID	Cond	Beh ID	Cond	Beh ID	Cond	Beh ID	Cond	Beh ID	Cond	Beh ID	Cond	Beh ID	Cond	Beh ID			
Internal Variable 1			P: a	B1	P: avail. O: avail.	B1							P: avail. O: avail.	B1					Start	B1	First state = B1
			P: a	B1	P: avail. O: avail.	B1							P: avail. O: avail.	B1					T: B5 success	B1	B5 transitions to B1 when it succeeds
Internal Variable 2		B2																	Ongoing	B2	B2 always running
External Variable 2					P: avail. O: avail.	B3	P: avail. O: avail.	B3			P: avail. O: avail.	B3			P: avail. O: avail.	B3	P: avail. O: avail.	B3	Ongoing	B3	B3 always running
Comms Out			P: t	B3	P: avail. O: avail.	B3	T: <=t	B3			P: avail. O: avail.	B3			P: avail. O: avail.	B3	P: avail. O: avail.	B3	O: B1	B3	B3 only generates outgoing message when Internal Variable 2 <= t and B1 is active
Parameter Control			P: t	B4			T: >t	B4											B1	B4	B1 transitions to B4 when Internal Variable 2 > t
Hardware Control					P: avail.	B5			T: update	B5									B1	B5	B1 transitions to B5 when External Variable 3 received

Fig. 4.2 Basic structure of a Capability Analysis Table

are *any* or *none* so that the user knows that the state transition information has been explicitly considered even in the null case.

Information in the Old State column can take the same *T*, *P*, and *O* condition markers as the other Condition columns. In this context, *T* implies a particular condition or outcome of the state (e.g., "T: B5 success" indicates that state B5 must succeed in order for B1 to become active). Condition *P* means that the old state must be active in order for the new state to become active, but that the old state may continue running in parallel with the new state. Condition *O* means that the old state must continue running in parallel with the new state. The lack of a specific condition type indicator implies a transition from the old state to the new state that is driven by some other variable in the row (e.g.,"B1" means that state B1 transitions to state B5 when the conditions of the row are met. (There is no state information in the Braitenberg vehicle CAT because the vehicle has no state changes—the same two behaviors are always running.)

The behaviors in this CAT represent complex lower level behaviors rather than simple function calls like the ones in the Braitenberg vehicle CAT. Each of these lower level behaviors can be defined with their own CAT, with their own sub-behaviors and functions, but the externally facing information must be captured in both CATs.

The *A Priori* preconditions are constants that are defined by the system before the high-level behavior starts. In this case, the example behavior Behavior ID knows the values for *a* and *t* before any of the behaviors in the cells are called. The system does not generate *a* or *t*, while Behavior ID is active.

Behavior B3 is capable of producing two outputs: a variable visible outside Behavior ID (External Variable 2) and a message sent via outgoing communication. The external variable is produced constantly, while the outgoing communication only occurs when Internal Variable 2 is below or equal to the a priori threshold *t*. In order to run, behavior B3 also requires information from Sensor Input 1, Internal Variable 1, and access to incoming communications containing both environmental and task-related information.

Internal Variable 2 is automatically generated by behavior B2 without any additional input.

Behavior B1 only generates one output, Internal Variable 1, but it is initiated in two ways and therefore requires two rows:

- If the Old State is Start, the a priori value for *a* is known, and Internal Variable 1 and Sensor Value 2 are both present and continue to update, B1 activates.
- If the Old State is B5 and B5 has succeeded, the a priori value for *a* is known, and Internal Variable 1 and Sensor Value 2 are both present and continue to update, B1 activates.

The system will transition from B1 to B5 when External Variable 3 updates, and a value for Internal Variable 1 is available to initialize B5. We see in the first column that B5 is some type of hardware control. Since there are no variable- or sensor-based transitions out of B5, but "B5 success" is a trigger for B1 to restart, we can see that B5 must contain its own exit conditions. (It would then behoove us to carefully examine a CAT of B5 by itself, to ensure that it cannot enter an infinite loop state.)

The system will transition from B1 to B4 if Internal Variable 2 exceeds a threshold t. Because behavior B4 does not show up in the Old State column, it is a terminating state. Once the algorithm reaches B4, it will stay in that state until the behavior Behavior ID is manually shut down. Note that the system will *not* transition from B5 to B4, regardless of the value of Internal Variable 2.

Even with this abstract example, it is possible to identify potential disconnects between the desired capability and the provided capability.

For example, if Behavior ID is supposed to exit cleanly rather than concluding in an ongoing state, the lack of an Exit state in the New State column indicates that failure. If one of the populated Behavior ID cells was missing an associated Condition, that would be a violation of a valid capability specification. If any of the Condition cells referencing threshold t were ongoing conditions (O) rather than triggering conditions (T), the Internal Variable 2 column would warrant special consideration. If a behavior has an ongoing condition that a variable be within a specific range of values, then it is essential to understand what happens when that variable is outside that specific range of values. Similarly, transitions between old and new states must cover the full range of possibilities. The transition out of behavior B5 illustrates this problem: Because we have defined a row where B5 transitions only if it succeeds, there must also be a row defining what should happen if B5 fails. Because this row does not exist, we know that this table is incomplete.

From this CAT we can also identify specifications for the capabilities associated with behaviors B1, B2, B3, B4, and B5. These specifications can in turn be used to define lower levels CATs for each behavior. We know, for example, that B2 generates Internal Variable 2 without any inputs and that B3 and B4 depend on it. An example of a potential B2 is a random number generator. The assumptions that B3 and B4 make about the properties of B2's output define what B2 needs to produce, just as the output of B2 defines what is available to B3 and B4. Because of the Conditions columns, we know that B2's output must be able to be thresholded by t. If t is a vector, then Internal Variable 2 must be a vector of the same size and type.

4.3.6.2 Concrete CAT Example: System Checkout

Figure 4.3 shows an example CAT for the System Checkout scenario in Sect. 4.3.4.3. There are five behavior components in the design of the *System Checkout* behavior. Because the arm motor and the gripper motor are assumed to be the same model of motor, the *move-arm-motor* and *move-gripper-motor* behaviors use the same input variables. The *pan-tilt-move* behavior requires its own inputs to drive the camera motor, while the *camera-data* behavior sends a command to the camera to generate image data. The *comms-out-check* behavior takes the camera data and uses it as the body of its test message.

System Checkout / *Input*	None Req.	A Priori		Active Motor ID		Active Motor Current Pose		Goal Change In Active Motor Pose (+/- angle)		Camera Motor Current Pose		Goal Change In Camera Motor Pose [pan angle, tilt angle]		Camera Data		Old State	New State	Notes
Output		Cond	Beh ID	Cond	Beh ID	Cond	Beh ID	Cond	Beh ID	Cond	Beh ID	Cond	Beh ID	Cond	Beh ID			
Comms Output														P: avail. O: avail.	*comms-out-check*	*camera-data T: Success*	*comms-out-check*	Send camera data as test message when camera data check is done
Active Motor ID		P: Start-Pose_c and t_c	*pan-tilt-move*	O: Camera	*pan-tilt-move*					P: avail. O: avail.	*pan-tilt-move*	P: avail. O: avail.	*pan-tilt-move*			*move-gripper-motor T: Success*	*pan-tilt-move*	Run camera pan/tilt motor check when the gripper motor check is done
		P: Start-Pose_m and t_m	*move-gripper-motor*	O: Gripper	*move-gripper-motor*	P: avail. O: avail.	*move-gripper-motor*	P: avail. O: avail.	*move-gripper-motor*							*move-arm-motor T: Success*	*move-gripper-motor*	Run gripper motor check after arm motor check
		P: Start-Pose_a and t_a	*move-arm-motor*	O: Arm	*move-arm-motor*	P: avail. O: avail.	*move-arm-motor*	P: avail. O: avail.	*move-arm-motor*							Start	*move-arm-motor*	Run arm motor check
Camera Pan/Tilt Command		P: Start-Pose_c and t_c	*pan-tilt-move*	O: Camera	*pan-tilt-move*					P: avail. O: avail.	*pan-tilt-move*	P: avail. O: avail.	*pan-tilt-move*			*move-gripper-motor T: Success*	*pan-tilt-move*	Run camera pan/tilt motor check when the gripper motor check is done
Camera Data Generation Command		P: test image parameters	*camera-data*													*pan-tilt-move T: Success*	*camera-data*	Generate camera data when the pan/tilt check is done
Gripper Motor Command		P: Start-Pose_m and t_m	*move-gripper-motor*	O: Gripper	*move-gripper-motor*	P: avail. O: avail.	*move-gripper-motor*	P: avail. O: avail.	*move-gripper-motor*							*move-arm-motor T: Success*	*move-gripper-motor*	Run gripper motor check after arm motor check
Arm Motor Command		P: Start-Pose_a and t_a	*move-arm-motor*	O: Arm	*move-arm-motor*	P: avail. O: avail.	*move-arm-motor*	P: avail. O: avail.	*move-arm-motor*							Start	*move-arm-motor*	Run arm motor check
Fault Command																*Any T: Fail*	*fault-response*	Exit when any behavior fails
End																*comms-out-check T: Success*	Exit	Exit when comms test message sent

Fig. 4.3 CAT for the System Checkout case expressed in behaviors

Because the inputs are defined as variables in which information is stored rather than as values, the same input variables Active Motor Current Pose and Active Motor ID and Goal Change In Active Motor Pose can be used to drive different behaviors depending on their values.

The Active Motor ID column illustrates the use of the *O* marker in the Cond sub-column, where the motor ID is marked *O* as an ongoing input to the *move-arm-motor*, *move-gripper-motor*, and *pan-tilt-move* behaviors. Active Motor ID is also an internal variable, generated as part of the motor motion behaviors. The *move-arm-motor* behavior initializes its value to "Arm," the *move-gripper-motor* behavior takes it as "Arm" and changes it to "Gripper," and the *pan-tilt-move* behavior takes it as "Gripper" and changes it to "Camera."

The inputs in the A Priori column are all values that do not change while a behavior is running and thus are all marked *P* for precondition. They must be available in order for the relevant behaviors to become active, but at this level in the hierarchy, their values do not determine which behaviors are active. Information about how they are used to control the relevant behavior is contained in the CAT for that behavior.

It may seem redundant to have both an a priori specified *Start-Pose_a* and a column for Active Motor Current Pose, but they specify two variables that can have different values. These particular behaviors may require that the current pose match the a priori starting pose in order to initiate motion. As noted in the Braitenberg vehicle CAT discussion, active behaviors are defined by the union of the conditions across each row—when all the conditions are true, the behavior is active and generates the output defined in that row.

In this CAT, we can see that when the system starts, the Arm Motor Command row is the only potentially active row. The Old State input defines the prior state as a precondition on the New State (which is the current state). If the old state does not match, the current state cannot become active. The only behavior with *Start* as a precondition is the *move-arm-motor* behavior.

The system will not issue the Arm Motor Command until the Active Motor Current Pose is available, the Goal Change In Active Motor Pose is available, the Active Motor ID is set to "Arm," and the a priori parameters Start-Pose_a and threshold_a are known. When all of these conditions are met, the Arm Motor Command will be generated. As long as these conditions are true, it will continue to run. When this use of the *move-arm-motor* behavior finishes, the Old State precondition requires a transition to the *move-gripper-motor* behavior.

If any of the other behaviors fail rather than succeeding—for example, the Active Motor ID ceases to be Gripper during *move-gripper-motor*—that triggers the *fault-response* behavior to issue a Fault Command that shuts down the system gracefully.

Given this CAT, the *move-arm-motor*, *move-gripper-motor*, and *pan-tilt-move* behaviors all require the ID for the motor that is supposed to move, the current pose of that motor, and the desired change in pose of that motor as inputs, as well as the motor in question's starting pose and the accuracy threshold that determines when the motor has reached the desired pose. They must each generate a motor command suitable for the motor in question as defined by the platform performing this System Checkout operation. They must also each have a defined exit point so that the *System Checkout* decision-making mechanism managing transitions is able to determine when it has finished and to trigger the next action in the sequence.

The CAT in Fig. 4.4 shows the details of the *move-arm-motor* capability within the System Checkout capability in Fig. 4.3. This CAT terminates the hierarchical structure, as instead of capabilities or behaviors, the underlying mathematics defining the relationships is used to define what happens in the cells. If this level of implementation detail is known for all the components of a given CAT, then that CAT can provide a termination point for a more abstract capability representation.

We can map the Active Motor Current Pose and Goal Change In Active Motor Pose (+/- angle) inputs in the CAT for the *System Checkout* behavior to the Current Pose (angle) and Goal Change In Motor Pose (+/-angle) in the CAT for the *move-arm-motor* behavior. Similarly, we can map the Start-Pose_a and t_a inputs in the *System Checkout* CAT to the SP_a and t_a inputs in the *move-arm-motor* CAT. The Motor Command output is the same for both. The lower level *move-arm-motor* CAT contains an additional internal variable, Change in Pose Remaining, but because it is generated and used internally, this does not affect the mapping between the *move-arm-motor* behavior's definition and its use in the *System Checkout* behavior. The bottom row in the *move-arm-motor* behavior CAT defines the exit criteria required by *System Checkout.*

The Active Motor ID variable is set to "Arm" by this behavior as part of the initialization process. Because this function is called on start and does not have any input information, it is represented in the None Req. column.

4.4 Task and Behavior Capability Specification

Autonomous system capabilities and design are inherently linked with their task specifications—each capability drives the development of a lower level specification capturing the details of the assumptions necessary to provide the required lower level capability without which the higher level function will fail.

This is, of course, true for all systems—without assurance that the lower level components are correct, the higher level function is not assured—but it is complicated in autonomous systems by the tight environmental interactions that drive autonomous behavior. Because these environmental interaction assumptions are often implicit or undocumented, they have an outsize impact on the risk associated with autonomous system operation and thus must be included in the specification as constraints on reasonable expectation of correct operation. Similarly, internal optimization functions embedded in the behavior decision-making may not match the external task goals. The effects of these internal optimization functions must also be captured in the specification to ensure correct operation.

move-arm-motor / *Input*	None Req.	A Priori		Active Motor ID		Change In Pose Remaining [+/- angle]		Current Pose (CP) [angle]		Goal Change In Pose (GCIP) [+/-angle]		Old State	New State	Notes
Output		Cond	Beh ID	Cond	Beh ID	Cond	Beh ID	Cond	Beh ID	Cond	Beh ID			
Change In Pose Remaining		T: SP_a	CIPR = GCIP - (SP_a-CP)					P: avail. O: avail.	CIPR = GCIP - (SP_a-CP)	P: avail. O: avail.	CIPR = GCIP - (SP_a-CP)	*Update Motor ID T: Success* OR *Ongoing*	CIPR = GCIP - (SP_a-CP)	Generate the Change In Pose Remaining value from the Current Pose, the Goal Change In Pose Remaining, and the Change In Pose Remaining
Active Motor ID	Active Motor ID = Arm											T: Start	Active Motor ID = Arm	Set Active Motor ID to Arm (initialization)
Active Motor Command		P: t_a	MC = *f(t_a,CIPR)*	P: avail	MC = *f(t_a,CIPR)*	O: |*|>=t_a	MC = *f(t_a,CIPR)*					*Update Motor ID T: Success* OR *Ongoing*	MC = *f(t_a,CIPR)*	Generate the Motor Command based on the Change In Pose Remaining
End		P: t_a	Exit			T: |*|<t_a	Exit					MC = *f(t_a,CIPR)*	Exit	If the change in pose remaining is less than the defined threshold, stop

Fig. 4.4 CAT for the *move-arm-motor* behavior

The Capability Analysis Tables enable explicit documentation of these relationships as they are instantiated in a particular autonomous behavior design. The Capability Representation enables specific identification of particular sub-capabilities in terms of the constraints imposed by the context in which they are required and the designs that they are able to support. The Task/Behavior Duality Principle enables us to distinguish between desired and provided capabilities, while simultaneously enabling us to transition between the capability specifications and the implemented behavior in a common conceptual framework.

The Task/Behavior Duality Principle ensures a complete desired capability specification is sufficient to define the task that needs to be accomplished in enough detail for the autonomous behavior designer to know explicitly whether a given provided capability is adequate to accomplish the task—it enables assessment of the match between desired and provided capabilities, between the job to be done and the behavior proposed to accomplish it.

The Capability Representation implements the Duality Principle and enables evaluation of specific autonomous behaviors against detailed descriptions of specific jobs that must be done. It provides a translation point between the customer's needs and the designer's intent. It defines both the desired behavior and the provided behavior.

The Capability Analysis Tables take advantage of the Duality Principle to enable accurate documentation of the relationships between capabilities within the context of a specific task or behavior. They support the behavior designer and enable improved debugging when the provided behavior as designed fails to match the desired behavior [19].

4.4.1 System Checkout Example

Because of the Task/Behavior Duality Principle, each of the behaviors Bx in this example is associated with a task Kx. Both Bx and Kx are specified by the same capability representation C_x, which can be substituted in the table for the behavior identifiers B1 through B5. This enables the CAT to support specifications that include modular, pre-verified components.

Within the System Checkout example CAT, we can explicitly connect the *move-motor* capability C_{mmSC} with the *move-arm-motor* and *move-gripper-motor* behaviors by looking at their individual CRs and seeing whether they are compatible with the more generic *move-motor* CR.

We start by defining C_{mamSC} (the *move-arm-motor* CR) and C_{mgmSC} (the *move-gripper-motor* CR).

The base *move-motor* capability for the system checkout case is pulled from Sect. 4.3.4.3 and defined as follows:

$$CR_{mmSC} = \begin{cases} T_{mmSC} : \text{goal (waypoint (} \\ \qquad \text{start-value = any,} \\ \qquad \text{end-value = [(any - start-value) > threshold])} \\ \qquad \text{S-tag = any)} \\ A_{mmSC} : \text{motion (self in self-motion frame, rel. to} \\ \qquad \text{self-fixed frame)} \\ P_{mmSC} : \text{motor (def. self-motion frame, self-fixed frame,} \\ \qquad \text{motor-id),} \\ \qquad \text{sensor (:)} \\ E_{mmSC} : \text{temperature } (T_L < \text{temperature} < T_H, \text{ S-tag = all)} \end{cases} \tag{4.6}$$

In order to define the capability representation for the *move-arm-motor* behavior, we need to modify the input parameters and add detail. In this case, we define the `start-value` as *Start-Pose_a*, the `threshold` as *t_a*, and the `motor-id` as *arm*, and we refine the `sensor` category to ensure that the *current pose* information is available. We define the *Goal Change In Pose* input from the *move-arm-motor* CAT as parameter *Y*, replacing the reference to *any* in the `end-value` equation. We specialize the `action` category by adding the actual equation used as a tag on the `motion` category.

$$CR_{mamSC} = \begin{cases} T_{mamSC} : \text{goal (waypoint (} \\ \qquad \text{start-value} = \textit{Start-Pose\_a}, \\ \qquad \text{end-value} = [(Y - \text{start-value}) > t\_a]) \\ \qquad \text{S-tag = any)} \\ A_{mamSC} : \text{motion (self in self-motion frame, rel. to} \\ \qquad \text{self-fixed frame, } MC = f(t\_a, CIPR), \\ \qquad CIPR = Y - (\textit{Start-Pose\_a} - \textit{current pose})) \\ P_{mamSC} : \text{motor (def. self-motion frame, self-fixed frame,} \\ \qquad \text{motor-id} = \textit{arm}), \\ \qquad \text{sensor (}\textit{current pose}) \\ E_{mamSC} : \text{temperature } (T_L < \text{temperature} < T_H, \\ \qquad \text{S-tag = all)} \end{cases} \tag{4.7}$$

We can use the same process for the *move-gripper-motor* behavior. The `start-value` becomes *Start-Pose_m*, `threshold` becomes *t_m*, and `motor-id` becomes *gripper*. Since we assume that the gripper motor is physically the same as the arm motor, we are able to reuse the control equations from the *move-arm-motor* case, and to assume that the same sensor, providing *current pose* is available.

$$CR_{mgmSC} = \begin{cases} T_{mgmSC} : \text{goal (waypoint (} \\ \quad \text{start-value} = \textit{Start-Pose\_m}, \\ \quad \text{end-value} = [(Y\text{- start-value}) > t\_m]) \\ \quad \text{S-tag = any)} \\ A_{mgmSC} : \text{motion (self in self-motion frame, rel. to} \\ \quad \text{self-fixed frame, } MC = f(t\_m, CIPR), \\ \quad CIPR = Y - (\textit{Start-Pose\_m} - \textit{current pose})) \\ P_{mgmSC} : \text{motor (def. self-motion frame, self-fixed frame,} \\ \quad \text{motor-id} = \textit{gripper}), \\ \quad \text{sensor } (\textit{current pose}) \\ E_{mgmSC} : \text{temperature } (T_L < \text{temperature} < T_H, \\ \quad \text{S-tag = all)} \end{cases} \tag{4.8}$$

We can generalize the C_{mgmSC} and C_{mamSC} capabilities into a parameterized version that is capable of accomplishing both tasks. This capability substitutes the parameter *Start-Pose_x* for the *Start-Pose_a* and *Start-Pose_g* inputs, parameterizes the *t_a* and *t_g* inputs with *t_x*, and parameterizes the motor-id input with *X*.

$$CR_{mXmSC} = \begin{cases} T_{mXmSC} : \text{goal (waypoint (} \\ \quad \text{start-value} = \textit{Start-Pose\_x}, \\ \quad \text{end-value} = [(Y \text{ - start-value}) > t\_x]) \\ \quad \text{S-tag = any)} \\ A_{mXmSC} : \text{motion (self in self-motion frame, rel. to} \\ \quad \text{self-fixed frame, } MC = f(t\_x, CIPR), \\ \quad CIPR = Y - (\textit{Start-Pose\_x} - \textit{current pose})) \\ P_{mXmSC} : \text{motor (def. self-motion frame, self-fixed frame,} \\ \quad \text{motor-id=}X), \\ \quad \text{sensor } (\textit{current pose}) \\ E_{mXmSC} : \text{temperature } (T_L < \text{temperature} < T_H, \\ \quad \text{S-tag = all)} \end{cases} \tag{4.9}$$

We have not changed the goal types or structure of the original capabilities, only parameterized some of the input variables. If we compare the *move-arm-motor* CAT in Fig. 4.4 to the CAT associated with the new capability in Fig. 4.5, they have almost exactly the same structure and cell contents. The input variable names in the A Priori column have changed, but that is not a structural change. The only structural change concerns the Active Motor ID variable. Instead of having a fixed value determined by the behavior and set during initialization, a new *Update Motor ID* behavior now uses the initial value of the variable to update it during initialization to the correct value. Instead of being placed in the None Req. column, the *Update Motor ID* behavior now resides in the Active Motor ID column.

<table>
<tr><th>Move-X-Motor: C_mXmSC
Input</th><th colspan="2">A Priori</th><th colspan="2">Active Motor ID (X)</th><th colspan="2">Change In Pose Remaining (CIPR) [+/-angle]</th><th colspan="2">Current Pose (CP) [angle]</th><th colspan="2">Goal Change In Pose (Y) [+/-angle]</th><th>Old State</th><th>New State</th><th>Notes</th></tr>
<tr><th>Output</th><th>Cond</th><th>Beh ID</th><th>Cond</th><th>Beh ID</th><th>Cond</th><th>Beh ID</th><th>Cond</th><th>Beh ID</th><th>Cond</th><th>Beh ID</th><th></th><th></th><th></th></tr>
<tr><td>Change In Pose Remaining</td><td>T: SP_x</td><td>CIPR = Y - (SP_x-CP)</td><td></td><td></td><td></td><td></td><td>P: avail.
O: avail.</td><td>CIPR = Y - (SP_x-CP)</td><td>P: avail.
O: avail.</td><td>CIPR = Y - (SP_x-CP)</td><td>Update Motor ID (2)
T: Success
OR Ongoing</td><td>CIPR = Y - (SP_x-CP)</td><td>Generate the Change In Pose Remaining value from the Current Pose, the Goal Change In Pose Remaining, and the Change In Pose Remaining</td></tr>
<tr><td>Active Motor ID</td><td></td><td></td><td>P: avail.</td><td>Update Motor ID</td><td></td><td></td><td></td><td></td><td></td><td></td><td>T: Start</td><td>Update Motor ID</td><td>Make sure Active Motor ID is updated during initialization; if X=empty -> Arm; if X=Arm -> Gripper</td></tr>
<tr><td>Motor Command (Active Motor ID)</td><td>O: t_x</td><td>MC = f(t_x,CIPR)</td><td>P: avail.</td><td>MC = f(t_x,CIPR)</td><td>O: |*|>=t_x</td><td>MC = f(t_x,CIPR)</td><td></td><td></td><td></td><td></td><td>Update Motor ID (2)
T: Success
OR Ongoing</td><td>MC = f(t_x,CIPR)</td><td>Generate the Motor Command based on the Change In Pose Remaining</td></tr>
<tr><td>End</td><td>P: t_x</td><td>Exit</td><td></td><td></td><td>T: |*|<t_x</td><td>Exit</td><td></td><td></td><td></td><td></td><td>MC = f(t_x,CIPR)</td><td>Exit</td><td>If the change in pose remaining is less than the defined threshold, stop</td></tr>
</table>

Fig. 4.5 CAT for the parameterized C_{mXmSC} capability

Because the new capability representation C_{mXmSC} has the same inputs and outputs and uses the same function calls with the same conditions, it can accomplish the tasks defined by both the *move-arm-motor* and *move-gripper-motor* cells in the original *System Checkout* CAT in Fig. 4.3.

In Fig. 4.6 we see the result of replacing each individual behavior with its corresponding Capability Representation. Because the parameterized C_{mXmSC} capability satisfies both the *move-arm-motor* and *move-gripper-motor* capability requirements, these two behaviors are now correctly replaced with the single parameterized capability. Two rows are required to capture the different calls for the Arm and Gripper functions and the associated motor commands. However, only a single row is required to capture generation of the Active Motor ID, because the output is an update to the same variable in both cases.

The *move-motor* capability and the camera motor's *pan-tilt-move* command are distinguished by a different set of input and output variables, so they require different Capability Representations C_{mXmSC} and C_{ptSC}. The camera motor command still requires the same two rows to capture its motor command and Active Motor ID outputs.

The None Req. column in Fig. 4.3 is empty; since no behaviors run without inputs, we can eliminate that column from our final CAT in Fig. 4.6.

4.4.2 Implications for Verification

The act of verification assumes the presence of sufficient and necessary specifications. To verify, one must compare actual properties to desired properties *as defined by the specification*. Specifications themselves must be testable in order to be of value. By explicitly linking the desired capability defined in the higher level CAT to the provided capability described in the lower level CAT via the structured Capability Representation, we provide a mechanism to identify the sufficient and necessary properties to specify that lower level capability. Looking at the context in which the provided capability will function, we can assess whether it actually meets the needs of that context or not. That context defines the evaluation mechanisms that must be used to determine whether or not the provided capability is adequate or appropriate.

As we have seen in Sect. 4.4.1, using a consistent representation for specifications can simplify the verification problem by identifying modular components and reducing the number of components that need to be verified.

The CR specifications also ensure that the evaluation metric and method for every component of the behavior represented in the CAT is clearly defined based on the context in which the CAT defines that it is to be used.

This approach further simplifies the verification problem by enabling support for some portions of the assurance case development based on the basic analysis of CAT structure. It defines a clear hierarchy of CR specifications based on the relationships between higher and lower level CATs.

System Checkout: *C_SC* / *Input*	A Priori		Active Motor ID (*X*)		Active Motor Current Pose		Goal Change In Active Motor Pose (*Y*) (+/- angle)		Camera Motor Current Pose		Goal Change In Camera Motor Pose [pan angle, tilt angle]		Camera Data		Old State	New State	Notes
Output	Cond	Beh ID	Cond	Beh ID	Cond	Beh ID	Cond	Beh ID	Cond	Beh ID	Cond	Beh ID	Cond	Beh ID			
Comms Output													P: avail. O: avail.	*C_comms*	*C_cd* *T: Success*	*C_comms*	Send camera data as test message when camera data check is done
Active Motor ID	P: Start Pose_c and t_c	*C_ptSC*	O: Camera	*C_ptSC*					P: avail. O: avail.	*C_ptSC*	P: avail. O: avail.	*C_ptSC*			*C_mXmSC* *T: Success*	*C_ptSC*	Run camera pan/tilt motor check when the gripper motor check is done
	P: Start Pose_m and t_m	*C_mXmSC*	O: Gripper OR Arm	*C_mXmSC*	P: avail. O: avail.	*C_mXmSC*	P: avail. O: avail.	*C_mXmSC*							*Start* OR *C_mXmSC* *T: Success*	*C_mXmSC*	Run gripper motor check after arm motor check
Camera Pan/Tilt Command	P: Start Pose_c and t_c	*C_ptSC*	O: Camera	*C_ptSC*					P: avail. O: avail.	*C_ptSC*	P: avail. O: avail.	*C_ptSC*			*C_mXmSC* *T: Success*	*C_ptSC*	Run camera pan/tilt motor check when the gripper motor check is done
Camera Data Generation Command	P: test image parameters	*C_cd*													*C_ptSC* *T: Success*	*C_cd*	Generate camera data when the pan/tilt check is done
Motor Command (Active Motor ID)	P: Start Pose_m and t_m	*C_mXmSC*	O: Gripper	*C_mXmSC*	P: avail. O: avail.	*C_mXmSC*	P: avail. O: avail.	*C_mXmSC*							*C_mXmSC* *T: Success*	*C_mXmSC*	Run gripper motor check after arm motor check
Motor Command (Active Motor ID)	P: Start Pose_a and t_a	*C_mXmSC*	O: Arm	*C_mXmSC*	P: avail. O: avail.	*C_mXmSC*	P: avail. O: avail.	*C_mXmSC*							Start	*C_mXmSC*	Run arm motor check
Fault Command															*Any* *T: Fail*	*C_fr*	Exit when any behavior fails
End															*C_Comms* *T: Success*	Exit	Exit when comms test message sent

Fig. 4.6 CAT for the *move-motor* component of the *system-checkout* behavior expressed in capabilities

The CAT itself defines the required input information and control outputs that the platform must provide to the behavior—any violation of these is easy to identify via a simple check of platform properties against CAT row and column elements.

Some inconsistencies in the CAT can be found using formal tools. For example, if there are rows that define valid ranges of input variables, it is easy to check that there is a definition for correct system behavior outside of those ranges by scanning that column and checking for those conditions. Because threshold requirements can only be specified in the particular context of the behavior that uses them, assessment of their consistency and validity must be made at that level in the CAT hierarchy.

The redundancy in the CAT representation is intentional—it provides the user with a manual mechanism to perform basic consistency checks and to understand the relationships between the hardware and the behavior requirements. Eliminating the redundancy would force users to rely on a computer to interpret the tables and would require them to trust the interpretation software.

The CR is set up to allow the specifier and the designer to explicitly call out platform constraints, but taxonomy like that suggested in [16] is needed to support clear communication between stakeholders. This approach supports a standardized representation of goals, evaluation mechanisms, and context without implying or enforcing specific design architecture.

Because we define a specification using a CR and define a design using a CAT, it is possible to clearly demarcate where a specification ends and a design begins. If information is in the CR, it is *by definition* part of the specification and not part of the design. If the CR defines the contents of the cells of its corresponding CAT, it has devolved into behavior design rather than autonomy specification.

This approach also potentially allows feedback between the physical/ground truth layer and the formal methods verification tools by providing structured representations for both the specification and the design that are easily translated to existing formal modeling tools while reducing the human burden of capturing the system needs, constraints, and goals.

4.5 Conclusion

This chapter set out to address the problem of specification development as a necessary precondition for autonomy verification through tools and processes that clearly define the boundary between the specification and the design, which generate a testable specification at the desired level of abstraction without requiring the people generating the specifications to design the autonomous behaviors.

Fundamentally, the core problem with specification of autonomous behaviors is not only that it is extremely difficult to generate the specification in the first place, but also that once we have the specification it is extremely difficult to evaluate whether the system meets it or not. Together, the Task/Behavior Duality Principle, the Capability Representation, and the Capability Analysis Tables support both specification development and assessment.

The Task/Behavior Duality Principle enables us to address the challenge of determining at what level we are transitioning from specifying requirements to designing the system and how we can develop principled requirements that do not devolve into designing the solution. Because the Capability Representation specifically allows representation of uncertainty (which defines the level at which the autonomous behavior is expected to handle things) and abstraction (which defines the domain within which the designer must operate), it supports comparisons between desired and provided capabilities. The level at which we transition from specifying requirements to designing the solution is the level at which that uncertainty and abstraction are defined in the desired capability.

Capability Representation provides a descriptive framework that does not constrain the developer to specific tools but is flexible enough to capture specific cases. It provides a mechanism for capturing the specification of what is to be analyzed (the behavior) and how it is to be evaluated (the task). It explicitly supports representation and analysis at both the specific and broad levels through the use of abstraction and uncertainty representations. Because the Capability Representation is built on a common classification system, it will define a standard approach to evaluation representation. As it matures it will enable disambiguation between cases where a specification is incorrect (the task description abstraction has failed to capture some required system action) and cases where the environmental model is incorrect (the environmental abstraction failed to capture some critical system-environment interaction). Task descriptions, system responses, and environmental assumptions are all captured explicitly, enabling future development of generalization relationships, as well as task/subtask relationships and evaluation of safety and functionality.

The Capability Representation approach also enables the user to focus on not only the explicit metrics but also elucidating the implicit metrics by which autonomous system behaviors are measured. It pushes the specification developer to require the user to specify particular elements of the problem in order to develop a complete and valid specification. Invalid task specifications that fail to include necessary information are obviously incomplete; where uncertainty and abstraction are used to broaden the desired capability definition, the representation must address all the relevant aspects, forcing the user to explore regions of the specification space where implicit requirements exist.

Capability Analysis Tables provide the context for a structured process allowing feedback between the physical layer and formal methods verification tools. The process of developing a CAT requires iterated sequences of requirements generation and behavior design and forces feedback between the physical layer and the formal representation. It provides a standard mechanism for system interface representation.

With regard to the specific problem outlined at the beginning of this chapter, the interaction between the Capability Representation and the Capability Analysis Tables supports the development of testable specifications at the desired level of abstraction. The specification is expressed in terms of a Capability Representation, while the design is expressed in terms of a corresponding top-level Capability

Analysis Table. The CR includes specific tags that define the testing necessary to determine whether the system meets its specified evaluation metrics or not. The CAT includes all the design elements and sublevel Capability Representations. The behavior specification defines the level of abstraction and the uncertainty. The CAT defines the details where the CR specifies abstraction.

References

1. Brooks R. 'A Robust Layered Control System for a Mobile Robot'. *IEEE Journal of Robotics and Automation*. 1986;**2**(1):14–23
2. Luckcuck M., Farrell M., Dennis L.A., et al. 'Formal specification and verification of autonomous robotic systems: A survey'. *ACM Computing Surveys (CSUR)*. 2019;**52**(5):1–41
3. Weyns D., Malek S., Andersson J. 'FORMS: Unifying Reference Model for Formal Specification of Distributed Self-Adaptive Systems'. *ACM Trans. Auton. Adapt. Syst.* 2012 May;**7**(1)
4. Abrial J.R. *Modeling in Event-B System and Software Engineering*. Cambridge University Press; 2010
5. Tarasyuk A., Pereverzeva I., Troubitsyna E., et al. 'Formal Development and Assessment of a Reconfigurable On-board Satellite System'. Ortmeier F., Daniel P., (eds.). *Computer Safety, Reliability, and Security. SAFECOMP*; vol. 7612 of Lecture Notes in Computer Science. Springer; 2012. p. 210–222
6. Finucane C., Jing G., Kress-Gazit H. 'LTLMoP: Experimenting with language, Temporal Logic and robot control'. *Proceedings of the 2010 IEEE/RSJ International Conference on Intelligent Robots and Systems (IROS)*; ; 2010. p. 1988–1993
7. Heitmeyer C.L., Leonard E.I. 'Obtaining Trust in Autonomous Systems: Tools for Formal Model Synthesis and Validation'. *2015 IEEE/ACM 3rd FME Workshop on Formal Methods in Software Engineering*; ; 2015. p. 54–60
8. van Lamsweerde A. 'Goal-oriented requirements engineering: A guided tour'. *Proceedings Fifth IEEE International Symposium on Requirements Engineering*; IEEE; 2001. p. 249–262
9. Goldsby H.J., Sawyer P., Bencomo N., et al. 'Goal-based modeling of dynamically adaptive system requirements'. *15th Annual IEEE International Conference and Workshop on the Engineering of Computer Based Systems (ECBS 2008)*; IEEE; 2008. p. 36–45
10. Vassev E., Hinchey M. 'Engineering Requirements for Autonomy Features'. *Software Engineering for Collective Autonomic Systems: Results of the ASCENS Project*; vol. 8998 of Lecture Notes in Computer Science. Springer Verlag; 2015. p. 379–403
11. Vassev E., Hinchey M. 'Knowledge Representation for Adaptive and Self-Aware Systems'. *Software Engineering for Collective Autonomic Systems: Results of the ASCENS Project*; vol. 8998 of Lecture Notes in Computer Science. Springer Verlag; 2015. p. 221–247
12. Smith B., Millar W., Dunphy J., et al. 'Validation and Verification of the Remote Agent for Spacecraft Autonomy'. *1999 IEEE Aerospace Conference. Proceedings (Cat. No.99TH8403)*; vol. 1. Snowmass at Aspen, CL, USA; 1999. p. 449–468
13. Brat G., Jonsson A. 'Challenges in verification and validation of autonomous systems for space exploration'. *Proceedings of the 2005 IEEE International Joint Conference on Neural Networks 2005 (IJCNN)*; ; 2005. p. 2909–2914
14. Balakirsky S., Schlenoff C., Rama Fiorini S., et al. 'Towards a Robot Task Ontology Standard'. *ASME International Manufacturing Science and Engineering Conference*; vol. 3, Manufacturing Equipment and Systems; 2017.
15. Olszewska J.I., Barreto M., Bermejo-Alonso J., et al. 'Ontology for Autonomous Robotics'. *2017 26th IEEE International Symposium on Robot and Human Interactive Communication (RO-MAN)*; ; 2017. p. 189–194

16. Redfield S.A. 'A review of robotics taxonomies in terms of form and structure'; 2021. ArXiv 2101.02659
17. Redfield S.A., Leonard E.I., Lennon J. 'Task specification and behavior verification for UUV behavior design'. Ehlers F., (ed.). *Autonomous Underwater Vehicles: Design and practice*. IET; 2020. p. 391–418
18. Braitenberg V. *Vehicles: Experiments in Synthetic Psychology*. Cambridge, MA: MIT Press; 1984
19. Edwards V., McGuire L., Redfield S. 'Establishing Reliable Robot Behavior using Capability Analysis Tables'. Cardoso R.C., Ferrando A., Briola D., et al., (eds.). *Proceedings of the First Workshop on Agents and Robots for Reliable Engineered Autonomy*; 319. EPTCS; 2020. p. 19–36. Proceedings https://doi.org/10.4204:EPTCS.319. arXiv DOI: 2007.11739

Chapter 5
A Philosophical and Mathematical Framework for Associated Problems of Hierarchical Verification of Autonomous Systems

Andrew Bouchard, Richard Tatum, Benjamin Hartman, and Demetrious Kutzke

Chapter 5 Editorial Note

This chapter describes a new conceptual and mathematical framework for comparative analysis of capabilities on the basis of mathematical functions describing their interfaces and a distance metric based on availability, connectedness, and performance. This new technique uses clear specifications of the system's capabilities to define its usefulness, functionality, and limitations and provides insight into one method for using analytic data to predict performance. It addresses technical challenge 2 (standardization of evaluation and interfaces through common models and frameworks) by providing a common method for analyzing compatibility and effectiveness of different combinations of autonomous system behavior modules as instantiated in similar systems. Defining capabilities as functions enables standardization of system interfaces. The introduction of the tatum, a measure for capturing the various attributes of the performance of a system, addresses challenges 16 (consistent measurement of autonomous behavior) and 17 (definition of metrics that enable performance comparison across platforms and behaviors). Further, by supporting analysis at an arbitrary level of abstraction, this chapter addresses both challenge 19 (how to generalize performance from a finite set of samples) by enabling comparison across systems and system components, specifically in terms of how performance generalizes across parameter value changes, as well as challenge 23 (tools that balance broad and specific representations) by providing a descriptive framework that functions at whatever level of abstraction desired, from the specific to the broad. Stakeholder concerns related to functionality, usefulness, and limitations are discussed in more

A. Bouchard · R. Tatum (✉) · B. Hartman · D. Kutzke
NSWC PCD, Panama City, FL, USA
e-mail: andrew.bouchard@navy.mil; richard.d.tatum@navy.mil; benjamin.hartman@navy.mil; demetrious.kutzke@navy.mil

S. Redfield et al. (eds.), *Verification of Autonomous Systems*,
https://doi.org/10.1007/978-3-031-88546-4_5

depth in the Chap. 1, Introduction. Technical challenges are introduced in Chap. 2, Verification Challenges.

Abstract The ubiquitous presence of autonomous systems has given rise to the need for verification to ensure transparency and trustworthiness of such systems. Traditional methods that treat autonomous systems as monolithic are not tractable given the complexity of in situ vehicle behavior in a stochastic environment. To assist with the problem of verification, we propose a new ontology, the form of a hierarchical abstract modeling language for autonomous system capabilities. This abstraction is useful in that it establishes a conceptual model describing capabilities of autonomous systems and the relationships between them to represent complete systems. Building upon this conceptual model, we develop a mathematical framework that describes autonomous systems as graphs of capabilities capable of analysis and develop a metric for the space of verified capabilities. These mathematical tools enable a formal, analytical comparison of multiple systems. With the mathematical framework in place, we demonstrate its use by considering and simulating a simple example involving an autonomous vacuuming robot.

5.1 Introduction

Historically, the widest use of robotic systems has been as stationary systems in structured environments performing limited and rigidly defined tasks for which specific constraints have provided a means of reducing the risk associated with their operation. However, recent years have seen a dramatic rise in the employment of mobile autonomous systems operating in unstructured environments and interacting directly with humans. The greater complexity of these systems and their environments has presented a significant challenge to providing assurances regarding performance and safety. While the risks associated with current household robots may be small, those associated with personal vehicles, weapons of war, and even financial management systems have proven to be much more significant, with impacts measured in millions of dollars and in human lives. These risks and their associated costs demand investment in test and evaluation of these systems to provide some assurances of performance, but previous work suggests that their inherent complexity and the stochastic nature of their operating environments makes traditional forms of verification intractable [1–6].

This complexity has stymied the development of verification solutions for monolithic autonomous systems [7], suggesting that a modular, compositional approach is applicable [7, 8]. Several solutions proposing such an approach have shown some isolated success, including compositional verification [7–10], verification by assume-guarantee contracts for probabilistic systems [11], automated proof verification for formally defined state space models [12], and natural deduction in minimal logic systems [13]. However, these techniques provide assurances limited

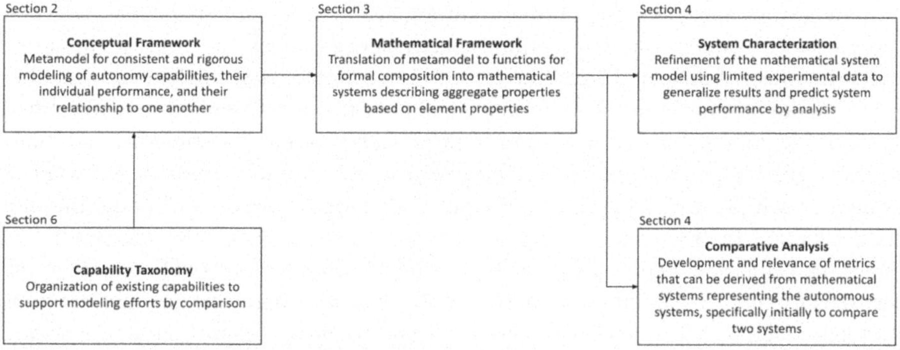

Fig. 5.1 Proposed verification process described in this work

to a particular technology and scope; what is needed is a general framework for analysis that is generic to the specific implementation or context used and which can compare similar systems to support decision-making. In fact, the definition of "similar" in this context is itself problematic given the lack of a coherent ontology or taxonomy upon which autonomous systems can be rigorously and formally described and compared. This lack of a consistent descriptive structure has limited efforts to develop mathematical structures to quantitatively analyze such systems more broadly.

To address these problems, we describe a process (shown in Fig. 5.1) for modeling and analysis of autonomous systems by considering them as systems composed of independently verified capabilities, develop a framework for their formal analysis, and then apply analytical tools and metrics to that framework in order to provide reliable estimates of full system performance based on limited experimental data to inform evaluation efforts. In Sect. 5.2 we introduce a new ontology for describing autonomous systems in the form of a hierarchical abstract modeling language for autonomous system capabilities. This language, defined using a metamodel, creates a consistent structure for the formal description of individual autonomy capabilities and definition of relationships between those capabilities to model complete autonomous systems. In Sect. 5.3, we extend this conceptual framework to develop a mathematical framework for formal analysis of collections of interrelated capabilities using category and sheaf theory. This framework provides the necessary structure to compose individual capabilities into a mathematical system that formally characterizes the functionality of an autonomous system relative to its inputs and outputs. We expand upon these frameworks in Sect. 5.4 to describe a metric that measures the differences between systems using system and capability attributes. While measuring the mathematical closeness of two systems is important, we need guarantees for how closeness in our metric space relates to similarity of system performance. Furthermore, we require a way to estimate system performance when system performance is not available. Therefore, we show how to use capability performance as an upper bound for system performance, as well as provide sufficient and necessary guarantees for when closeness of systems implies similar system performances. In order to illustrate the principles of this approach, in Sect. 5.5 we model and analyze a sweeping robot model to show several important aspects of this work applied to a simplified system. We demonstrate composition of the system from its constituent capabilities to show the efficacy of our metric with regard to scenarios in which we want to compare multiple real or hypothetical systems that may or may not meet our mission requirements. In Sect. 5.6, we lay out an approach for using the metamodel to support an emergent taxonomy of autonomy capabilities to aid modeling and outline the development of additional metrics for formally characterizing autonomous systems. Section 5.7 summarizes the work and identifies opportunities for future work to implement and develop these frameworks as part of a complete verification process and leverage them to develop new metrics.

5.2 Conceptual Framework

The foundation of our approach is a formal conceptual framework establishing an ontology for describing autonomous systems based on their capabilities. Definition of such an ontology is essential to any effort aimed at describing and assessing autonomous systems because it establishes the concepts and categories of these systems and therefore determines the philosophical basis for analytical tools and techniques. We define this ontology in the form of a metamodel that describes a language for modeling the capabilities of autonomous systems as hierarchical functions that can be linked to describe complete systems.

5.2.1 Capability Metamodel

Previous efforts to provide definitions and models for autonomous systems attempted to define a closed set of concepts and categories either for the systems as a whole or for specific components that compose them [3–5]. These efforts were hampered by the breadth and complexity of autonomous systems and their associated technologies; as quickly as these descriptive structures were created, counterexamples were identified that showed them insufficient for current applications. We argue that these previous efforts were hindered by five major weaknesses:

- **Restrictive Categories**: Whatever aspect of autonomous systems was used to define categories, these efforts generally attempted to completely define all categories a priori. We suggest that the multidisciplinary nature of autonomous systems and pace of innovation in the field makes it intractable for any reasonably sized group of experts to completely define all possible categories.

- **Implementation Assumptions**: Many of these efforts-based definitions or categories on the particular technology or technique were misused. We suggest that this created ambiguity and confusion when attempting to describe systems in which the same technique was used to implement different functions or to compare cases in which the same function was realized using different techniques.
- **Assumption of Access**: Likewise, many previous efforts [14–16] to define metrics for autonomous systems required information specific to algorithms, software code, or other details of implementation that in commercial contexts are generally protected as intellectual property. We suggest that a reasonable reluctance on the part of commercial companies to share this information proved an impediment to the use of such solutions.
- **Full System Characterization**: Past efforts [4, 17, 18] have focused on providing structures, definitions, and metrics to describe complete autonomous systems. We suggest that the complexity of these systems and the variety of attributes used to describe them made the development of meaningful standards for describing them as monolithic systems intractable.
- **Subjective Evaluation**: Most notably in the cases of the various "levels of autonomy" frameworks [4, 6, 19] and likely due to the complexity issue described above, many previous attempts to provide a means for verification for autonomous systems have relied on subjective evaluation by some expert, a qualification that may be defined in many ways. We suggest again that the complexity of these systems and the breadth of possible ways to characterize their behavior make them particularly susceptible to bias in such evaluations and made these efforts particularly ineffective. This can be seen in numerous examples in which experts evaluating the same systems with the same guidance rated systems very differently [14, 20–22].

Having identified these limitations to existing efforts, we sought a solution that, rather than imposing a complete listing of concepts and categories on the community, would define a descriptive structure and allow those attributes to develop organically from the community, growing as innovation and expertise bring forth new ideas. Additionally, an approach was desired that allowed autonomous systems to be decomposed such that their characterization was more tractable than as full systems. Therefore, what was desired was not a comprehensive model for all autonomous systems, but a consistent language for the community to formally describe the fundamental components that make up autonomous systems. Borrowing from concepts in psychology used to model and assess human behavior, this resulted in the identification of capabilities as the fundamental component of an autonomous system and the definition of a metamodel that addresses all of the weaknesses described above:

- **Defines Structure Rather Than Content**: The use of a metamodel is not restrictive in the types and attributes that can be defined, but only creates a common structure for their definition, facilitating the emergence of standards rather than their imposition.

- **Independent of Implementation**: Because this approach models abstract capabilities, the metamodel is agnostic to the implementation and allows different applications of the same techniques to be clearly distinguished and different implementations of the same purpose to be immediately compared.
- **Protects Intellectual Property**: The metamodel describes capabilities according to their interfaces with other capabilities or with the environment; therefore, no internal information is required that might violate intellectual property considerations.
- **Characterization by Parts**: The hierarchical nature of the metamodel allows an autonomous system to be modeled at whatever level of abstraction capabilities can be defined, making the effort more tractable than attempting to fully characterize a complete system.
- **Objective Data for Evaluation**: The metamodel describes capabilities as functions that operate on input data to provide output data; this data and the associated quality metrics provide objective criteria for evaluation of performance.

Assuming that the goal of verification of a system is only to assess the quality of its performance rather than to determine how a system accomplishes a given function, this approach is especially well-suited for the purpose.

5.2.2 Metamodel Description

The metamodel for the formal description of the functions of an autonomous system is founded on the identification and modeling of an individual **capability**, for which a formal definition is provided in Table 5.1 along with the definitions of the other elements of the metamodel. This formalism defines a system modeling language to describe the individual operations of an autonomous system; we later show that these operations can then be linked to describe the system as a whole. This approach assumes that any capability or function of an autonomous system can be considered as a collection of hierarchical sub-functions (sub-capabilities), each of which is necessary to accomplish the top-level function; each of these sub-functions can similarly be taken as a top-level function and decomposed into a collection of sub-

Table 5.1 Definitions of capability metamodel elements

Capability	"Defines what a program (e.g., a service, a business process, a sensor) does from a functional perspective" [23–25]
Component	"A modular part of a system that encapsulates its contents and whose manifestation is replaceable within its environment" [26]
Data	Information that is either provided to a component (input) or provided by a component (output) as a normal condition or product of its operation
Tatum	A characteristic of data that describes its quality
Constraint	A limiting factor that must be satisfied for the model to be valid

functions. Note that the distinction between a function and a sub-function or the separation between sub-functions is arbitrary and represents a choice made by the system modeler that can be used to select interfaces that can be verified.

Taking a capability as a function of the system, it may be considered as an operation of the system that takes in some input and generates some output. The metamodel represents these inputs and outputs generally as *data*, acknowledging that the data will in some instances describe a measurable aspect of either a phenomenon in the environment or the action of some effector. On this basis, we assert that the performance of an autonomous capability can be measured based on the quality of the output that it produces relative to the quality of the input that it requires. Broadly, the performance of a capability is proportional to the quality and quantity of the data it produces and inversely proportional to the quality and quantity of the data it requires for its operation. That is, a "better" capability will produce more and higher quality data given less and poorer quality data. We define the term *tatum* to refer generally to metrics that describe data quality, since the specific metrics will vary by data type.

Each capability must be realized by some *component* which instantiates the functionality described. Note that these components may be either software or hardware as long as they are modular and replaceable. Additionally, a component may instantiate multiple capabilities or single capabilities be instantiated by multiple components, which may provide a formal metric for system modularity.

Finally, the operation of a component or a capability may be described with respect only to certain assumptions, such as operating within some specific environmental constraints or requiring the presence of some other component or resource as a dependency. We model these assumptions as *constraints*, which may be specific to either the component (e.g., maximum operating temperature) or the capability (e.g., assumed underwater).

Figure 5.2 shows the resulting metamodel in UML format. Note that because this metamodel defines the structure of a capability rather than its content, it is not limited to any type of function or technical implementation. The utility of this approach to verification is based on the assumption that it is significantly easier to characterize and test the operation of a single function than it is a complete system,

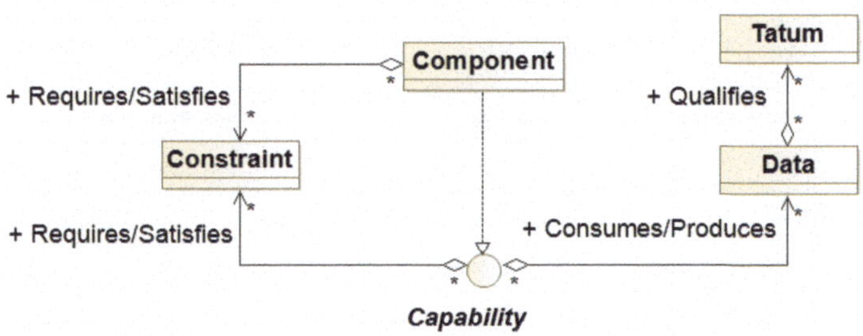

Fig. 5.2 Capability metamodel for system description, shown in UML format

and the characterization of that function can be expressed in terms of the outputs provided given required inputs under defined assumptions.

We define the metamodel broadly, recognizing that a complete implementation of this conceptual framework must be flexible to support broad definitions of these capability attributes. In practice, the tatums that describe the quality of output data may be multivariate and a function of the input data, environmental factors, or constraints. While we develop the framework generically, we focus on proving the simple case in which data inputs and outputs are scalar values described by tatums that define the allowable values for that data as an interval.

Example We illustrate the application of the metamodel with a sweeping capability used in the example in Sect. 5.5. The capability is instantiated by a sweeper component, whose performance is dependent on the type of floor (modeled as an input) and whose output is the removal of soil from the floor (modeled as an output). The performance characteristics of the sweeping function of the robot are based on the Neato D4 as characterized in [27], indicating that the attribute of the input that affects the capability performance is the floor type and the quality of cleaning is given as a percentage of soil removed based on that floor type; these performance attributes are modeled as tatums of the input and output. The resulting capability model is shown in Fig. 5.3.

Having established a metamodel capable of describing the individual capabilities of an autonomous system, we next consider whether the same metamodel can be leveraged to describe complete systems. We suggest that this can be accomplished by linking related capabilities where the output data of one capability is the input data of another and where the constraints required by one capability are satisfied by another. In order to show that this is possible, consider some autonomous system with capabilities defined according to the metamodel. The inputs for each of these capabilities must either be provided from outside the system or be the outputs of

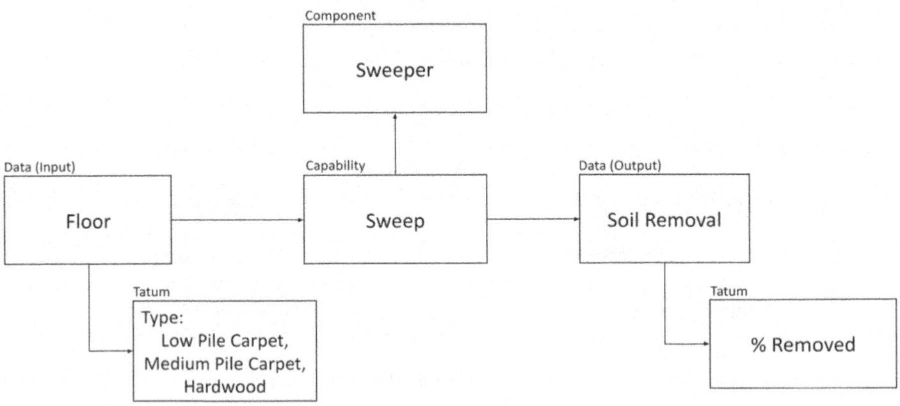

Fig. 5.3 Expression of the sweeping capability using the defined metamodel. Performance characteristics are based on the Neato D4 as published in a CNet assessment [27]

other capabilities within the system. Likewise, constraints for capabilities within the system must either be satisfied by other capabilities within the system or become constraints on the system as a whole. In this way, a complete autonomous system can be modeled by enumerating and connecting all of the capabilities that define its behavior.

The astute reader will note that this work has so far declined to hazard a definition for autonomy or to bound the application of this work to a particular subset of these systems. In addition to simply avoiding the controversy over that terminology, the key reason for this omission is that, because the proposed solution is a framework, it is limited in application not by a particular definition or level of autonomy, but by those capabilities that can be represented and their performance characterized using the proposed metamodel.

5.3 Mathematical Framework

The conceptual framework established above provides a consistent structure for the definition of the individual functions of an autonomous system and their characterization according to quantitative attributes of performance. It further provides a way to model an autonomous system as a whole by connecting related functions based on their inputs and outputs. Our intuition suggests that the attributes of these individual capabilities affect one another in such a way that, to the extent that they are verified, they can be somehow combined to describe and verify the attributes of the system as a whole. What is desired, therefore, is a mathematical formulation of the conceptual framework that matches this intuition.

In order to develop such an analytical framework for this application, we use category theory to formalize the representation of an autonomous system as a directed graph in which the capabilities of that system are the vertices of the graph and the data connections between capabilities are the edges. To supplement the capability graph, we employ sheaf theory to combine simultaneous capabilities together along common input and output data types to produce shared requirements and performance metrics [28]. The result is a model of capabilities as functions with clear metrics and constraints, definitions of the necessary conditions of a capability for this model to be valid, and a set of rules for the composition of capabilities.

5.3.1 Translation of Metamodel

We begin by representing conceptual definitions defined in Sect. 5.2.1 in a mathematical framework whose mechanics match our intuitions for the process of verification. As it is central to the verification philosophy we adopt, we will first discuss the definition of a capability within this context, the mechanisms by which capabilities come together to model the system, and the assumptions we make when adopting this framework. Recall that we would like to use capabilities to represent a broad range of possible behaviors of the system that its components can have. We would also like to ensure that the intuitive steps a designer would take to extrapolate system performance from the performances of its capabilities are reflected in the way capabilities are defined and allowed to be combined within the mathematical framework. To that end, we treat capabilities as functions while making the data types and constraints of a capability explicit, so interactions between them can be clearly resolved. By defining tatums as below, we also ensure that the relevant behavior of a capability can be modeled even when detailed performance data is not available. Finally, a clear, intuitive definition of a verified capability is necessary to ensure that combining capabilities produces verified composite behaviors of the system.

Definition: A *capability* C is a function $f_C : D_C \rightarrow R_C$ that represents an ability of any number of components of a system. The capability has a domain of states of the world D_C over which the capability will function and affect the state of the world, resulting in a number of range states R_C. Note that the state of the world with respect to a given capability includes the states of other components of the system. As a convention, when we define a new capability function f_C, we define a capability's domain D_C and range R_C so that f_C is surjective. It is useful to describe a capability's domain and range in terms of its *input and output data types*. Let D and R be the total span of possible states the system and environment can take at the moments a capability's input and output are processed, respectively. We refer to a capability's data types by the units over which D and R must be restricted to produce D_C and R_C, and consequently the units for the input and output data need not be the same. A capability also comes equipped with constraints N_C, a set of conditions that are required for the capability to function that are distinct from the domain and range.

Definition: The *tatums* t_C of a capability C are an approximate representation of the requirements in the domain that one needs to meet, in practice, for the capability to produce an expected guarantee in the range. Applying a tatum t to the capability function f_c converts the functional language into contract language. That is, $t_c = t(f_c) : t(D_C) \rightarrow t(R_C)$, where $t(D_C)$ is a set of assumptions, and $t(R_C)$ is a corresponding set of guarantees if those assumptions are met.

Definition: A capability C is *verified* if:

- f_C is well defined and nontrivial (consistently inputting the same data will generate the same expected results).
- t_C, if defined, has been tested and is confirmed to be a standard the capability holds to in practice.
- Any behavior encoded in f_C has been sufficiently confirmed through testing and/or analysis.
- The constraints N_C are confirmed to accurately represent the limitations of C. C will otherwise be verified if N_C are met and will not be verified if the constraints are not met.

A point of tension between the abstract definition of a capability and the pragmatic considerations of verifying an actual system is determining the relevant input and output data types. Even when a verifier is able to identify the primary data types involved in a capability's function, they cannot expect to completely ignore the effect of unmeasured variation in the remaining state of the world on its output performance. However, if the capability models the performance of the component(s) it represents well enough, all of the data types which significantly affect performance will be represented in its domain and range. A well-defined capability is, intuitively, one which does not display unexpected behavior over a series of similar expected operating scenarios, although applying this definition too surgically can lead to practical standards that are impossible to meet. From another perspective, a behavior of the system which is not well defined is likely not properly classified as a capability if there is no clear relationship between inputs and outputs.

The standards for verification of individual capabilities are left intentionally broad to allow solutions to vary based on application. This is important because the methods available to verify a given capability may depend on its implementation and technical details of the application; this approach allows experts in those details to set the standards within their domain rather than attempting to broadly impose a one-size-fits-all approach. Implicit in this understanding is the idea that the stronger the evidence confirming the model of each capability, the stronger the claim that the system is verified becomes. Through the remainder of this primer, when we speak of a capability or a system being verified, understand that this means it is verified at a probability which meets the standards of the verifier.

5.3.2 Composition of Capabilities

Definition: The *capability graph* is a directed graph in which the vertices are the capabilities composing a system and the edges from the source capability to the sink capability represent transfer of data in an autonomous system, i.e., points in the system where the range of the sink capability intersects nontrivially with the domain of the source capability.

The capability graph paints a picture of the autonomous system, but it does not contain all the structure necessary to build it from its components. The graph indicates which capabilities occur in sequence, but it does not represent which ones occur simultaneously. Because the processes associated with an autonomous system are complicated such that capabilities may be used multiple times or never used at all during its operation, we cannot easily use a time variable as a way to manage the structure of the capabilities. We assume that the capabilities can be sensibly classified into groups that represent their overall function in an autonomous system (e.g., sense, orient, plan, and execute). The purpose of these groups is to provide a baseline to guide the verifier in checking that the requirements for multiple capabilities are consistent with each other; capabilities in the same group draw on the same pool of data and output a pool of data which is itself used by another group of capabilities. More specialized frameworks that focus on classifying capabilities by their function, such as the OODA loop, are well-accepted examples of such groupings and have the advantage of being highly intuitive. By classifying capabilities into groups, we can keep some of that intuition while ensuring that the metamodel is as general as possible in describing many different types of autonomous systems.

With this structure imposed, we now describe the process of collapsing the capability graph into a representation as a mathematical system suitable for analysis. This simultaneously formalizes the steps to verify the system and provides a way to determine if two capability graphs are equivalent, representing the same system.

Theorem 5.3.1 (Composition of Sequential Capabilities) *If A and B are separate, verified capabilities, the composition of capabilities $C = B \circ A$ may be defined such that $f_c = f_B \circ f_A : D_C \to R_C$ and $N_C = N_A \cup N_B$, where $D_C = f_A^{-1}(R_A \cap D_B)$ and $R_C = \{f_B(f_A(x)) | x \in D_C\}$. It is verified if $R_A \cap D_B \neq \emptyset$ and the conditions in N_C may be met simultaneously. If $t(D_B) \subseteq t(R_A)$, t_A and t_B may be composed of t_C; when the assumptions of A are met, the guarantees of B are delivered. Otherwise, new tatums will have to be defined to capture the new metrics of C. Physically speaking, this composed capability represents the result when two capabilities are applied in sequence.*

Proof With the definition for f_C and N_C above, we have all the information necessary to define C as a capability. If $R_A \cap D_B \neq \emptyset$, f_C is nontrivial, and $f_B \circ f_A$ is well defined if f_A and f_B are. The behavior of f_C is defined by the behaviors of f_A and f_B and is supported by the empirical tests which support the models of $f_A$and f_B. If the conditions in N_C may be collectively met, then both N_A and N_B are met, and f_A and f_B are verified. Otherwise, either one of f_A and f_B is not verified, and therefore f_C is not verified. □

Theorem 5.3.2 (Further Composition of Capabilities) *Let $f_1, \ldots, f_n$ be a series of capabilities for which $f_1 \circ \cdots \circ f_i$ and f_{i+1} meet the conditions of Theorem 5.3.1 for all i. Then there exists some f which represents the composition of all n capabilities in sequence.*

Proof *True by Induction* □

Because capabilities are broadly represented as functions and contracts, composing them in sequence is not difficult to manage. Combining capabilities that are meant to operate at the same time, however, requires a bit more formalism. Essentially, we want to create a set of restriction maps that map from the shared environmental data seen by all capabilities to every possible snapshot of shared data that would need to be consistent at each simultaneous interaction. For this, the problem is best formulated in sheaf theoretic language, as in [29].

Definition: A *scenario space* is a set of all data types that are relevant for the operation of the system within its specified goals, together with a topology on that set generated by a basis $\mathscr{B}$, where a subset of data types is in $\mathscr{B}$ if those data types are the ones operated on by either the inputs or outputs of a capability in the system.

Broadly, it represents the data types an environment comprises at a specific moment during the system's operation.

Definition: The *attributes* of each data type are the values that it takes. The presheaf of a system at a particular state is a topology of data types together with a set of possible attributes for those data types as determined by the relevant capabilities at that particular state. The relevant capabilities are the ones which either output to or take input from this particular state of the world.

Definition: The *global sections* of a scenario space are attributes of the set of all data types in a scenario space. They represent a snapshot of an entire state of the world that is consistent with reality, a single instance of all relevant data which is acted on by multiple capabilities or is the result of action by multiple capabilities. Note that this snapshot may not necessarily be a single instant in time. For example, multiple sensing capabilities might analyze different areas of an environment at three different times before they each process the input and output the result data at a fourth time. In this case, the global sections of the shared input scenario space are the states of the environment at any three times it could be measured by the capabilities.

The goal in defining these topologies of data types is to create a mathematical framework by which we can discuss the simultaneous operation on or simultaneous effect of two or more capabilities on a state of the world. In so doing, we gain the ability to answer the following questions:

- Can the requirements of multiple capabilities be met simultaneously? $\rightarrow$Do global sections exist which, when restricted to the open sets corresponding to each capability's data types, meet that capability's requirements?
- If a capability existed which represented the simultaneous combination of other capabilities, what tatums would it have? $\rightarrow$What is the set of all global sections which meet all requirements when restricted as above?
- Can the output thresholds be met simultaneously? What performance metric can we assign to a combined output of capabilities? $\rightarrow$Similar to the above using the scenario space of the output environment as opposed to the input environment.
- If there are capability gaps, can we construct a theoretical system with capabilities that fill in those gaps? $\rightarrow$Can we determine a sheafification of a scenario space?

Lemma 5.3.3 (Composition of Parallel Capabilities) *Let capabilities $C_1, \ldots, C_N$ take input from one scenario space S_1 and output into another scenario space S_2, and assume their constraints may be met simultaneously. A candidate capability C_S which represents their combined performance is verified over domain D_S and range R_S if and only if the following holds:*

- *Locality: Across all global sections residing in D_S, the data restricted to each open set matches the requirements for each capability whose domain intersects with the data types represented in the open set. Similarly, the data for the performance metrics in R_S properly restricts to meet the performance requirements of all capabilities intersecting each open set.*

- *Consistency: If a candidate compound capability $C_{S'}$ is equal to C_S in performance when restricted to each smaller open set in the domain and range, then C_S is equal to $C_{S'}$ over the global environment.*

Proof When the assumptions hold, the scenario spaces defined by the inputs and all corresponding outputs of the capabilities are sheaves. The data from the capabilities can be glued together at the intersections of their domains in a unique fashion. To show the resulting capability is verified, we need to prove that it is well defined and performs to the standard set by the previous testing. C_S is well defined because each global section in the input scenario space maps to a unique global section in the output space, by the Consistency condition. By the Locality condition, C_S is also guaranteed to meet the performance standards set by testing; restricted to reach set of relevant data types, the function meets every standard the composing capabilities do over those domains.

The conditions are not just sufficient, but necessary. Without Locality, there are no states of the world where all the requirements of all capabilities are met simultaneously. Without Consistency, the resulting capability is not well defined, since the same input can produce various outputs represented in the various formulations of the compound capability. □

With these tools at our disposal, a fully verifiable system can be sequentially collapsed into a representation of the complete system given external factors. We now also have a method to determine if two capability graphs that organize their capabilities in a different fashion but represent the same system are equivalent.

Theorem 5.3.4 *The act of performing simultaneous composition followed by sequential composition of two capability graphs and checking if they collapse to the same function is an equivalence relation.*

Proof

- $x \backsim x$: The same capability graph is clearly equivalent to itself, as the same set of operations will be taken and result in the same resulting graph in each case.
- $x \backsim y$ implies $y \backsim x$: This process is completely independent of any symmetry considerations.
- $x \backsim y$ and $y \backsim z$ imply $x \backsim z$: Because the steps of this operation are clearly ordered and each step produces a unique result, the capability graph y reduces to in each equivalence is the same, so $x \backsim z$ because x and z reduce to the same capability graph.

□

5.3.3 Holism and Emergent Capabilities

Like all modeling frameworks, this approach to describing an autonomous system assumes that the model accurately captures all of the interactions that affect its operation. Therefore it is critical that the system modeler describes boundaries between capabilities such that they can be reliably characterized and capture all connections between those capabilities, so all interactions are included in the analysis. While such explicit modeling is a challenge when approximating natural systems, since autonomous systems are designed and constructed by humans this is assumed to be tractable as we know what inputs the capability understands and the outputs it generates. However, even when we assume that the parts of the system are properly modeled, it is necessary to address the possibility that some aspects of the system may only be understood holistically. If there are distinct behaviors that cannot be separated out as capabilities and can only be described by referencing the whole system, they may not arise even when all well-modeled behaviors are properly combined.

Ongoing research into autonomous systems describes the existence of such emergent capabilities or qualitatively more sophisticated capabilities of a system that are not explicitly designed but emerge as a result of interactions between lower level capabilities. A classic example is a Braitenberg vehicle, an early research robot with no explicit cognition but simply a circuit link between two right-left light sensors and two right-left wheel motors such that a wheel will drive forward proportional to the amount of light sensed by the opposite sensor. The consequence of these simple rules is that the robot follows light, which it is not explicitly programmed to do, but a behavior that emerges out of the dynamics of the system.

While anticipation of such emergent capabilities would obviously be of value, our current understanding of these emergent capabilities is that they are additional capabilities that emerge as a result of the autonomous system correctly operating within the prescribed performance of the explicit capabilities. Because the focus of verification is to confirm the functionality of those explicit capabilities, this challenge of holism and the possibility of emergent capabilities do not interfere with that objective. Therefore, such emergent capabilities would not violate performance guarantees based on this analysis, and therefore the framework remains valid for verification of autonomous systems even in the presence of emergent capabilities.

5.4 Comparative Analysis

In this section, we expand the previous mathematical framework to build tools to assist with problems associated with comparing multiple systems. Effective comparisons of systems can assist with choosing between systems to best meet the mission requirements. In some cases, systems may be actual systems which have verified system performances in the specified environments, while in others there are no actual systems, but only non-integrated capabilities with independent verified capability performances. In either case, we want to save time and resources by allowing for the comparison of systems in a principled way. In this section, we describe a method to compare systems by measuring the distance between systems in a suitable metric space. The metric developed for system comparison is sensitive to system differences including the presence or lack of different capabilities, the difference in the connectedness of capabilities, and the performance of the capabilities. For connectedness, we are referring to how well the outputs of one capability will match the inputs of an associated, ensuing capability. Our metric will not necessarily guarantee agreement in system performance if two systems are close. This can be further exacerbated by the lack of system performance data. For the lack of system performance data, we show that capability performances act as a worst-case predictive surrogate for system performance when no system tests have been performed or no system performance data is available. We also give theoretical conditions for which closeness of systems will necessarily imply agreement with system performance.

5.4.1 Metric for Space of Verified Capabilities

We begin by comparing systems based upon the capabilities that each one has. The system and its capabilities are represented as a directed graph, which are the result of the composition operators as described in Sect. 5.3. We define the space of verified capabilities as $\mathcal{S}_V$ and note that all capabilities in this space have been independently verified. The capabilities can be atomic or composed of other capabilities; in either case, we assume they have been verified as standalone capabilities. A system $\mathbb{S}$ has connected capabilities, each with a defined set of constraints and a measurement of performance that varies over the environment external to the system. If we want to compare multiple systems, then we need to measure how each of these features of capabilities differs.

Capabilities of Systems For a system $\mathbb{S}$, its capabilities are given as

$$\alpha_C(\mathbb{S}_i) = \{C_1^{(i)}, \cdots, C_{n_i}^{(i)}\}, \tag{5.1}$$

which maps system $\mathbb{S}_i$ to capabilities $\{C_1^{(i)}, \cdots, C_{n_i}^{(i)}\}$. We can measure the difference between two systems by determining which systems have which capabilities. For systems $\mathbb{S}_1$ and $\mathbb{S}_2$, we can use the Jaccard dissimilarity metric and compute

$$d_C(\mathbb{S}_1, \mathbb{S}_2) = 1 - \frac{|\alpha_C(\mathbb{S}_1) \cap \alpha_C(\mathbb{S}_2)|}{|\alpha_C(\mathbb{S}_1) \cup \alpha_C(\mathbb{S}_2)|}. \tag{5.2}$$

See [30] for a discussion on the use of Jaccard dissimilarity. If $\alpha_C(\mathbb{S}_1) = \alpha_C(\mathbb{S}_2)$, then note that $d_C(\mathbb{S}_1, \mathbb{S}_2) = 0$, meaning that all capabilities are present in both systems with no extraneous capabilities present in either. On the other hand, if $\alpha_C(\mathbb{S}_1) \cap \alpha_C(\mathbb{S}_2) = \emptyset$, then $d_C(\mathbb{S}_1, \mathbb{S}_2) = 1$, meaning that the two systems have no capabilities in common. For the remaining cases, d_C will give values between 0 and 1.

System Connectedness Once we know that both systems have the same capabilities, we need to compare the connectedness of the two systems. To measure connectedness , for $1 < k < n$, we first assume that capability $C_k^{(i)}$ potentially accepts input from $C_{k-1}^{(i)}$ and potentially provides output to $C_{k+1}^{(i)}$. We use the word potential since it is expected for these capabilities to produce and consume data in this manner, but we do not know for sure until we determine if the inputs from one capability are necessary and sufficient for the next, intended capability. As discussed in the previous section, the inputs and outputs for a capability $C_k^{(i)}$ of system $\mathbb{S}_i$ are related as $f_k^{(i)} : D_k^{(i)} \to R_k^{(i)}$, which allows us to compute connectedness across all capabilities as

$$\beta(\mathbb{S}_i) = \frac{1}{n_i - 1} \sum_{k=1}^{n_i - 1} \left(\frac{|f(D_k^{(i)}) \cap R_{k+1}^{(i)}|}{|f(D_k^{(i)}) \cup R_{k+1}^{(i)}|} \right). \tag{5.3}$$

Note that $D_k^{(i)}$ is a restriction of the domain of $f_k^{(i)}$ due to the capability constraints. If $f(D_k^{(i)}) = D_{k+1}^{(i)}$ for all $1 < k < n$, then $\beta(\mathbb{S}_i) = 1$, meaning the system has capabilities that are completely connected in the sense that no capability is missing an input and not producing an unnecessary output. On the other hand, if $f(D_k^{(i)}) \cap D_{k+1}^{(i)} = \emptyset$ for all $1 < k < n$, then $\beta(\mathbb{S}_i) = 0$, which means every capability is missing all of its outputs and producing unnecessary outputs. For cases in between complete connectedness and disconnectedness, the metric will give values between 0 and 1. Comparing the connectedness of the two systems is achieved by computing

$$d_\beta(\mathbb{S}_1, \mathbb{S}_2) = |\beta(\mathbb{S}_1) - \beta(\mathbb{S}_2)| \,. \tag{5.4}$$

While we are primarily interested in comparing systems, the above metric is useful for locating breaks in the system data stream during the design stage for very complex systems; any value less than 1 means that there is a connection problem between two capabilities.

Capability Performance To evaluate capability performance, we need to evaluate how well a capability will produce a set of expected outputs, given a set of inputs; the probability of success of a capability measures how well the pairing of inputs to expected outputs occurs. Therefore, for system S_i, each capability $C_k^{(i)}$ has an associated performance function

$$\psi_k^{(i)} : D_k^{(i)} \to R_k^{(i)} \subset \mathbb{R}, \tag{5.5}$$

which measures how likely a capability will match inputs to the expected outputs. An example of such a function would be one that shows that under harsh environmental conditions, a perception algorithm is able to correctly identify 25% of the items in an image, while it can identify 75% of the items in the image under less difficult environmental conditions. For the sake of exposition, we will remove the superscript (i) when it is clear as to which system we are referring. If we fix the inputs into $\psi_k^{(i)}$ but change the parameters of the goal, it is possible to define a probability density function. For our performance function $\psi_k^{(i)}$, different input values can occur due to different environments, and as such, we group these inputs into the set $E_k^{(i)} \subset D_k^{(i)}$. The environment will consist of those external factors that can affect the performance of a system. We therefore identify the performance interval

$$I_E = [\psi_{min}, \psi_{max}], \tag{5.6}$$

which represents the range of performance values for the capability $C_k^{(i)}$ with inputs in $E_k^{(i)}$. Thus, we need a metric that can compute the difference between intervals or sets in general. We can use the Hausdorff metric and compute the capability performance difference between two systems as

$$d_p(\mathbb{S}_1, \mathbb{S}_2) = \sum_{\substack{C_i^{(1)} \in \alpha_C(\mathbb{S}_1) \\ C_j^{(2)} \in \alpha_C(\mathbb{S}_2) \\ C_i^{(1)} = C_j^{(2)}}} H\left(R_i^{(1)}, R_j^{(2)}\right). \tag{5.7}$$

The metric describes how to estimate the difference between systems in fixed environment E. This is an important yet subtle point that needs to be clear. **Extrapolating the capability performance from one environment to another environment without any testing or additional information is beyond the scope of this work and should be avoided if at all possible.** Future efforts will need to address the incurred risk associated with extrapolation with little relevant information available.

Metric for $\mathcal{S}_V$ The next lemma shows us how to combine all of these metrics for the comparison of two systems.

Lemma 5.4.1 *Suppose that* $\mathbb{S}_1$ *and* $\mathbb{S}_2$ *are systems with verified capabilities, and let*

$$d_C(\mathbb{S}_1, \mathbb{S}_2) = 1 - \frac{|\alpha_C(\mathbb{S}_1) \cap \alpha_C(\mathbb{S}_2)|}{|\alpha_C(\mathbb{S}_1) \cup \alpha_C(\mathbb{S}_2)|}, \quad \textit{(System Capabilities)} \tag{5.8}$$

$$d_\beta(\mathbb{S}_1, \mathbb{S}_2) = |\beta(\mathbb{S}_1) - \beta(\mathbb{S}_2)|, \quad \textit{(System Connectedness)} \tag{5.9}$$

$$d_p(\mathbb{S}_1, \mathbb{S}_2) = \sum_{\substack{C_i^{(1)} \in \alpha_C(\mathbb{S}_1) \\ C_j^{(2)} \in \alpha_C(\mathbb{S}_2) \\ C_i^{(1)} = C_j^{(2)}}} H(R_i^{(1)}, R_j^{(2)}). \quad \textit{(Capability Performance)}, \tag{5.10}$$

where H is the Hausdorff metric between two intervals. If

$$d(\mathbb{S}_1, \mathbb{S}_2) = \kappa_C d_C(\mathbb{S}_1, \mathbb{S}_2) + \kappa_\beta d_\beta(\mathbb{S}_1, \mathbb{S}_2) + \kappa_p d_p(\mathbb{S}_1, \mathbb{S}_2), \tag{5.11}$$

where $\kappa_C, \kappa_\beta, \kappa_p \in [0, 1]$, *then d is a metric.*

Proof Since d_C is a dissimilarity metric [31], d_β is the L_1 metric over $\mathbb{R}$, and d_p are also metrics [32], d must be a metric as it is a finite sum of metrics. □

The presence of a metric for $\mathcal{S}_V$ is an important step for verification of autonomous systems as it can be used to eliminate unnecessary testing in the following way. In the case where someone desires a new hypothetical system but has to choose from a large number of actual systems, the metric can be applied to identify those systems that are close to the desired system. While the selected system may need to be tested, it is clearly unnecessary to test those systems that are not close to the desired system. However, a critical aspect to consider is how the total system performance varies from system to system, which is discussed in the next section.

5.4.2 Capability and System Performance

Capability performance is paramount to verification. If we do not know capability performance because we do not understand the goal of the capability, we are unable to evaluate how well the capability accomplished the goal in an environment, and then there is nothing, by definition, that we can verify. Furthermore, in the absence of system performance data, we need some bounds or estimate on potential system performance, even if the system is hypothetical and does not exist. The next theorem describes how capability performance relates to system performance.

Theorem 5.4.2 (System and Capability Performance Relationship) *Suppose a system* $\mathbb{S}$ *has capabilities* $C_1, \cdots, C_n$ *that are independently verified in the desired or expected environment E with corresponding performance evaluation functions* ψ_i. *For fixed inputs* $\delta_i \in D_i$, *if the probability of a successful system in the expected environment and capability performance is related by*

$$p(C_1, \cdots, C_n | \mathbb{S}) \geq p(\mathbb{S} | C_1, \cdots, C_n), \tag{5.12}$$

then the probability of a successful system is bounded by

$$p(\mathbb{S}) \leq \prod_{i=1}^{n} p(C_1). \tag{5.13}$$

Proof Using Bayes' theorem, we compute the joint probability distribution as

$$p(C_1, \cdots, C_n) = \frac{p(C_1, \cdots, C_n|\mathbb{S})}{p(\mathbb{S}|C_1, \cdots, C_n)} p(\mathbb{S}). \tag{5.14}$$

By assumption, we see that

$$\frac{p(C_1, \cdots, C_n|\mathbb{S})}{p(\mathbb{S}|C_1, \cdots, C_n)} \geq 1, \tag{5.15}$$

which implies that

$$p(C_1, \cdots, C_n) \geq p(\mathbb{S}). \tag{5.16}$$

Since the capabilities can be independently verified, we have

$$p(\mathbb{S}) \leq \prod_{i=1}^{n} p(C_i). \tag{5.17}$$

□

We now make a few observations of the previous theorem. Note that we assume that $p(C_1, \cdots, C_n|\mathbb{S}) \geq p(\mathbb{S}|C_1, \cdots, C_n)$, which means that we are more likely for a working system to imply the capabilities are working than working capabilities are to imply a working system, which seems intuitively to be the case for most scenarios. Observe that strict equality, namely,

$$p(\mathbb{S}) = \prod_{i=1}^{n} p(C_i), \tag{5.18}$$

occurs when $p(C_1, \cdots, C_n|\mathbb{S}) = p(\mathbb{S}|C_1, \cdots, C_n)$, which means the system is strictly defined in terms of its capabilities with no system-level issues such as integration issues or latency between capabilities, for example. We can also find a more conservative bound for system performance by computing the following:

$$p(\mathbb{S}) \leq \prod_{i=1}^{n} \max_{\delta_i \in E_i} \psi_i. \tag{5.19}$$

5.4.3 Closeness and System Performance

Even with estimates for system performance, closeness of systems does not necessarily guarantee agreement with respect to system performance. In particular, if we fix a desired system and find a sequence of systems that get arbitrarily close to our desired system, we have no guarantee that the system performance of these closer systems will approach that of the system in question. In other words, we need some type of guarantee that says if two systems are close with respect to our metric, then their respective performances are also close. To address this, we have the following theorem.

Theorem 5.4.3 *Suppose that* $f_V : \mathcal{S}_V \to [0, 1]$ *is a continuous function with respect to any metric* d *that measures the distance between systems, where* $\mathcal{S}_V$ *is the space of systems of independently verified capabilities. If* $\mathbb{S}^*$ *is a limit point of composite systems with independently verified capabilities, then as* $d(\mathbb{S}_j, \mathbb{S}^*) \to 0$, $f_V(\mathbb{S}_j) \to f_V(\mathbb{S}^*)$ *if and only if* f *is continuous.*

Proof Since f_V is continuous and d is a metric by Lemma 5.4.1, the result is true. However, we include the following details to explain why this is the case. Suppose $\mathbb{S}^*$ is a limit point of the set of composite systems with independently verified capabilities. Thus, there exists a sequence $\mathbb{S}_j$ such that as $j \to \infty$, $d(\mathbb{S}_j, \mathbb{S}^*) \to 0$. Since f_V is continuous, $f_V(\mathbb{S}_j) \to f_V(\mathbb{S}^*)$.

If f_V is not continuous, then there must exist a point $\mathbb{S}^*$ such that for a sequence $\mathbb{S}_j \to \mathbb{S}^*$, $f_V(\mathbb{S}_j)$ does not converge to $f_V(\mathbb{S}^*)$, which implies that $f_V(\mathbb{S}_j)$ does not converge to $f_V(\mathbb{S}^*)$. □

The assumption of a continuous performance function is a very weak, only surpassed by the assumption of f_V being a function itself. In other words, the assumption is not so restrictive to lose much in terms of pragmatic application to problems related to verification of autonomous systems. As this speaks to the utility and applicability of the result, continuity of the performance function is essential and must be verified, at least in neighborhoods around the desired system. This is powerful and can save time in the following scenario. Suppose that we have a system that has been verified and has an acceptable system performance value. Now suppose that we want to replace one of the capabilities with potentially one of many cheaper implemented capabilities that offer the same capability. If our system performance function is continuous, then all of those new hypothetical systems that are close with respect to the metric d to our original system are guaranteed to have a close system performance. Furthermore, we know this without having to build and test potentially expensive system prototypes. Note that we are not advocating for the elimination of testing, but rather, by eliminating many alternatives by analysis, this allows the system builder to focus his or her engineering and testing efforts on fewer, more relevant systems.

Observe also that the continuity of f_V does not tell us how small the neighborhoods around our desired system need to be to guarantee the closeness of system performance. This largely relates to how well-behaved the function f_V is and is the subject of our next lemma.

Lemma 5.4.4 *Suppose that* $f_V : \mathcal{S}_V \subset \mathcal{S} \to [0, 1]$ *is a Lipschitz continuous function with Lipschitz constant* k. *Consider the system* $\mathbb{S}^*$ *with system performance* $f(\mathbb{S}^*)$, *and let* t *be a desired tolerance of* $f_V(\mathbb{S}^*)$. *All systems* $\mathbb{S}$ *within* *the neighborhood* N_δ *where* $\delta \leq t/k$ *have a system performance such that* $|f_V(\mathbb{S}^*) - f_V(\mathbb{S})| < t$.

Proof Choose $\delta = t/k$, where k is the Lipschitz constant of f_V. Since f_V is Lipschitz continuous, we have that $|f_V(\mathbb{S}^*) - f_V(\mathbb{S})| < kd(\mathbb{S}^*, \mathbb{S})$ for all $\mathbb{S} \in N_\delta$. Thus, $|f_V(\mathbb{S}^*) - f_V(\mathbb{S})| < t$. □

The point of Lemma 5.4.4 is to show with a slightly more restrictive assumption about f_V than continuity, we can say something about how small to make our neighborhoods around our desired system to find those systems with comparable system performances. Without this lemma, we may have to choose very restrictive neighborhoods to find comparable system performances.

5.5 Example Application to a Simple Sweeping Robot

In order to demonstrate the principle of the frameworks and metrics that we have developed in this chapter, we apply them to a simple notional sweeping robot. We first model the system using the metamodel for individual capabilities, show the connections between capabilities, and model the whole system. We next use the mathematical framework to compose the model into a system that can be used for analysis to characterize its performance. Finally, we use the metrics developed in Sect. 5.4 in concert with limited simulation to describe the performance of the system, demonstrating how the result can be used to generalize the results to system configurations not explicitly tested.

We consider an autonomous mobile sweeping robot designed to cover and sweep a rectangular area. The capabilities of the robot have been deliberately simplified to facilitate the goal of demonstrating the principles of the method rather than rigorous analysis of a realistic robot.

5.5.1 System Modeling

The system as described above has four capabilities that must be modeled, which we detail below.

Path Planning For the purposes of this example, we specify a simple path planner capability implemented by a lawnmower coverage algorithm which assumes a rectangular space to cover with no obstacles. We assume the vehicle starts in one corner of the room aligned with the wall such that this movement pattern can be defined as driving forward for the length of the room, turning 90°, driving for the effective width of the sweeper, turning 90°, and repeating turning the opposite direction. In order to generate this pattern, the capability requires the dimensions of the room and the effective width of the sweeper, modeled as data inputs. The

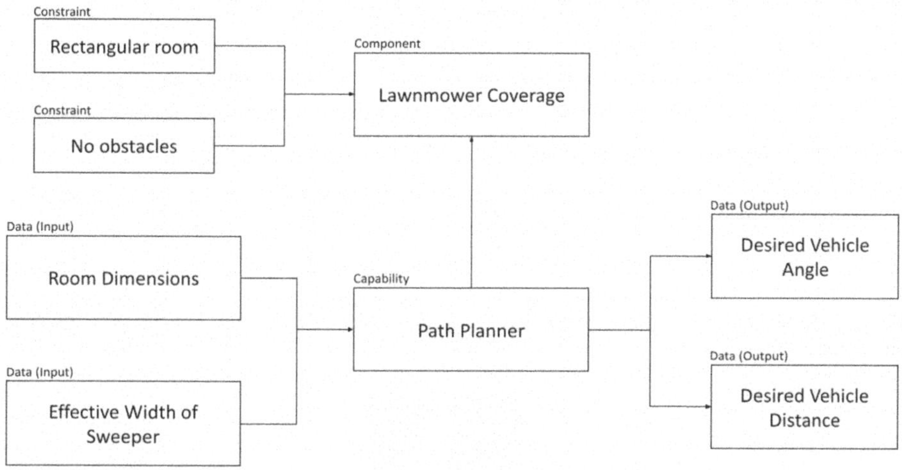

Fig. 5.4 Simple lawnmower path planner capability, expressed according to the defined metamodel

planner data outputs are commands to turn and move forward that generate the desired sweeping path. The resulting capability model is shown in Fig. 5.4.

Translation Our performance model for the robot translational movement is based on the characterization of the Roomba platform in [33]. According to this characterization, given a commanded distance to drive d_d cm, the robot will actually drive a distance d cm, expressed by $d = d_d * 0.9284 - 3.5583$ with an error of $d_d * 0.0028 + 0.886$ cm. We describe this as a translation capability associated with the drivetrain component having input data that is the desired distance for the robot to travel and an output of the actual vehicle distance traveled using the above equation with the error captured as a tatum.

Rotation Like the translation model, the rotation model of the sweeping robot is based on [33]. Based on this analysis, given a commanded desired angle θ_d for the robot, the actual angle achieved θ is expressed by $\theta = 0.916899\theta_d$ with an error of 0.0034159 radians. Therefore, we describe this as a rotation capability associated with the drivetrain component with an input of the desired rotation of the robot and an output of the actual vehicle rotation using the above equation with an accuracy of 0.0034 radians captured as a tatum.

Sweeping The sweeping capability model was already described in 5.2.2 to provide an illustration of the application of the metamodel.

We connect the capabilities as shown in Fig. 5.7, starting with the path planning capability and connecting it to the translation and rotation capabilities that act upon the data output by the path planner. The input and output data corresponding to the individual capability diagrams can be seen connecting the capabilities. Components, constraints, and tatums are omitted for readability. The desired system performance

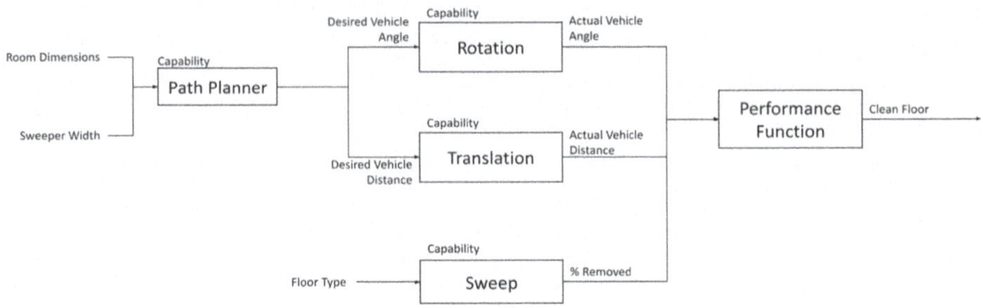

Fig. 5.5 Model of the complete sweeping robot system, composed by linking capabilities by input and output data

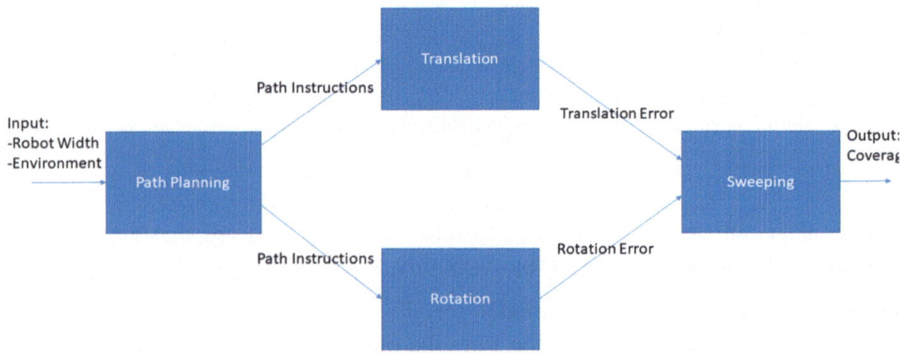

Fig. 5.6 Capability model of sweeping robot

measure is a cleaned floor, for which we use the tatum of cleaning effectiveness, or overall percentage of soil removed. Because this measure is not provided directly by any capability, we describe it as an output of an implied system performance capability, which will be composed into the system with the explicit capabilities (Fig. 5.5).

5.5.2 *Capability Composition of Sweeping Robot*

Using the mathematical framework, we next model the sweeping robot system as a directed graph of its capabilities. In Fig. 5.6, we delineate the connections between capabilities and note the outputs of source capabilities which are taken as relevant inputs for the target capabilities. We then apply Lemma 5.3.3 and Theorem 5.3.1 in the following figures to construct composite capabilities ultimately representing the performance of the whole system. Figure 5.7 demonstrates combining functions that act simultaneously by determining how they act on shared input and output data types. Note that in this case there is not much to check when combining

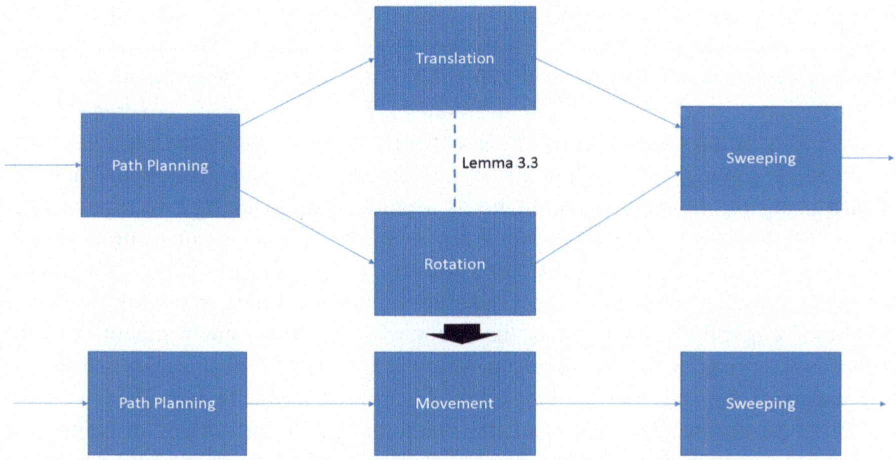

Fig. 5.7 Composition of the translational and rotation capabilities into one move capability by virtue of Lemma 5.3.3

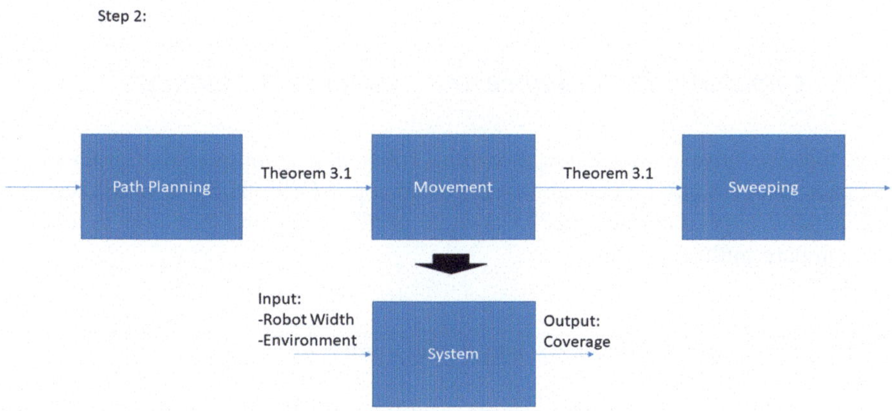

Fig. 5.8 Composition of the entire system by virtue of Theorem 5.3.1

the translation and rotation capabilities, because they do not share any data types. Consequently, there is nothing surprising that happens when they are combined since the resultant capability represents what happens when a translation, then a rotation is processed in the system. They are essentially combined into a "move" capability which takes $\{d_d, \theta_d\}$ as input and outputs $\{d, \theta\}$, which in this example are equal to $\{d_d * 0.9284 - 3.5583, 0.916890_d\}$. Figure 5.8 demonstrates composition of the resulting functions in the usual sense, and now that the functional flow of the system is simplified. While in our capability graph we have confirmed the way that each individual capability's inputs and outputs line up and how they connect, there are still some assumptions we must make about the scenario in order to determine system performance. The width of the robot and the dimensions of the environment are necessary inputs for the planning capability, and the floor type of each tile in the environment is necessary for the sweeping capability. This means that our resulting system-level capability is a function which takes environment and robot width as inputs and produces our ultimate performance metric, coverage of the whole room, as an output. As it is currently formulated, this function determines coverage by generating the whole list of movement commands, following each one in turn using the movement command as a guide, and checking coverage at each step using the sweep command. However, as in the following subsection, we can represent this system using a different intuitive understanding of the capabilities. If the planning and both movement capabilities are combined into a single "move coverage" capability, the input is the robot width and the environment, and the output is the areas of the environment which are visited when the plan is followed and executed by the movement capability. This then feeds into the sweeping capability as before. The power of the mathematical framework is in being able to show that this understanding of the system is equivalent to the one where the capabilities are all individually defined because when combined, they collapse to the same system performance capability.

5.5.3 Capability Performance and System Performance

Now that we have modeled the sweeping robot and composed the system as a compact, directed graph of capabilities, we can apply Theorem 5.4.2 to first see how well capability performance predicts system performance and then how to analyze and compare multiple systems.

5.5.3.1 Prediction of Single System Performance

We first show the ability of this approach to analytically predict the performance of the simulated robot under different environmental conditions by specifically varying room size. Because the selected performance measure of overall cleaning effectiveness is not a verified output of any of the capabilities, we first use sample cases to characterize the relationship between this measure of performance and the rest of the system. We then use that relationship to predict performance and run the corresponding simulation to confirm the results.

The resulting data is shown in Fig. 5.9. The blue asterisks show the results of the initial samples, describing the space of performance results given a room size. Note that results vary due to the error in robot translation and rotation performance, which is allowed to vary in the simulation. The trend shows that smaller rooms have a greater range of possible outcomes, but while the results over a larger room are more predictable, they also tend to be worse.

The predicted performance based on the mathematical framework developed falls within the bounds found by simulation, showing that the analysis is consistent with

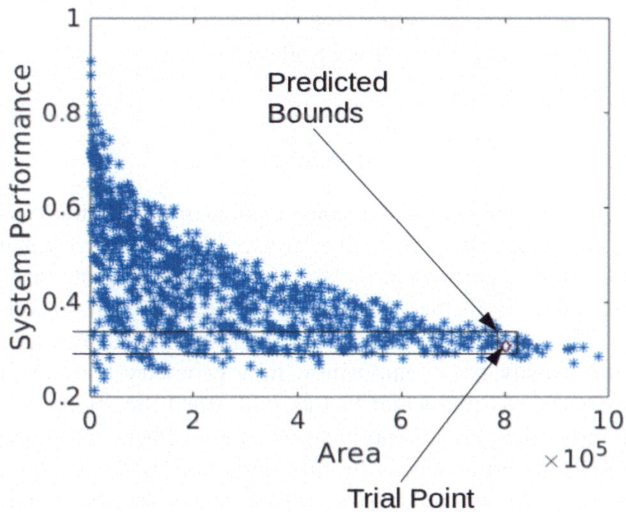

Fig. 5.9 Simulation results from a fixed sweeping robot while varying the area of the robot. Sampling over the parameter space yields the blue asterisks. The purple diamond is a test point, which corresponds to 80,000 cm^2 (800 cm × 100 cm). For this value of 80,000 cm^2, we examined the nearby blue points and estimated a lower bound of approximately 0.3 and upper bound of approximately 0.35 before running the simulation for 80,000 cm^2 case. The resulting simulated system performance value associated with the purple diamond falls within this bound as prescribed by the blue points in the plot

the experimental data. Further work is ongoing to more rigorously develop this process of performance prediction using fewer test samples to generate more precise results.

5.5.3.2 Generalization of Results for System Exploration

The single vehicle case shows the application of the analytical framework to assess a fixed robot against a variety of scenarios and accurately predict performance. We next extend the work to show its ability to assess a variety of theoretical robot designs against a set of scenarios.

We will generate several hundreds of system, each with system performance, move capability performance, and sweep performance. We assume that we have an $L \times L$ grid of discrete tiles. For the sweeping robot, we have three performance values that we want to compute:

$$p_t = \frac{\text{Area of Tiles Cleaned}}{\text{Total Area}}, \tag{5.20}$$

$$p_s = \frac{\text{Soil Extracted from Tile}}{\text{Total Soil on Tile}}, \tag{5.21}$$

$$p_m = \frac{\text{Number of Tiles Visited}}{\text{Total Number of Tiles}}, \tag{5.22}$$

where p_t, p_s, and p_m are the performance estimates for the total system, sweep capability, and move capability. We will compare the system performance to that of the individual capability performances later in this section. Note that p_t measures the total dirt removed from the entire grid, p_s measures how much dirt will be extracted from a tile, and p_m measures the fraction of total tiles that are visited. The dirt type, dirt density, floor composition, floor geometry, humidity, temperature, and all other relevant environmental factors will affect the ability of a vacuum to successfully remove dirt. As a simplification, these effects are aggregated into a single probability measuring successful dirt extraction, which we have denoted as p_s. Our numerical results will refer to three distinctive probabilities of dirt extraction that we refer to as three unique floor types.

Move Capability To model the move capability of a sweeping robot, we use the linear error models as found in [33]

$$d = m_d d_d + j_d, \tag{5.23}$$

where d is the actual distance traveled, and d_d is the planned distance traveled. The terms m_d and j_d are fixed, but arbitrary constants that are sampled from the uniform distributions $U[m_d - \sigma_d, m_d + \sigma_d]$ and $U[c_d - \sigma_j, c_d + \sigma_j]$. The values σ_d and σ_j are the standard deviations of the respective distributions. To measure the angle turned we use

$$\theta = m_\theta \theta_d + k_\theta, \tag{5.24}$$

where θ is the actual angular distance traveled, and θ_d is the planned angular distance traveled. The terms m_θ and k_θ are fixed, but arbitrary constants that are sampled from the uniform distributions $U[m_\theta - \sigma_\theta, m_\theta + \sigma_\theta]$ and $U[k_\theta - \sigma_k, k_\theta + \sigma_k]$. The values σ_θ and σ_k are the standard deviations of the respective distributions. Thus, a unique move capability is created when values for the means and standard deviations are selected. The better move capabilities will have means closer to 1, with very small standard deviations.

To estimate the move capability, we simulate several runs over the floor and compute p_m. This results in a set of values, which we denote as $[p_{min}, p_{max}]$ for each system.

Table 5.2 Simulation parameters

(a) Parameters that vary

Parameter	Values
σ_d, σ_θ	$\{.1, .01, .001, 0\}$
p_1, p_2, p_3	$\{.7, .8, .9, 1.0\}$

(b) Constant parameters

Parameter	Values
N_1,N_2,N_3	$\frac{1}{3}$
m_d, m_θ	.97
j_d, k_θ	0 cm, 0 rad
L	500 cm
w	33 cm

Sweep Capability To generate a unique sweep capability, we use the following weighted average as described in [27]:

$$p_s = \frac{1}{N}(p_1 \cdot N_1 + p_2 \cdot N_2 + p_3 \cdot N_3), \tag{5.25}$$

where p_1, p_2, and p_3 model the effectiveness of sweeping a floor in which N_1, N_2, andN_3 are the number of tiles of floor type 1, 2, and 3, respectively, and

$$N = N_1 + N_2 + N_3. \tag{5.26}$$

Once we set the floor type, unlike the move capability for the system, we have only a single point estimate for the sweep performance. Observe that during runs of our simulation, we will randomly generate tiles of type 1, 2, and 3 numbering close to the values of N_1, N_2, and N_3, respectively.

System Generation In general we may not know the actual system performance. However, one of our main purposes of this simulation is to demonstrate how knowledge of the individual capabilities performances can serve as a surrogate for the total system performance. Therefore, we will simulate several systems with varying move and sweep capabilities that will allow us to explore the relationship to the total system performance. We will also explore additional uses of our metric that will readily allow for a trade-off analysis that can assist during the design phase, which can reduce the number of total system test that may have to be conducted.

For each simulation run, we randomly create a room with the same floor type composition, which we set to always be of equal proportions of type 1, type 2, and type 3 floor types. Table 5.2 summarizes our simulation runs with respect to constant and varied parameters. For each value of varied parameters in the table, we will generate a minimum and maximum probability, $[p_{min}, p_{max}]$, that measures how much of the floor we visit. This range, coupled with the probability of sweeping defined by Eq. 5.25, will define a unique system. Thus, we will consider 1024 unique systems in this section.

Example paths are given in Fig. 5.10. Observe that in Fig. 5.10 we are showing the trajectories of two unique systems, one that has a very good move capability and one that has a poor move capability. As an approximation to computing the

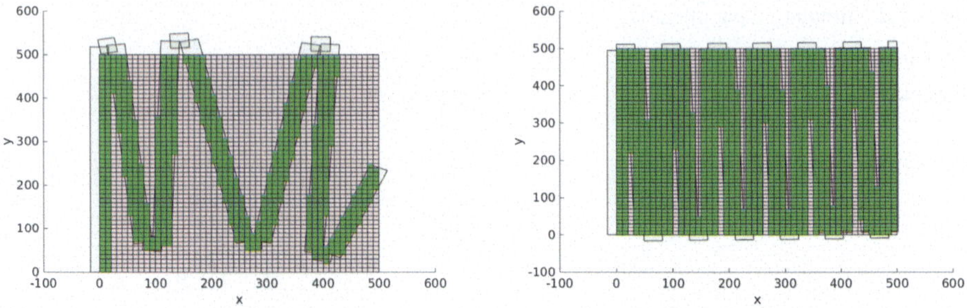

Bad path. Good path.

Fig. 5.10 Examples of trajectories. The slim rectangular areas are areas that have been visited and swept

area swept, we determine the centroids of the tiles that are contained within the sweep area. If that centroid is within the sweep area, the tile is then the percentage of dirt removed as dictated by the associated probability of dirt extraction. Observe that smaller tiles will give more accurate results. For our simulation, the sweep area covers roughly about eight tiles.

Results Let $\bar{p}$ be the average of $[p_{min}, p_{max}]$. For our purposes, we estimate the performance of the move capability, p_m with $\bar{p}$ so that $p_m \approx \bar{p}$. Although we used the mean for p_m and not the largest value of its performance range, Fig. 5.11 shows that the product $p_m p_s$ is an upper bound for the overall system performance and also a good estimate for system performance in this example dataset. While we will claim that the conservative bound established by Eq. 5.19 is always true, we have no guarantee that the product of the capability performances will always have good agreement with the actual system performance. Our simulation experiment is ideal as we did not model integration issues, latency issues, or other issues that can occur once a system is assembled and consequently degrade system performance. However, for the case in which we want to compare hypothetical systems using only capability performances, the comparison is at least based upon a consistent system performance estimate. In the next section, we will discuss these comparisons.

5.5.4 Comparing Multiple Systems

Based upon our results comparing capability performance and system performance, we are able to entertain very interesting questions. First, suppose that we have access to only capability performance and no system performance, but we have a hypothetical system in mind, $\mathbb{S}^*$ with a sweep performance of $p_s^* = .8$, move

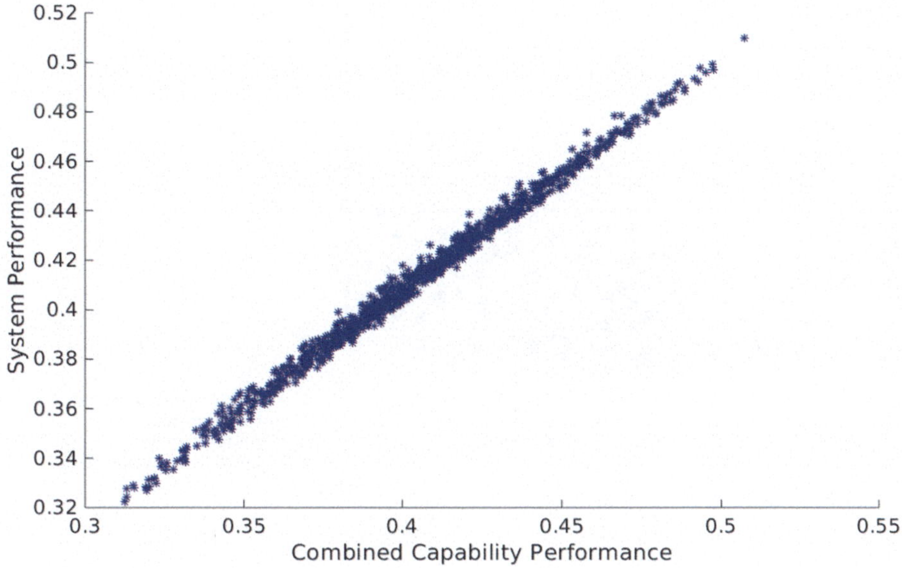

Fig. 5.11 Comparing the product of capability performances to the system performance of 1024 unique systems

performance $p_m^* \in [.5, .6]$, and an estimated system performance of $P^* = .45$. Which combination of capabilities do we choose? We can use

$$\frac{\Delta P_j}{P^*} = \frac{P^* - P_j}{P^*} \tag{5.27}$$

and the distance between system $\mathbb{S}_j$ and $\mathbb{S}^*$ as

$$\Delta S_j = d(\mathbb{S}_j, \mathbb{S}^*), \tag{5.28}$$

to choose which systems are potentially viable. Note that for this example we are assuming that all systems will have a move and sweep capability and that those capabilities will be perfectly connected. Thus, our metric collapses to

$$d(\mathbb{S}_j, \mathbb{S}^*) = d_p(\mathbb{S}_j, \mathbb{S}^*), \tag{5.29}$$

meaning that we are looking for differences only in capability performances in this example.

In Fig. 5.12, we can quickly identify which systems are more likely to achieve our desired results and which systems we can ignore. We observe that systems with $\Delta S_j \leq .01$ do not meet the desired system performance. In fact, Fig. 5.12 shows that we would have to look at systems such that $\Delta S_j \geq .04$. We see that the first system, as indicated by the magenta asterisk, may meet the desired system performance by

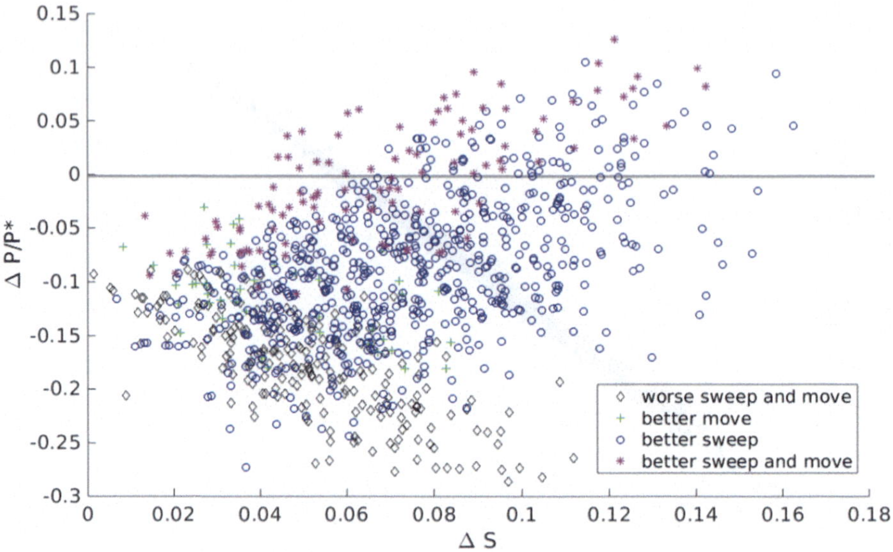

Fig. 5.12 Comparing the system performance of a hypothetical system to other systems when only capability performance is known (no system performance is available). Systems above the gray line have improved system performance over our desired system but may be too far away if factors such as cost or availability correlate to distance, for example

using an improved move capability, as well as a better sweep capability. On the other hand, we may want to consider systems within $\Delta S_j \leq .01$ as these capabilities may be cheaper to acquire if the cost is correlated with the distance between systems. Nonetheless, we are now able to make more reasoned decisions about which capabilities we want to use for our new system without necessarily performing a single integrated test. Furthermore, we can quickly winnow the number of systems by significant numbers if we consider the relative distance and estimated system performance. Of course, this represents a potential significant savings in time and money with respect to verification as we know exactly which combinations of systems will not satisfy our requirements, meaning we can rigorously justify not spending resources on testing combinations of capabilities that we now know are not useful. Most importantly, we can use the results of the metric to justify our rationale.

For the problem of comparing a known system to other known systems, we assume that we have performance estimates for the system and individual capabilities. As an example, we chose one of the systems that we generated and compared it to the remaining systems. Figure 5.13 shows the results. Here, we can see which systems are close to our system, which is represented as the asterisk at the origin. Observe that systems that are with $\Delta S_j \leq .01$ are very comparable to our own system. Furthermore, we can see how far we would have to go to find a system that is better than our own system and why it is better. As with our previous

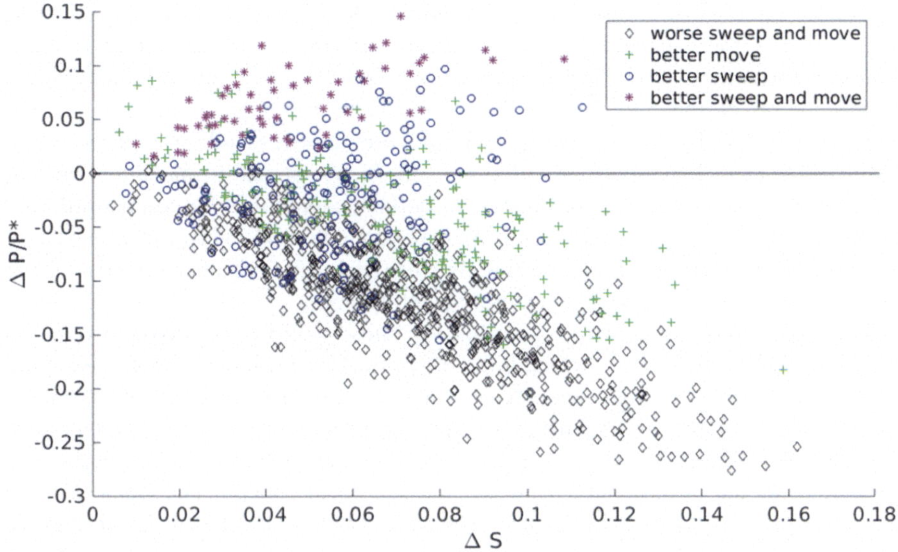

Fig. 5.13 Comparing the system performance of a known system to other known systems. The nominal system is represented as the diamond that appears at the origin. Systems above the gray line have increased system performance

problem, the point is that we can make rational decisions based upon the data as measured with our metric and save time and money with respect to verification by not considering clearly bad or impractical choices.

5.6 Future Work

This chapter establishes rigorous philosophical and mathematical frameworks for the description and analysis of autonomous systems. While the metrics already developed are already useful to the challenge of verification, further research is needed to develop more comprehensive metrics and tools that can apply to these systems more broadly. We describe here some of these avenues for further work already identified.

5.6.1 Taxonomy and Clustering

The ontology described by our metamodel is useful for the description of individual capabilities, and the structure can be used to build a repository of these autonomy capabilities. Much as companies such as NewEgg or McMaster-Carr provide well-organized catalogs of components according to their use and attributes, it would be useful for these capabilities to be organized and sorted according to their common attributes [34, 35]. Such organization requires a taxonomy or a hierarchical representation of the relationships between capabilities defined in terms of the ontology created by the metamodel. While previous research has proposed to characterize the whole space of autonomy [3–6], we appeal to machine learning techniques to identify groups or clusters of defined capabilities based on similarity. There may be a temptation here to appeal to a Delphi method and rely on some panel of "experts" to categorize capabilities, but research has indicated crowdsourcing methods also pose challenges to accurate labeling [36].

For inspiration, we consider perhaps the most well-known taxonomy is the Linnaean Taxonomy of Life, which in biology provides a framework for classifying all life on Earth. Any new species of plant or animal discovered is organized according to similarity of its attributes, generally genetic or physical characteristics [37, 38]. An alternative view is a phylogenetic tree, which is a representation derived from an evolutionary view of species development. The book *On the Origin of Species*, written by Charles Darwin in 1859, introduced the alternative view of generating clusters and groups of "taxa," showing a temporal classification of species [39]. These views, collectively, have shaped the way that biologists classify and establish the interconnectedness of life. While physical and genetic characteristics offer objective criteria to establish similarity between species, the determination of which criteria to use to distinguish or group a given species is still ultimately left up to biologist, sometimes creating controversy [40]. However, this method is essential for resolving ambiguities that may otherwise arise from the attempt to categorize complex entities with many attributes that may align with multiple possible classifications.

Therefore, while the metamodel provides a consistent structure of objective criteria for comparison, we suggest that the human input is still necessary to interpret these distinctions and apply meaningful labels. What is desirable, therefore, is some tool that will intelligently exploit the objective criteria available through the metamodel to minimize (rather than eliminate) the amount of subjectivity required in the labeling of capabilities and their categories. We propose that capabilities be grouped using a learning clustering algorithm according to "likeness" based on their metamodel representation. Once capabilities are clustered, the robotics and autonomous systems community can apply the appropriate labels for the emergent clusters of capabilities, without the need to predict the groupings a priori.

Let us imagine that a library of autonomy capabilities exists, each expressed using the metamodel as a template, providing a body of candidates for classification into an organized taxonomy. Whereas there is an established hierarchy of Kingdoms, Phyla, Orders, etc. in the Taxonomy of Life, currently we lack the same taxonomic structure in the autonomy community. Therefore, we require a solution that not only categorizes capabilities into a hierarchical structure of categories but also aids in the definition of those categories themselves. To address both of these challenges, we propose a machine learning-based approach to capability categorization, providing suggestions based on metamodel attributes regarding which capabilities are similar.

This technique can be used to identify clusters to create new categories or to identify which category best corresponds to a new capability. This approach avoids bias from naming conventions or category names and evaluates likeness exclusively based on objective attributes and values alone. Once clusters are formed, then the "taxonomy" is defined by the application of meaningful labels by experts, who are able to do so based on what is common to the cluster rather than imposing an arbitrary outside structure.

To facilitate the clustering, we propose a machine learning workflow to cluster common capabilities. We choose a combination of an unsupervised, offline algorithm, with a supervised, online algorithm. The unsupervised algorithm is an adaptation of a DBSCAN for the abstract capability space, where the distance metric and neighborhood are defined by the similarity metric established in Sect. 5.4.1. The supervised learning component utilizes labels to create general filter categories for searching a catalog of capabilities but does not represent a strict labeling of individual components. By utilizing the secondary supervised learning component, we create category classes that can be used by system designers and project managers seeking to filter search results via common criteria. Moreover, the establishment of category labels gives us a convenient mechanism for performing compositional analysis and verification.

5.6.2 Refinement of Data and Tatums

As noted, for the purpose of this initial work, we have defined the data and tatums in the metamodel generically but have focused particularly in the application on simple scalar data values with tatums that represent the allowable range for those values. While this is useful as a proof of concept, in reality the data handled by autonomous systems is more complex, with a variety of tatums describing its attributes.

We propose to refine the metamodel and the definitions of data and tatums through application to several real autonomous systems and their capabilities.

5.6.3 Modularity Metric

While modularity has been identified as a desirable feature for software-intensive systems, no quantitative measure has yet been identified for this property, making it difficult to require. We propose to use the relationship between capabilities and components to develop such a metric that will serve as a measure of the separability of the essential functions of systems.

5.6.4 *Cost Analysis*

The distance metric developed in Sect. 5.4 provides a means of comparing systems that correlates a single quantitative measure to anticipated performance. A natural and potentially useful extension of this analysis is to incorporate cost considerations alongside performance. The cost may represent any number of types of resources which a designer would like to minimally consume, a consideration that is often difficult in practice when higher cost designs exhibit better performance. One way to resolve this issue in cases where the cost correlates with the distance of a system from ideal is by employing a measure of how close one can afford to get to a desired performance. In cases in which the cost is independent of the distance between systems, we can imagine metrics for the value of a system to optimize performance for dollar. Such applications are of clear value to the same stakeholders interested in performance verification and will be the subject of the future study.

5.7 Conclusion

We have developed both conceptual and mathematical frameworks which are coupled together by the metamodel, an abstract hierarchical model of autonomous systems, and their respective capabilities. The metamodel allows autonomous systems to be modeled agnostically with respect to implementations. Our mathematical framework describes how we can compose capabilities through the use of category and sheaf theory. We developed and described techniques for the comparative analysis of systems that uses a distance metric that we also developed. The metric measures differences with respect to capability availability, capability connectedness, and capability performance. Capability performance indicates how well the capability maps input data to expected output and is measured as a probability, which is very general and necessary for verification. Using capability performance, we have shown how to estimate system performance for cases in which system performance is not available. Furthermore, we have given conditions for which closeness of systems implies the existence of similar system performances. While testing is very important, our comparative analysis techniques are expected to reduce the number of systems required to test, based upon the desired requirements and capability performances. These results were shown through application to a simple sweeping robot. Future work will apply these frameworks and metrics to real systems with the intention of developing a formal process for the verification of autonomous systems. We also described a way to generate a taxonomy using machine learning techniques that attempt to remove human bias, which will be applied as the body of modeled capabilities grows.

Acknowledgments We thank Dr. Marc Steinberg at the Office of Naval Research (ONR) Code 31 for his support of this research under the Science of Autonomy program. We also thank Patrick Tam for his help in the editing and refinement of this document.

References

1. A. Bouchard and R. Tatum, "Verification of autonomous systems: Challenges of the present and areas for exploration," in *OCEANS 2015-MTS/IEEE Washington*. IEEE, 2015, pp. 1–10.
2. A. Bouchard, R. Tatum, and S. Horan, "Verification of autonomous systems by capability verification composition (CVC)," in *OCEANS 2017-Anchorage*. IEEE, 2017, pp. 1–7.
3. H.-M. Huang, E. Messina, and J. Albus, "Toward a generic model for autonomy levels for unmanned systems (ALFUS)," National Institute of Standards and Technology, Gaithersburg, MD, Tech. Rep., September 2003. [Online]. Available: https://apps.dtic.mil/sti/citations/ADA515323
4. H.-M. Huang, E. Messina, and J. Albus, "Toward a generic model for autonomy levels for unmanned systems (ALFUS)," in *Proceedings of the 2003 Performance Metrics for Intelligent Systems (PerMIS) Workshop*. Gaithersburg, MD: National Institute of Standards and Technology, August 2003.
5. H.-M. Huang, J. Albus, E. Messina, R. L. Wade, and R. W. English, "Specifying autonomy levels for unmanned systems: interim report," in *Proceedings of the SPIE Volume 5422, Unmanned Ground Vehicle Technology VI*, vol. 5422. Orlando, FL: SPIE, September 2004.
6. P. J. Durst and M. W. Gray, "Levels of autonomy and autonomous system performance assessment for intelligent unmanned systems," Geotechnical and Structures Laboratory (U.S.) and Engineer Research and Development Center (U.S.), Special Report ERDC/GSL SR ; 14-1, April 2014. [Online]. Available: https://hdl.handle.net/11681/3284
7. G. Brat, E. Denney, D. Giannakopoulou, J. Frank, and A. Jónsson, "Verification of autonomous systems for space applications," in *2006 IEEE Aerospace Conference*. IEEE, 2006, pp. 11–pp.
8. S. Ben-Rayana, M. Bozga, S. Bensalem, and J. Combaz, "RTD-Finder: A tool for compositional verification of real-time component-based systems," in *International Conference on Tools and Algorithms for the Construction and Analysis of Systems*. Springer, 2016, pp. 394–406.
9. H. Giese, S. Burmester, W. Schäfer, and O. Oberschelp, "Modular design and verification of component-based mechatronic systems with online-reconfiguration," in *Proceedings of the 12th ACM SIGSOFT twelfth international symposium on Foundations of software engineering*, 2004, pp. 179–188.
10. S. Bensalem, M. Bozga, J. Sifakis, and T.-H. Nguyen, "Compositional verification for component-based systems and application," in *International Symposium on Automated Technology for Verification and Analysis*. Springer, 2008, pp. 64–79.
11. M. Kwiatkowska, G. Norman, D. Parker, and H. Qu, "Assume-guarantee verification for probabilistic systems," in *International Conference on Tools and Algorithms for the Construction and Analysis of Systems*. Springer, 2010, pp. 23–37.
12. R. Gerth, D. Peled, M. Y. Vardi, and P. Wolper, "Simple on-the-fly automatic verification of linear temporal logic," in *International Conference on Protocol Specification, Testing and Verification*. Springer, 1995, pp. 3–18.
13. L. Warren, "Minimal logic and automated proof verification." 2019.
14. D. F. Woerner, "The selection and infusion of autonomy for Mars rovers," in *Proceedings First International IEEE Symposium Intelligent Systems*, vol. 1. IEEE, 2002, pp. 86–91.
15. A. Lampe and R. Chatila, "Performance measure for the evaluation of mobile robot autonomy," in *Proceedings 2006 IEEE International Conference on Robotics and Automation, 2006. ICRA 2006*. IEEE, 2006, pp. 4057–4062.
16. G. S. Sukhatme, G. A. Bekey *et al.*, "An evaluation methodology for autonomous mobile robots for planetary exploration," in *Proc. of the First ECPD International Conference on Advanced Robotics and Intelligent Automation*. Citeseer, 1995, pp. 558–563.
17. B. T. Clough, "Metrics, schmetrics! how the heck do you determine a UAV's autonomy anyway," Air Force Research Lab Wright-Patterson AFB OH, Tech. Rep., 2002.

18. H. Chen, X.-m. Wang, and Y. Li, "A survey of autonomous control for UAV," in *2009 International Conference on Artificial Intelligence and Computational Intelligence*, vol. 2. IEEE, 2009, pp. 267–271.
19. B. Hasslacher and M. W. Tilden, "Living machines," *Robotics and autonomous systems*, vol. 15, no. 1–2, pp. 143–169, 1995.
20. G. T. McWilliams, M. A. Brown, R. D. Lamm, C. J. Guerra, P. A. Avery, K. C. Kozak, and B. Surampudi, "Evaluation of autonomy in recent ground vehicles using the autonomy levels for unmanned systems (ALFUS) framework," in *Proceedings of the 2007 Workshop on Performance Metrics for Intelligent Systems*, 2007, pp. 54–61.
21. P. J. Durst, W. Gray, A. Nikitenko, J. Caetano, M. Trentini, and R. King, "A framework for predicting the mission-specific performance of autonomous unmanned systems," in *2014 IEEE/RSJ International Conference on Intelligent Robots and Systems*. IEEE, 2014, pp. 1962–1969.
22. D. Gertman, C. McFarland, T. Klein, A. Gertman, and D. Bruemmer, "A methodology for testing unmanned vehicle behavior and autonomy," in *Proceedings of the 2007 Workshop on Performance Metrics for Intelligent Systems*, 2007, pp. 62–69.
23. S. Baccar, W. Derguech, E. Curry, and M. Abid, "Modeling and querying sensor services using ontologies," in *International Conference on Business Information Systems*. Springer, 2015, pp. 90–101.
24. S. Bhiri, W. Derguech, and M. Zaremba, "Modelling capabilities as attribute-featured entities," in *International Conference on Web Information Systems and Technologies*. Springer, 2012, pp. 70–85.
25. W. Derguech, S. Bhiri, S. Hasan, and E. Curry, "Using formal concept analysis for organizing and discovering sensor capabilities," *The Computer Journal*, vol. 58, no. 3, pp. 356–367, 2015.
26. O. A. Specification, "OMG unified modeling language (OMG UML), superstructure, v2. 1.2," *Object Management Group*, vol. 70, 2007.
27. "Best robot vacuum for 2020: Electrolux, iRobot Roomba, Neato, Eufy and more," http://cnet.com/news/best-robot-vacuum-for-2020-irobot-Roomba-Electrolux-neato-eufy-and-more, accessed: 2020-10-02.
28. R. Hartshorne, *Algebraic Geometry*. Springer-Verlag, New York, 1977.
29. M. Robinson, "Sheaves are the canonical data structure for sensor integration," *Information Fusion*, vol. 36, pp. 208–224, 2017.
30. S. Kosub, "A note on the triangle inequality for the Jaccard distance," *Pattern Recognition Letters*, vol. 120, pp. 36–38, 2019.
31. G. Gilbert, "Distance between sets," *Nature*, vol. 239, no. 5368, pp. 174–174, 1972.
32. M. Kumbar, U. Ingalagi, T. Patel, P. Marigoudar, H. Nadaf, and S. Narasimhamurthy, "Geometric approach for Banach space using Hausdorff distance," *International Journal of Mathematics And its*, vol. 55, p. 95, 2017.
33. B. Tribelhorn and Z. Dodds, "Evaluating the Roomba: A low-cost, ubiquitous platform for robotics research and education," in *Proceedings 2007 IEEE International Conference on Robotics and Automation*. IEEE, 2007, pp. 1393–1399.
34. [Online]. Available: https://www.newegg.com/
35. [Online]. Available: https://www.mcmaster.com/
36. L. B. Chilton, G. Little, D. Edge, D. S. Weld, and J. A. Landay, "Cascade: Crowdsourcing taxonomy creation," in *Proceedings of the SIGCHI Conference on Human Factors in Computing Systems*, ser. CHI '13. New York, NY, USA: Association for Computing Machinery, 2013, p. 1999–2008. [Online]. Available: https://doi.org/10.1145/2470654.2466265
37. C. v. Linné and L. Salvius, *Caroli Linnaei ... Species plantarum: exhibentes plantas rite cognitas, ad genera relatas, cum differentiis specificis, nominibus trivialibus, synonymis selectis, locis natalibus, secundum systema sexuale digestas...* Holmiae: Impensis Laurentii Salvii, 1753, vol. vol. 1, https://www.biodiversitylibrary.org/bibliography/669—Pages 483, 638, 639, and 674 misnumbered 481, 938, 939, and 774, respectively. — Soulsby | 480 — Stafleu (2nd) | 4769 — Pritzel (2nd) | 5427. [Online]. Available: https://www.biodiversitylibrary.org/item/13829

38. C. v. Linné and L. Salvius, *Caroli Linnaei ... Species plantarum: exhibentes plantas rite cognitas, ad genera relatas, cum differentiis specificis, nominibus trivialibus, synonymis selectis, locis natalibus, secundum systema sexuale digestas...* Holmiae: Impensis Laurentii Salvii, 1753, vol. 2, https://www.biodiversitylibrary.org/bibliography/669 — Pages 483, 638, 639, and 674 misnumbered 481, 938, 939, and 774, respectively. — Soulsby | 480 — Stafleu (2nd) | 4769 — Pritzel (2nd) | 5427. [Online]. Available: https://www.biodiversitylibrary.org/item/13830
39. C. Darwin and L. Kebler, *On the origin of species by means of natural selection, or, The preservation of favoured races in the struggle for life*. London: J. Murray, 1859. [Online]. Available: https://www.loc.gov/item/06017473/
40. B. Jones, "A few bad scientists are threatening to topple taxonomy," *Smithsonian Magazine*, Sep 2017. [Online]. Available: https://www.smithsonianmag.com/science-nature/the-big-ugly-problem-heart-of-taxonomy-180964629/

Chapter 6
Structuring Learning from Demonstration to Support Verifiable Robot Control

Subramanian Ramamoorthy, Craig Innes, and Yordan Hristov

Chapter 6 Editorial Note

This chapter focuses on the development of a mathematical representation connecting learning from demonstration with linear temporal logic constraints and autoencoder-based representations of the relevant environment to define verifiable controllers for learning systems using reinforcement learning with deep neural networks, increasing our understanding of how to verify these complex systems. This addresses the portion of technical challenge 3 (development of models for debugging and formal analysis) that is concerned with how mathematical/logical models suitable for formal analysis may be produced from empirical observations, since the learning component provides the empirical observations and the mathematical formalism provides the connection to formal analysis. It provides a technique for developing a formal mapping between learning systems and formal analysis, partially addressing challenge 21 (identifying needed tools and techniques). And this chapter provides one possible answer to challenge 27 (verification of learning systems) by defining an approach that enables verification of systems that are designed to learn as they operate. The stakeholder question of understanding how to verify an autonomous system is discussed in the Chap. 1, Introduction. Technical challenges 3, 21, and 27 are introduced in Chap. 2, Verification Challenges.

Abstract With the increasing level of adoption of autonomous systems in our daily lives, as robots make their way out of the restricted factory floors into our hospitals and shopping malls, there is now a need to consider how they should be designed so that they may be trusted. Achieving trustworthiness involves several complex criteria, but a useful description of the desiderata in a form that is compatible with the engineering design process is to ask whether a system will function as

S. Ramamoorthy (✉) · C. Innes · Y. Hristov
School of Informatics, University of Edinburgh, Edinburgh, UK
e-mail: s.ramamoorthy@ed.ac.uk; craig.innes@ed.ac.uk; yordan.hristov@ed.ac.uk

S. Redfield et al. (eds.), *Verification of Autonomous Systems*,
https://doi.org/10.1007/978-3-031-88546-4_6

we say it should, whether we know how it will fail and if we have evidence regarding how it continues to function after it has failed to satisfy some criteria. Learning and adaptation are central features of many modern autonomous systems. In particular, the paradigm of Learning from Demonstrations is a widely used approach in numerous systems being deployed in fielded applications. Establishing whether these desiderata are satisfied by a controlled autonomous system based on learning from demonstrations can be particularly difficult. We argue that this is best achieved when models and representations used by the learning algorithms are suitably structured and when the learning paradigms explicitly incorporate input from human experts or end users. In this chapter, we outline an approach to achieving this, and we describe some techniques that aid in such a design process. This is grounded in examples taken from the domain of assistive operations with a humanoid robot, but similar principles apply also in most other applications of robotics and autonomous systems.

6.1 Introduction

The field of robotics, since its inception in the 1950s, has made sustained progress through the confluence of advances in hardware, software and electronic systems designs. Even by the standards of such a vibrant field, the past decade has seen an unprecedented level of advance and excitement powered by the advent of modern data-driven machine learning methods that have unlocked new levels of skill in perception and adaptation of behaviour. This has enabled the use of robots in environments beyond the controlled domains of factory shop floors and specialised teleoperation setups. We now have autonomous robots performing deliveries in our neighbourhoods [43], flying medicines to remote parts of the world [46], assisting nurses in hospitals [39] and driving on our public roads [1, 44]. Learning plays a crucial role in facilitating such autonomy in open worlds, with all of the variability implied in task contexts, environments and sources of disturbances. In particular, fielded robotics systems often use the paradigm of Learning from Demonstrations, to help end users directly modify and re-target the tasks being performed autonomously by the robot.

This raises the crucial question concerning us in this chapter: to what extent and how do we achieve assurances regarding the behaviour of such robots and autonomous systems. We will outline a methodology which is presented through the medium of examples from the domain of manipulation tasks involving humanoid robots. The underpinning ideas are fairly general and could be applied in many other domains beyond the ones used as examples here.

6.1.1 Autonomous Systems and Specifications of Behaviour

Robots are ultimately mechatronic systems subject to design processes akin to many other engineered products. What distinguishes them from other embedded systems we might find in our daily lives? The answer lies in the way they *interact* with their environment.

To understand this concept, consider the example of an automated driving system such as a car that must navigate urban environments. The problem being solved by this autonomous system is not just the control problem of following a lane but also issues of collision avoidance with respect to a dynamic and interactive environment with other decision-making agents. So, the model of the world and the context defined by the sensory inputs play a crucial role in defining the inputs, hence determining what the system will ultimately do.

The notion of correctness of behaviour of such a system is very difficult to make precise without also enumerating all variables and dynamics variations implied by the world model and other drivers. As an example, consider a requirement such as that a good driver should never pull out from a parked position in such a way as to cause another driver to slow down [12]. Determining the satisfaction of this requirement by an autonomous controller is crucially not only dependent on the details of the input-output relations encoded in the controller but also contingent on the validity of the models and beliefs about others' behaviours. Furthermore, in a learning-based solution to the generation of such driving behaviour, the need to query variables required to describe such attributes of observed behaviour places specific requirements on the models being used.

Crucially, in many of these applications, many aspects of what would conventionally be viewed as functional specifications are not made explicit. So, while certain elements of safe behaviour (say, collision avoidance) could be made explicit, other user-defined aspects of the task may be implicit in the example data and not made explicit as a formal specification. An extreme example of this phenomenon would be so-called end-to-end learning architectures [4], where the task is never made formally explicit and the details of the task requirements are really left implicit in the training data and processes for collecting them. Even when the entire system is not monolithically opaque in this way, many modern robotic systems involve task requirements being provided only through the nature of the diversity in the data. In keeping with this, our use of the term 'specifications' in this chapter will cover both what one conventionally understands by formal specifications of behaviour and this form of implied specification in systems that are allowed to learn from the end user to adapt their behaviours.

In the same vein, the line normally drawn between implementation of the controller and the concerns of verification and validation may begin to become blurred in practice. In this chapter, we begin from considerations of how learning from demonstration can be constrained when learning is used to *implement* the controller, motivated by considerations of making the implementation *verifiable*, and then end with the considerations of how the controller can actually be verified under different environmental conditions.

6.1.2 Learning Paradigms and Verification Needs

A common categorisation of machine learning is in terms of unsupervised, supervised and reinforcement learning methods. In modern practice of deep learning methods, one sees further specialisation in the form of methods for self-supervision, meta-learning and related approaches to devising architectures. From the point of view of robotics in human-centred environments, the focus of attention shifts slightly to methods including learning from demonstration, interactive task learning and machine teaching.

Giving physical demonstrations is an established method of teaching a robot complex movements [3]. One could consider a more extensive notion of interactions between the human user and the robot, involving a more sustained dialogue or mechanisms for allowing the human user to shape the models. This is the concept of interactive task learning [32]. This is also related to the idea of machine teaching [42].

While much of the motivation for the original development of these methods was based on a desire to enable end-user programming, we argue that these methods also lend themselves to improving the verifiability of systems. Our approach is to use these interactive paradigms to both inject knowledge into the models and structure them for querying in order to engender trustworthiness. This is aligned with the recognition that all uses of AI systems in safety critical systems must lend themselves to scrutiny in ways that remain challenging for the current state of the art machine learning systems [16, 20].

6.2 An Approach to Structuring the Design of Robot Control Strategies

In this chapter, we describe three techniques that constitute aspects of a coherent methodology for structuring models used in robot control through a process of interaction with the human expert or end user. The methods we describe are fully developed in [22–24, 27] and [28]. The aim of this chapter is to bring the key results together in order to expose the synergies.

We start by asking how the Learning from Demonstration (LfD) paradigm might be modified to take explicit logical specifications of behaviour in conjunction with the demonstration data [27]. Next, we ask how the representations within neural network based models might be constrained so as to map to high-level concepts that might appear in instructions from a human end user [22–24]. Finally, we describe a language for specifying models of task environments that are crucial for exploring when and how policies for action might fail [28]. In this chapter, we will not explicitly address techniques for verification of the models and policies for robot action. Instead, these techniques address the gaps between conventional tools for verification and the design processes required to create cyber-physical systems that could cope with open environments and evolving requirements. Indeed, verification of neural network-based models is an active area of investigation, see, e.g. [7, 25]. However, on the one hand, these methods are still far from addressing the scale of models being deployed in practice within robotics domains, and they do not yet address the issues of specification gaps and open environments we have identified as being crucial in this domain.

6.3 Incrementally Elaborating Task Constraints in the Learning from Demonstration (LfD) Paradigm

Consider the setting where the robot is provided with data for several demonstrations $D = \{D_0 \ldots D_M\}$ in the form $D_i = \langle \mathcal{I}_i, y_i \rangle$, where $\mathcal{I}_i$ denotes the starting input for the demonstration, which may include relevant objects positions, the initial joint angles/pose of the robot or many other sensory inputs. The term $y_i = [y_{i,0} \ldots y_{i,T}]$ represents the *trajectory* of the robot, where $y_{i,j}$ is the recorded pose at time j of demonstration i.

The objective of LfD is to learn a function that maps $\mathcal{I}_i$ to output trajectories similar to the demonstrations, which might be parameterized in different ways, e.g., as Dynamic Movement Primitives [26] which may be described by a series of differential equations of the form:

$$\ddot{y} = \alpha_y(\beta_y(y_{goal} - y) - \dot{y}) + f(x) \tag{6.1}$$

$$\dot{x} = -\alpha_x x \tag{6.2}$$

In this description, the function $f(x)$ expresses complex non-linear motion as a combination of radial basis functions $\psi_{0\ldots N}$, weighted by parameters $w_{0\ldots N}$:

$$f(x) = \frac{\sum_{i=0}^{N} w_i \psi_i(x)}{\sum_{i=0}^{N} \psi_i(x)} x(y_{goal} - y_{start}) \tag{6.3}$$

Given these equations, one could define the LFD problem: Given a demonstration D_i, we wish to learn a parameterized function g_θ which takes $\mathcal{I}_i$ as input and produces weights w as outputs such that the following loss function is minimized:

$$\mathcal{L}_d(\theta, \mathcal{I}_i, y_i) = \frac{1}{T} \sum_{t=0}^{T} \|\text{DMP}_{\theta,\mathcal{I}_i}(t) - y_{i,t}\|^2 \tag{6.4}$$

Here, $\text{DMP}_{\theta,\mathcal{I}_i}(t)$ denotes the value at time t of a DMP defined by $y_{start} = y_{i,0}$, $y_{goal} = y_{i,T}$, and weights learned via a neural network function parameterised by θ: $w = g_\theta(\mathcal{I}_i)$

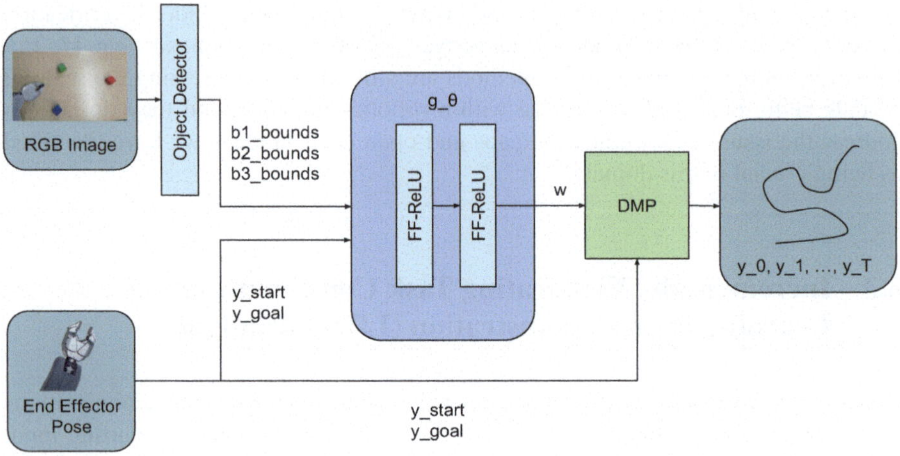

Fig. 6.1 Architecture overview. Inputs are the gripper pose and object descriptions (either given or extracted from RGBD with an off-the-shelf object detector). We run these through two feed-forward layers with ReLU activations and output weights w for Eq. (6.3). To calculate the trajectory, we roll out Eqs. (6.1–6.2)

In words, (6.4) gives the average distance between the learned and demonstrated trajectories. We call (6.4) the *imitation loss*.

Figure 6.1 shows an architecture overview. Like other works [34, 37], we represent g_θ as a feed-forward network, allowing us to optimize (6.4) using stochastic gradient-descent.[1]

While this basic procedure is remarkably versatile and used extensively in robotics, there are a few reasons why this might be insufficient in the field. Firstly, the task may be difficult to demonstrate, especially when teleoperation is cumbersome for the performance of the task, which may require subtle modulation of sensorimotor variables that are otherwise hard to measure. Crucially in such domains, simply copying a movement trace is rarely sufficient, and a closed-loop control policy may need to be induced. A second source of difficulties may be in the complexity of the specification, making it easier to specify at a high level but hard to cover fully in explicitly generated examples. There is also the issue of potential overtraining with limited training data, which calls for additional guidance.

An alternative is to provide the specification at a high level, such as in Linear Temporal Logic (LTL) [38], a language that extends traditional propositional logic with modal operators. With these additional operators, we can specify robotic constraints that must hold *through time*. The syntax for constructing an LTL

[1] For simple DMPs, we can approximate w in closed form [26]. However, such methods do not accommodate additional sensory inputs, nor the modifications we make to the loss function in the next section.

constraint φ is given by (6.5)

$$\begin{aligned} \varphi &:= p \mid \neg\varphi \mid \varphi_1 \wedge \varphi_2 \mid \mathcal{N}\varphi \mid \Box\varphi \mid \Diamond\varphi \mid \varphi_1 \mathcal{U} \varphi_2 \\ p &:= T_1 < T_2 \mid T_1 \leq T_2 \mid T_1 = T_2 \mid T_1 \geq T_2 \mid T_1 > T_2 \end{aligned} \tag{6.5}$$

Here, $\mathcal{N}\varphi$ means that φ must hold at the *next* time-step, $\Box\varphi$ means that φ must hold at *every* time-step, $\Diamond\varphi$ means that φ must *eventually* hold at some future time-step and $\varphi_1\mathcal{U}\varphi_2$ means that φ_1 must remain true *until* φ_2 holds. Atomic comparisons p may be made up of *static* terms (which do not depend on time) or *dynamic* terms (which do depend on time, such as the value of a DMP). Terms are denoted by T_i.

We wish to make use of the LTL definition given above in the optimization process of LfD, as outlined above. In order to do so, we need to be able to calculate continuous gradients. Towards this end, we introduce a *constraint loss* $\mathcal{L}_c$ which takes as input an LTL statement φ and time-step t and outputs a real-valued, differentiable loss. This loss measures quantitatively how close φ is to being satisfied.

Equation (6.6) gives the quantitative semantics for the atomic comparison terms. These build on the work by Fischer et al. [14] who define a differentiable quantitative semantics for propositional logic:

$$\begin{aligned} \mathcal{L}_c(t_1 \leq t_2, i) &= max(t_1(i) - t_2(i), 0) \\ \mathcal{L}_c(t_1 \neq t_2, i) &= \zeta[t_1(i) = t_2(i)] \\ \mathcal{L}_c(t_1 < t_2, i) &= \mathcal{L}_c(t_1 \leq t_2 \wedge t_1 \neq t_2, i) \\ \mathcal{L}_c(t_1 = t_2, i) &= \mathcal{L}_c(t_1 \leq t_2 \wedge t_1 \geq t_2, i) \end{aligned} \tag{6.6}$$

Here, $[a = b]$ is a function which evaluates to 1 when $a = b$ and 0 otherwise. The constant ζ is a small positive scalar.

Equation (6.7) gives the quantitative semantics for the modal and composition operators. It builds on quantitative measures of robustness from signal temporal logic [13]:

$$\begin{aligned} \mathcal{L}_c(\varphi_1 \wedge \varphi_2, i) &= max^\gamma\,(\mathcal{L}_c(\varphi_1, i), \mathcal{L}_c(\varphi_2, i)) \\ \mathcal{L}_c(\varphi_1 \vee \varphi_2, i) &= min^\gamma(\mathcal{L}_c(\varphi_1, i), \mathcal{L}_c(\varphi_2, i)) \\ \mathcal{L}_c(\mathcal{N}\varphi, i) &= \mathcal{L}_c(\varphi, i+1) \\ \mathcal{L}_c(\Box\varphi, i) &= max^\gamma(\mathcal{L}_c(\varphi, i), \ldots, \mathcal{L}_c(\varphi, T)) \\ \mathcal{L}_c(\Diamond\varphi, i) &= min^\gamma(\mathcal{L}_c(\varphi, i), \ldots, \mathcal{L}_c(\varphi, T)) \\ \mathcal{L}_c(\neg\mathcal{N}\varphi, i) &= \mathcal{L}_c(\mathcal{N}(\neg\varphi), i) \\ \mathcal{L}_c(\neg\Box\varphi, i) &= \mathcal{L}_c(\Diamond(\neg\varphi), i) \\ \mathcal{L}_c(\neg\Diamond\varphi, i) &= \mathcal{L}_c(\Box(\neg\varphi), i) \end{aligned} \tag{6.7}$$

$$\begin{aligned}
\mathcal{L}_c(\varphi_1 \mathcal{U} \varphi_2, i) = max^{\gamma}(\\
& min^{\gamma}(\mathcal{L}_c(\varphi_1, i), \mathcal{L}_c(\varphi_2, i)), \\
& min^{\gamma}(\mathcal{L}_c(\varphi_1, i+1), \mathcal{L}_c(\varphi_2, i), \mathcal{L}_c(\varphi_2, i+1)), \\
& \ldots, \\
& min^{\gamma}(\mathcal{L}_c(\varphi_1, T), \mathcal{L}_c(\varphi_2, i), \mathcal{L}_c(\varphi_2, i+1), \ldots, \mathcal{L}_c(\varphi_2, T)))
\end{aligned}$$

Because maximization and minimization functions are non-differentiable, we use the soft approximations max^{γ} and min^{γ} (inspired by the work of [10] on soft approximations for dynamic time warping). These are defined in Eqs. (6.8) and (6.9) below:

$$max^{\gamma}(x_1, \ldots, x_n) = \gamma \ln \sum_{i=0}^{n} e^{x_i/\gamma} \tag{6.8}$$

$$min^{\gamma}(x_1, \ldots, x_n) = -\gamma \ln \sum_{i=0}^{n} e^{-x_i/\gamma} \tag{6.9}$$

With the above properties, we can now augment our LFD loss from (6.4) with LTL constraints φ to give Eq. (6.10):

$$\mathcal{L}_{full}(\theta, D, \varphi) = \frac{1}{M} \sum_{i=0}^{M} \mathcal{L}_d(\theta, \mathcal{I}_i, y_i) + \eta \mathcal{L}_c(\varphi(\theta, \mathcal{I}_i), 0) \tag{6.10}$$

What does this enable us to achieve in terms of robot control? Consider dexterous manipulation tasks that could be performed by a humanoid robot, in our case the PR-2 robot. The robot might wish to learn (from a combination of RGB-D camera images of the scene and traces of joint angle movement data) to pour from one container to another. In this setting, the end user might wish to elaborate, with additional LTL constraints, to say for instance, 'Don't tip the cup until you are close to the container', or stated in LTL:

$$\begin{aligned}
&\Box(\|p_{xyz} - o_1\|^2 \geq 0.1 \ \wedge \ p_z \geq o_{1,z}) \\
\implies &(\langle 0, 0, -1.0\rangle \leq p_{rpy} \leq \langle 0.2, 0.2, 0.0\rangle)
\end{aligned}$$

Figure 6.2 shows the setup for the task. Similarly, one could specify goals for a reaching task, asking to reach from the edge of the table to the red cube. After the demonstration was recorded, we then incrementally add two additional constraints: 'Avoid the purple bowl' and 'Visit the green cube'. The LTL for this additional constraint is

$$(\Box \|p_{xyz} - o_2\|^2 \geq 0.2) \wedge (\Diamond p_{xyz} = o_3)$$

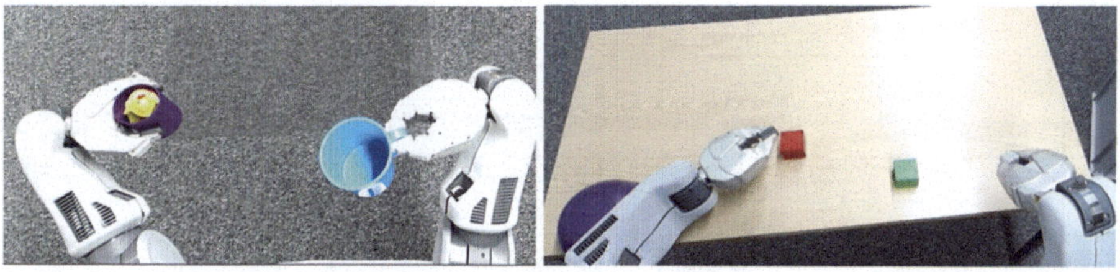

Fig. 6.2 Setup for the pour (left) and reach (right) tasks

There are by now numerous works combining symbolic logic with neural networks. Xu et al. [45] derive a 'semantic loss' for logical constraints, which can be used when a network's output layer represents a distribution over a set of Boolean variables. Similarly, [14] use their Deep Learning with Differentiable Logic framework to transform propositional logic into differentiable costs for use in classification tasks. They also use an adversarial algorithm to search for constraint-violating counterexamples. As we show in [27], the method described here can similarly accommodate adversarial training, thus extending the expressiveness to temporal logic for robotic control.

6.4 Achieving Alignment Between Concepts in High-Level Specifications and Variables in Data-Driven Models

When the objectives of a robot's task are being elicited from domain experts who are not otherwise experts in computational methods, it is inevitably the case that they describe the task in terms of high-level domain-specific terms that can be vague but at the same time content-rich. One way to conceptualise the relationship between these forms of specifications and the corresponding concrete representations within robot programs is as depicted in Fig. 6.3.

Coping with this variety of representations requires a collection of learning problems to be solved, intermediating between the different levels of description, as shown in Fig. 6.4.

In this section, we describe a method for addressing how a machine learning model could be structured in order to enable high-level concepts to be represented in an interpretable manner within the data-driven neural network model. This is based on the family of models known as auto-encoders.

An auto-encoder is typically comprised of two main parts—an encoder p_θ and a decoder q_ϕ, both represented as deep neural networks with stacked nonlinear layers. The task of the encoder is to translate a given data point $\mathbf{x}$ to a feature vector $\mathbf{z} = p_\theta(\mathbf{x})$. The task of the decoder is then to try and reconstruct the original input from just observing a given feature vector—$\hat{\mathbf{x}} = q_\phi(\mathbf{z})$. For a given dataset of N observations $\mathcal{D} = \{x^{(1)}, x^{(2)}, \ldots, x^{(N)}\}$, the parameters of the encoder p_θ and

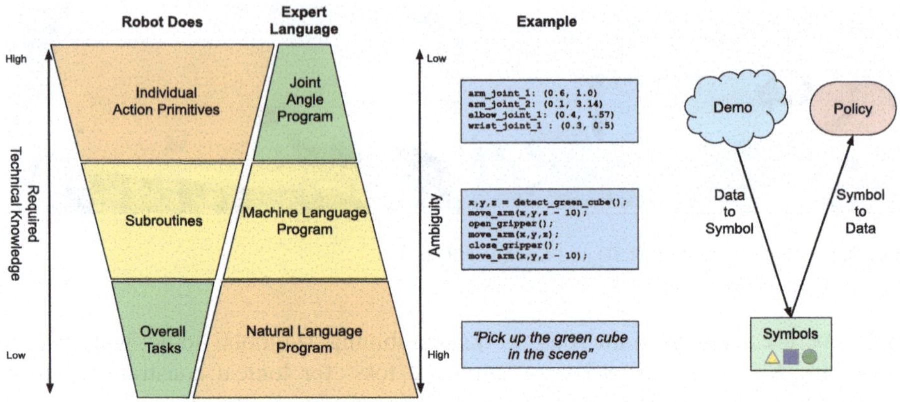

Fig. 6.3 Natural ways to program a robot. The same underlying task—grasping and picking up a coloured object—can be expressed through commands with different levels of abstraction. The more concrete and 'closer to the metal' the commands are, the less ambiguous they are, but more technical prior knowledge is required

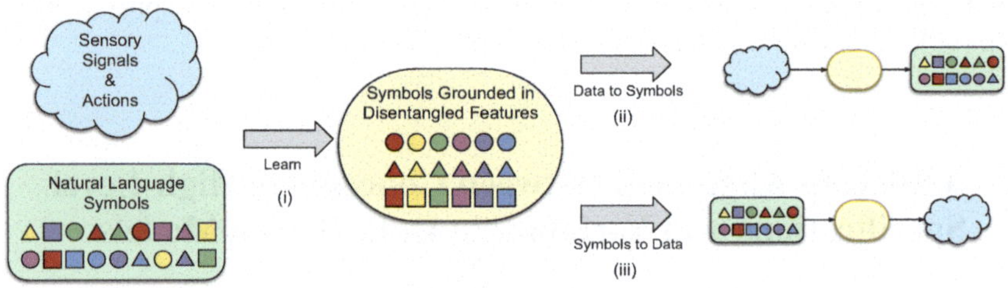

Fig. 6.4 A pipeline of learning problems that must be solved in order to go between the levels identified in Fig. 6.3. These include **(i)** learning interpretable features for symbol grounding, **(ii)** explaining high-dimensional observations through grounded symbols and **(iii)** generating high-dimensional actions conditioned on grounded symbols

decoder q_ϕ are jointly optimised by minimising the following objective:

$$\mathcal{L}_{AE} = \frac{1}{N}\sum_{i}^{N}(x^{(i)} - q_\phi(p_\theta(x^{(i)})))^2 \tag{6.11}$$

Minimising the mean squared distance between observed and reconstructed data points is a necessary but not sufficient condition for $\mathbf{z}$ to be a *good* representation. Another crucial factor is its dimensionality—it should be much lower than the dimensionality of $\mathbf{x}$—$|\mathbf{x}| >> |\mathbf{z}|$. Thus, $\mathbf{z}$ plays the role of an *information bottleneck* which only passes information *crucial* for reconstructing the original input. This implicit regularisation encourages the encoder to be efficient and to learn hierarchical and reusable feature extractors in its layers. Alternatively, if $\mathbf{z}$

and $\mathbf{x}$ have similar dimensionalities, the most optimal solution for p_θ and q_ϕ, which would drive $\mathcal{L}_{AE}$ to 0, is to be the identity function.

Minimising $\mathcal{L}_{AE}$ is essentially equivalent to maximising the data likelihood $p(\mathbf{x})$ under the model parameters θ, ϕ. The network as a whole can be perceived as modelling the data distribution in a non-parametric fashion. Thus, while useful in producing *good* low-dimensional representations, auto-encoders on their own do not lend themselves to reasoning about the relationship between $\mathbf{x}$ and $\mathbf{z}$.

Variational Autoencoders, VAEs [31], are a type of auto-encoder model, where additional constraints are introduced over the information bottleneck. The encoder p_θ and decoder q_ϕ are designed to predict the parameters of upfront-chosen parametric distributions, e.g. Gaussian, rather than point predictions for $\mathbf{z}$ and $\hat{\mathbf{x}}$. After predicting the parameter values, samples for the feature vectors or the reconstructions can be drawn in a Monte Carlo fashion. Introducing stochastic operations like sampling in the computation graph of the model makes gradient computation non-trivial, and training becomes sample inefficient. Thus, utilisation of the Parametrisation Trick [31] is of utmost importance.

Moreover, vanilla auto-encoder models have been shown to memorise different data points as individual modes in the latent space, without reliably being able to interpolate between them. Thus, additional regularisation of the latent space $\mathbf{z}$ can be achieved by imposing a prior distribution over it, which the posterior $p(\mathbf{z}|\mathbf{x})$ is optimised to follow. This additionally constrains the information bottleneck, forcing it to be smoother, capturing the underlying generative factor distribution more closely (up to a transformation of it) [5].

The overall loss function, with respect to which these models are optimised, is

$$\mathcal{L}_{VAE} = D_{KL}(q_\phi(\mathbf{z}|\mathbf{x})||p(\mathbf{z})) - \mathbb{E}_{q_\phi(\mathbf{z}|\mathbf{x})}(\log p_\theta(\mathbf{x}|\mathbf{z})) \tag{6.12}$$

Minimising it results in (i) bringing the approximating distribution p_θ as close as possible to the chosen prior $p(\mathbf{z})$ by minimising the Kullback-Leibler divergence term D_{KL} and (ii) maximising the data log-likelihood of the observed data $\mathbf{x}$ or in other words minimising the reconstruction error. The prior $p(\mathbf{z})$ is typically chosen such that the D_{KL} can be analytically computed—e.g. an isotropic Gaussian—making the training procedure more efficient.

Let us now consider how the human expert and a robotic agent, who share their environment and who can both observe and interact with objects in the surrounding world, could arrive at a common grounding of concepts regarding these objects. We assume that the agent can extract images of distinct objects from the world and receive labels for them from the expert—this being the interaction paradigm for our modelling.

In our experiments, we could work with any data that aligns a sequential abstract plan for a task (perhaps parsed from a linguistic instruction) with a temporal trace of expert movements and actions in the external environment—e.g. movement tracking [2] and eye-tracking [35]. So, specific symbols from the plan can be related to regions of interest in the environment, where the object, encoding the label's meaning, resides.

Let $O = \{\mathbf{x}_1, \ldots, \mathbf{x}_N\}, \mathbf{x} \in \mathbb{R}^D$, be a set of unlabelled observations, which have been passively gathered prior to the expert demonstration. An expert demonstration introduces a set of L conceptual groups $\mathcal{G} = \{g_1, \ldots, g_L\}$ (e.g. colour, shape, size etc.), where each group is a set of mutually exclusive discrete labels: $g_j = \{y_{j1}, \ldots, y_{jn}\}, n = |g_j|$ (e.g. colour can be red, blue, yellow etc.). The demonstration also provides labels, one drawn from each concept group, to M of the observations: $O = \{(\mathbf{x}_1, \mathbf{y}_1), \ldots, (\mathbf{x}_M, \mathbf{y}_M), (\mathbf{x}_{M+1}, \emptyset) \ldots (\mathbf{x}_N, \emptyset)\}$. $\mathbf{y} = \{y_1, \ldots y_L\}$, is a set of labels that have been attached to an image $\mathbf{x}$—one per concept group.

The crucial task of interest in this section is to learn a mapping from $\mathbf{x} \in \mathbb{R}^D$ to $\mathbf{Z} \in \mathbb{R}^C, C \ll D$, where the space of $\mathbf{Z}$ possesses the following properties:

Axes Alignment Guarantees a one-to-one mapping from the concept groups $\mathcal{G}$ to the orthogonal basis vectors that span $\mathbf{Z}$. This would guarantee that independent concepts in image space are kept independent in $\mathbf{Z}$—e.g. colour does not depend on the size, orientation or shape of an object.

Intra-Group Linear Separability The latent clusters in $\mathbf{Z}$, corresponding to the labels in each concept group g_j, are linearly separable across the basis vector which has been aligned with that concept group.

These two properties allow for $\mathbf{Z}$ to be used as a feature space for performing probabilistic symbol grounding with the ability to recognise *unknown* objects. For that we use 1-NN classification with 1-dim normal distributions (one per label) for each concept group in $\mathcal{G}$—see Fig. 6.5c, d. The distributions are estimated after optimising $\mathbf{Z}$.

In learning $\mathbf{Z}$, we explore the effects of adding an auxiliary classification loss to a Variational Autoencoder [31] as a base architecture, specifically the β-VAE [21]. Through weak supervision, in the form of partially labelled data, the auxiliary loss influences the latent space of the model to exhibit properties which make it suitable for robust symbol classification. The model consists of a convolutional encoder network q_ϕ, parametrised by $\boldsymbol{\phi}$, a deconvolutional decoder network p_θ, parametrised by $\boldsymbol{\theta}$ and a set of j linear classifiers parametrised by $\mathbf{w}_j \in \mathbb{R}^{|g_j|}$, one for each group $g_j \in \mathcal{G}$. Additional parameters—α, β, γ—are added on the three terms of the overall loss function—see (6.13)—in order to leverage their importance. The values for all three coefficients are chosen empirically such that the values for all the loss terms have a similar magnitude, and thus none of them overwhelms the gradient updates while training.

$$\min_{\theta,\phi,\mathbf{W}} \mathcal{L}(\mathbf{x}, \mathbf{y}, \theta, \phi) = \beta D_{KL}(q_\phi(\mathbf{z}|\mathbf{x})||p(\mathbf{z})) - \alpha \mathbb{E}_{q_\phi(\mathbf{z}|\mathbf{x})}(\log p_\theta(\mathbf{x}|\mathbf{z}))$$

$$+ \gamma \sum_{j}^{|\mathcal{G}|} H\left(z_j \mathbf{w}_j^T, y_j\right) \tag{6.13}$$

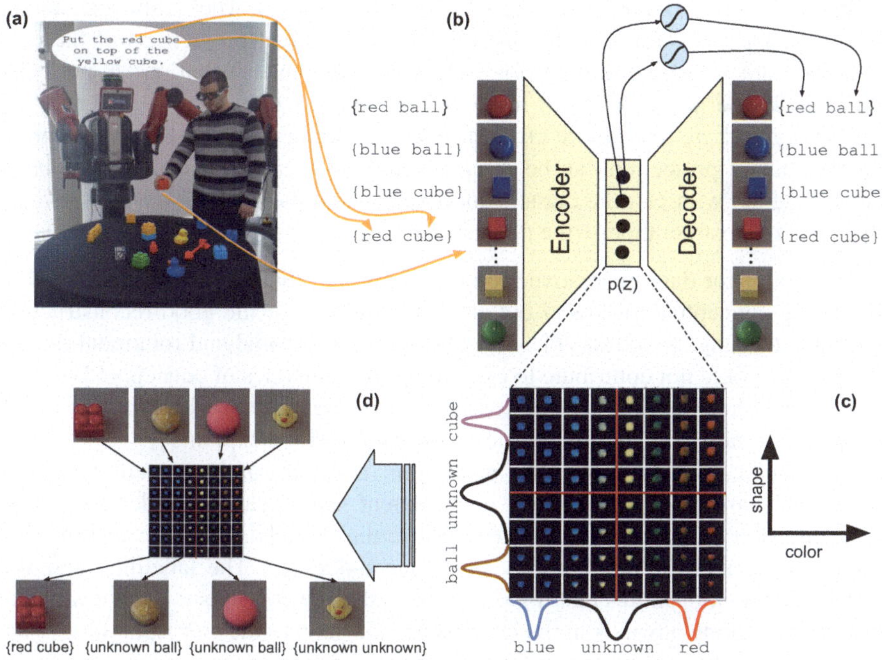

Fig. 6.5 Overview of the pipeline for a system that performs symbol grounding. An expert is demonstrating a task at hand while uttering symbols which are grounded according to their attention in the environment **(a)**. The conceptual grouping of the symbols is given. The gathered dataset is used to train a VAE with a set of auxiliary classification losses—one per conceptual group. Each classifier takes information from a single latent dimension **(b)**. The training procedure guarantees that the semantic meaning of each group will be encoded in a separate latent dimension with linearly separable classes **(c)**. Using the per-class estimated 1D normal distributions, together with in-between distributions for unknown labels, 1-NN classification can be performed **(d)**

Classification Term (Weighted by γ) To force the learnt latent space to explain the variations in the data, we add a linear classifier for each concept group. Each classifier has to predict the set of labels for its assigned concept group using information only from a single latent dimension. No two classifiers have access to the same latent dimension. That forces each dimension to only explain labels from the particular concept group. A softmax cross-entropy term is used over the predictions of each classifier.

Reconstruction Term (Weighted by α) The reconstruction loss is a standard Mean Squared Error (MSE), which is used to predict the pixel values across the three RGB channels. The motivation behind this term is that we do not assume all data points in the dataset are labelled. Thus the reconstruction loss would force data points that look similar in image space to be projected close to each other in the latent space.

Kullback-Leibler Divergence Term (Weighted by β) The Kullback-Leibler divergence term ensures that the distribution of the latent projections of the data in $\mathbf{Z}$ does not diverge from a prior isotropic Gaussian distribution. A perfectly optimised KL term would result in all latent projections to be 0. This forces the encoder network q_ϕ to be more efficient when encoding the image observations so that their latent projection can be discriminated from each other across the basis vectors aligned with $\mathcal{G}$—the classification term—and the decoder network p_θ can efficiently reconstruct them—the reconstruction term.

To account for data-generative factors of variation that might not be needed to encode the conceptual groups $\mathcal{G}$ but are still essential for the good reconstruction and stable training, we allow $|\mathbf{Z}| > |\mathcal{G}|$. For example, spatial and rotational factors of variation would not contribute to explaining the semantics of concepts like shape and size but should still be accounted for so that the subset of basis vectors in $\mathbf{Z}$ which are aligned with $\mathcal{G}$ only encodes information which explains $\mathcal{G}$.

The results of such a learning procedure, in terms of how well it enables interpretable and aligned labelling in the latent space, can be understood from inspecting the visualisation in Fig. 6.6. Here, the model is learning to represent images that contain items that vary in colour and shape. The labelling achieved by the model demonstrates that it has represented these concepts in such a way that classification along any axis in the latent space maps on to the corresponding visual concept we would expect to use to describe that variation.

The above description in this section is the core of an idea that has been extended in [24] and [23] to address much more sophisticated notion of spatial relationships between objects and concepts associated with gradation of continuous variables such as applied forces, speeds and distances.

6.5 A Specification Language for Robotic Task Environments

A typical task one might expect robots to be performing in the field today is pick and place in a factory or warehouse setting. For instance, there might be in- and out-trays containing parts, with the robot being asked to automatically transfer parts from one to the other. In current development practice, designers would be expected to make use of simulators to test such functionality. Examples of simulators popular within the robotics community include PyBullet [9] or CoppeliaSim [40]. A crucial gap in this pipeline is that while there exist numerous methods for specifying what *task* a robot must accomplish, few options exist for specifying the *range of environments* the robot is expected to accomplish that task in. This is despite the fact that adequately defining the intended operational environment of a robot is a crucial component of system validation [17].

In current practice, testing whether a robot works in multiple environments requires us to explicitly construct each simulation scene. So, for instance, the user may manually place the robot, table, trays and other objects and then run the

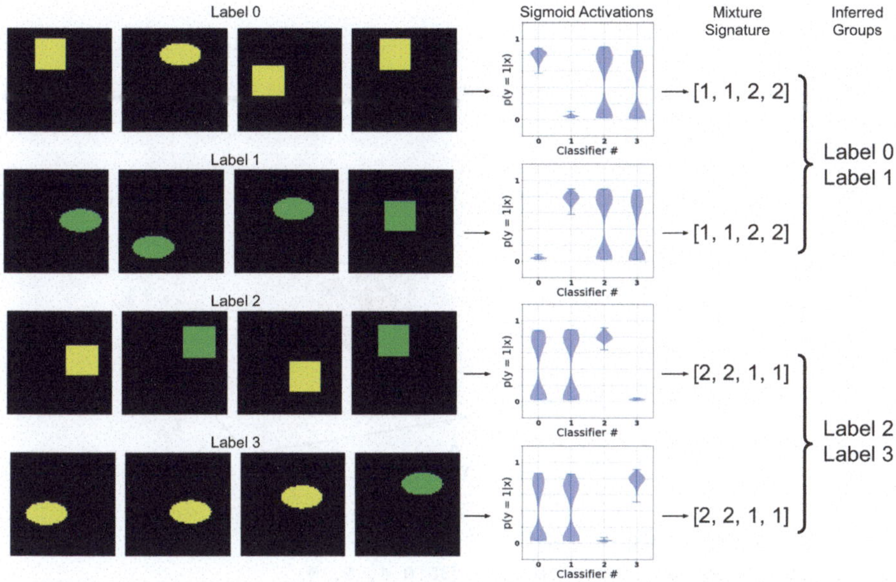

Fig. 6.6 Group inference over four labels, with two ground truth groups—colour and shape. Curated data batches with the corresponding labels: 0 ≡ yellow, 1 ≡ green, 2 ≡ square, 3 ≡ ellipse. A binary classifier is trained for each label **(left)**. Distribution of sigmoid activations for all four binary classifiers for all labels. For example classifier 0 on the first row predicts all presented observations are 'yellow', while classifier 2 predicts some are 'square' and some are not **(middle)**. A mixture of 1D Gaussians is fit over the sigmoid activations using a Bayesian Information Criterion (BIC). Labels with similar number of mixtures (*mixture signature*) are grouped together **(right)**

simulation using their robot controller. Subsequently, this will require a campaign of ad hoc changes to the scene in order to convince oneself that the controller can indeed handle variations. This is the approach of *exploratory testing* in traditional programming [29].

An alternative methodology would be to automate the process of randomized testing [19]. In this approach, input constraints correspond to constraints on valid environment setups in which our robot is expected to operate. For robotic manipulation tasks, these environmental constraints typically take the form of *geometric relations* between objects. Following the goals of End User Programming, we would like users to be able to easily express their desired environmental constraints without deep knowledge of the internals of a system. A small number of such specification languages exist already, for domains such as 2D autonomous vehicle simulation [18]. ProbRobScene [28] is a probabilistic specification language specifically aimed at three-dimensional, inherently relational domain of robotic manipulation. Figure 6.7 shows an example specification for a table-top environment, along with a sample from this specification. By combining techniques from probabilistic programming and convex computational geometry, ProbRobScene can efficiently

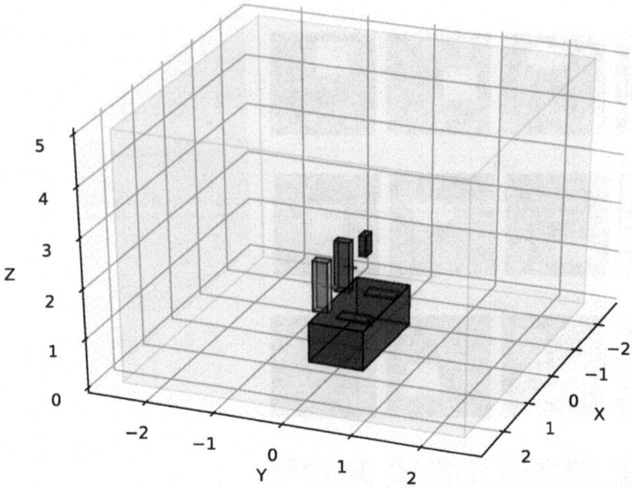


```
t = Table on V3D(0,0,0)
r1 = Robot on (top back t) - V3D(0.4, 0, 0)
r2 = Robot on (top back t) + V3D(0.4, 0, 0)

tr_1 = Tray completely on t,
            ahead of r1,
            left of t

tr_2 = Tray completely on t,
            ahead of r2,
            right of t

Cube completely on tr_1

Camera at V3D((-0.1, 0.1),
              (-0.1, 0.1),
              (1.9, 2.1)),
       facing V3D(0, 0, -1)
```

Fig. 6.7 'Table-Cube' specification—two robots at either side of a table, two trays with a cube in one and a ceiling camera facing downwards

generate samples environments consisting of complex relations between objects in three dimensions.

Table 6.1 summarizes the specifiers available in ProbRobScene. The left column shows the syntax of each specifier, the object property it applies to (e.g., position, orientation or any) and the data type of the arguments it takes. The right column shows the type of value the given specifier evaluates to. For most simple specifiers, this value is a three-dimensional vector (V3D). We summarize the semantics of each of these simple specifiers below:

Table 6.1 Specifier syntax and return assignment type

Specifier syntax	Return assignment type
Position specifiers	
at *V3D* [relative to (*V3D* | *Object*)]	*V3D*
beyond (*V3D* | *Object*) by (*Float* | *V3D*) from (*V3D* | *Object*)	*V3D*
in *Region*	*Region-Dist*
(left | right | ahead | ...) of (*Object* | *V3D*)	*Halfspace-Dist*
[completely] on (*Object* | *Rectangle-Region* | *V3D*)	*V3D* | *Rectangle-Dist*
aligned with (*Object* | *V3D*) on (*x* | *y* | *z*)	*Rectangle-Region*
Orientation specifiers	
facing *V3D*	*V3D*
facing Toward (*Object* | *V3D*)	*V3D*
Generic specifiers	
with *String Any*	*Any*

Fig. 6.8 Region constructors

```
Cuboid(origin, orientation, dims)
Rect3D(origin, orientation, dims)
Halfspace(origin, normal)
ConvexPolygon3D(hsi, origin, normal)
ConvexPolyhedron(hsi)
All()
Empty()
```

with, general purpose direct assignment. For example, `Cube with color "green"` sets the cube's `color` property to `"green"`.

facing / facing towards rotates the object to face either a given normal vector, or towards another object. For example `Cube facing V3D(1, 0, 0)` rotates the cube 90° to face forwards along the x-axis.[2]

at sets the position of an object to a given vector.

beyond, best explained by example—the statement `Cube beyond X by d from Y` sets the cube position `d` metres from `X` along the direction vector from `Y` to `X`.

Rather than specifying a property to be a single vector, the remaining specifiers assign a property to a *distribution* of points over an entire geometric region. Users can define custom regions explicitly using one of several `Region` constructors, given in Fig. 6.8.

The most direct way to set an object's position to a region distribution is with the **in** specifier. This specifier simply produces a uniform distribution over all points contained within the given geometric region.

However, rather than defining their own regions explicitly, an easier way for the user to specify their environment is by producing regions *implicitly* via other relations.

[2] This assumes the canonical 'forward' direction is the positive y-axis.

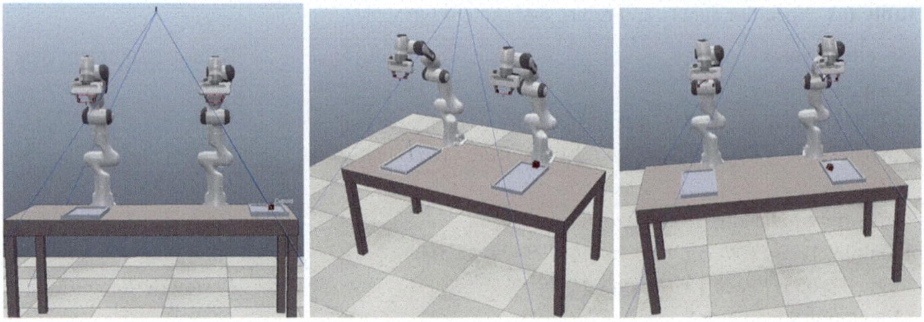

(a) (b) (c)

Fig. 6.9 Samples from 'Table-Cube' specification simulated using Coppelia-Sim. One of each of the identified failure types is shown. (**a**) Perception fault. (**b**) No path. (**c**) Poor grip

First, consider `Cube ahead of X`. This allows the cube to be placed anywhere in the world, as long as this position is front of X. This is geometrically equivalent to saying position that must lie within the *half-space* with an origin and normal given by the front face of X.

The next is `Cube on X`. It says that the cube is placed somewhere on space on top of X. Geometrically, this corresponds to a rectangular region defined by the top face of X. Notice though that there are points on the surface of X for which the edges of the cube might hang off the edge of X. So optionally, the user can use the `completely` keyword to assert that the full bounding box of the cube must be inside the surface rectangle of X.

Finally, let us look at `Cube aligned with X along z`. Geometrically, this restricts the z component of our cube to equal the z component of `X`. Thus the available positions of our cube are given by a plane with origin $\langle 0, 0, X.z \rangle$ and normal $\langle 0, 0, 1 \rangle$.

Reference [28] describes an implementation of efficient methods for sampling scenes described by this specification language. The outcome of using such methods is the identification of failure cases such as described in Fig. 6.9.

6.6 Discussion and Opportunities for Future Work

In the foregoing discussion, we have presented a few tools that are aimed at alleviating problems arising in the verification and validation of learning-enabled robot control systems. This is an area requiring much more extensive study and the development of a significantly more mature toolchain that can support industrial-scale deployments of autonomous systems. Here we identify some areas that we believe to be particularly ripe for immediate further development:

- **Finding edge cases**: What distinguishes robots and autonomous systems from most other embedded systems is the diversity of factors that can change the task context in an open environment. As we have discussed, this calls for systematic exploration of the task environment and the cross product of the robotic task variables and the environment variables. In a setting where the task is determined by a combination of spatial relational, temporal and concurrency constraints, with the added feature that the variables involved in these relationships are under the control of an external decision-maker in an active environment, there is a need for exploration methods that are able to cover the semantic space in an efficient manner. In this sense, current neural network verification methods, e.g., [25], and methods for guarding against adversarial examples, e.g., [7], are still in need of a significant extension and scale-up.
- **Challenges with simulation and data augmentation**: In the domain of autonomous vehicles, there has been long-standing discussion regarding the extent to which physical testing in the field is required to establish safety of the overall system, alongside the question of the balance between simulation testing and field testing [30]. In that domain, it is clear that achieving full coverage through physical testing alone may not really be economical. Yet, achieving similar coverage through simulation methods rests on the circularity implied by the need for the simulator to be designed to capture the unanticipated reality in the first place. An answer may be the use of generative models [11] that are able to support coverage in the same ways as formal methods from the automated verification literature. However, there remains the need for significant new research to scale up such approaches. Furthermore, there remains the question of emergent dynamics in complex interactive systems, as identified in [36]—phenomena that may only be made more acute by the black box nature of many components in the systems we are discussing.
- **Missing specifications**: It is well known in cognitive science and social studies of technology that much of what we know we cannot say [8]. This is especially the case when the information we seek is regarding rare events and edge cases that determine safe operation of complex systems in open environments. In computational terms, this translates to problems of many essential specifications being missing from the original requirements capture processes. In this chapter, we have sketched aspects of a methodology that allows for interaction and dialogue to fill in these gaps. However, any such interaction may need to be steered efficiently in order for it to be tractable. One might approach this in deployed systems through learning from incident reporting and other field data [33, 41], although much needs to be understood about how such qualitative information can be factored into the technical aspects of AI-based system design.
- **How will we capture responsibility and socio-technical factors?**: Ultimately, robots and autonomous systems will be tasked with serving people as part of complex socio-technical domains where there are numerous considerations beyond the technical aspects of functionality. There is a need for us to understand these gaps [6], to identify factors of accountability, responsibility and data ethics that also imply requirements and specifications that must guide technical system

design. Significant new work may be required to fully address these issues that are becoming increasingly more urgent with the rapid adoption of autonomous systems in our daily lives.

6.7 Conclusions

Techniques from machine learning and the data-driven methodology of modern AI have enabled significant new advances with robots and autonomous systems, spawning whole new industry segments where robots have come to be deployed in the field. With these rapid advances, we find new challenges for assuring the safety and trustworthiness of these autonomous systems. Indeed, leading edge deployments in domains ranging from autonomous vehicles to warehouse automation have made clear that identification of edge cases and efficient forms of weak and high-level supervision of learning models are essential ingredients of a pipeline for systems engineering. This conversation around the pressing need for new methods has even reached the highest levels of policymaking [15]. In this chapter, we describe a set of techniques that could support a holistic interactive approach to elicitation of task specifications, incorporation of these into machine learning models and finally searching for violations of these specifications in efficient ways. We have tried to elucidate these concepts using basic tasks in a robotic manipulation pipeline. In current work, we are actively engaged in applying such methods to demanding application domains including robots in healthcare and other safety critical domains.

Acknowledgments The work described in this chapter has benefitted from support of the Alan Turing Institute, through the project Safe AI for Surgical Assistance, and the UKRI Research Node on Trustworthy Autonomous Systems Governance and Regulation, through EPSRC Grant EP/V026607/1.

References

1. Five AI. Home Page. https://five.ai/, 2021.
2. Muhannad Al-Omari, Paul Duckworth, David C Hogg, and Anthony G Cohn. Natural language acquisition and grounding for embodied robotic systems. In *AAAI*, pages 4349–4356, 2017.
3. Brenna D. Argall, Sonia Chernova, Manuela Veloso, and Brett Browning. A survey of robot learning from demonstration. *Robotics and Autonomous Systems*, 57 (5): 469–483, 5 2009.
4. Mariusz Bojarski, Davide Del Testa, Daniel Dworakowski, Bernhard Firner, Beat Flepp, Prasoon Goyal, Lawrence D Jackel, Mathew Monfort, Urs Muller, Jiakai Zhang, et al. End to end learning for self-driving cars. *arXiv preprint arXiv:1604.07316*, 2016.
5. Christopher P Burgess, Irina Higgins, Arka Pal, Loic Matthey, Nick Watters, Guillaume Desjardins, and Alexander Lerchner. Understanding disentangling in beta-VAE. *arXiv preprint arXiv:1804.03599*, 2018.
6. Simon Burton, Ibrahim Habli, Tom Lawton, John McDermid, Phillip Morgan, and Zoe Porter. Mind the gaps: Assuring the safety of autonomous systems from an engineering, ethical, and legal perspective. *Artificial Intelligence*, 279: 103201, 2020.

7. Nicholas Carlini and David Wagner. Towards evaluating the robustness of neural networks. In *2017 IEEE symposium on security and privacy (SP)*, pages 39–57. IEEE, 2017.
8. Harry Collins. *Tacit and explicit knowledge*. University of Chicago Press, 2010.
9. Erwin Coumans and Yunfei Bai. Pybullet, a python module for physics simulation for games, robotics and machine learning. http://pybullet.org, 2016–2019.
10. Marco Cuturi and Mathieu Blondel. Soft-DTW: A differentiable loss function for time-series. In *34th International Conference on Machine Learning*, volume 2, pages 1483–1505, 3 2017.
11. José Rafael Valle Gomes da Costa. *Data Hallucination, Falsification and Validation using Generative Models and Formal Methods*. PhD thesis, University of California, Berkeley, 2018.
12. UK DoT. The Highway Code. https://www.gov.uk/guidance/the-highway-code, 2021.
13. Georgios E. Fainekos and George J. Pappas. Robustness of temporal logic specifications for continuous-time signals. *Theoretical Computer Science*, 410 (42): 4262–4291, 2009.
14. Marc Fischer, Mislav Balunovic, Dana Drachsler-Cohen, Timon Gehr, Ce Zhang, and Martin Vechev. DL2: Training and Querying Neural Networks with Logic. *International Joint Conference on Artificial Intelligence*, pages 1532–1538, 2019.
15. M Flournoy, Avril Haines, and Gabrielle Chefitz. Building trust through testing, 2020.
16. Scott Fouse, Stephen Cross, and Zachary J Lapin. Darpa's impact on artificial intelligence. *AI Magazine*, 41 (2): 3–8, 2020.
17. Frank Fratrik. How many operational design domains, objects, and events? In *SafeAI@ AAAI*, 2019.
18. Daniel Fremont, Xiangyu Yue, Tommaso Dreossi, Shromona Ghosh, Alberto L Sangiovanni-Vincentelli, and Sanjit A Seshia. Scenic: Language-based scene generation. *arXiv preprint arXiv:1809.09310*, 2018.
19. Patrice Godefroid, Adam Kiezun, and Michael Y Levin. Grammar-based whitebox fuzzing. In *Proceedings of the 29th ACM SIGPLAN Conference on Programming Language Design and Implementation*, pages 206–215, 2008.
20. David Gunning and David Aha. Darpa's explainable artificial intelligence (XAI) program. *AI Magazine*, 40 (2): 44–58, 2019.
21. Irina Higgins, Loic Matthey, Arka Pal, Christopher Burgess, Xavier Glorot, Matthew Botvinick, Shakir Mohamed, and Alexander Lerchner. beta-VAE: Learning basic visual concepts with a constrained variational framework. 2016.
22. Yordan Hristov, Alex Lascarides, and Subramanian Ramamoorthy. Interpretable latent spaces for learning from demonstration. In *Conference on Robot Learning*, pages 957–968. PMLR, 2018.
23. Yordan Hristov, Daniel Angelov, Michael Burke, Alex Lascarides, and Subramanian Ramamoorthy. Disentangled Relational Representations for Explaining and Learning from Demonstration. In *Conference on Robot Learning*, 2019a.
24. Yordan Hristov, Daniel Angelov, Michael Burke, Alex Lascarides, and Subramanian Ramamoorthy. Disentangled relational representations for explaining and learning from demonstration. In *Conference on Robot Learning (CoRL)*, 2019b.
25. Xiaowei Huang, Marta Kwiatkowska, Sen Wang, and Min Wu. Safety verification of deep neural networks. In *International conference on computer aided verification*, pages 3–29. Springer, 2017.
26. Auke Jan Ijspeert, Jun Nakanishi, Heiko Hoffmann, Peter Pastor, and Stefan Schaal. Dynamical Movement Primitives: Learning Attractor Models for Motor Behaviors. *Neural Computation*, 25 (2): 328–373, 2 2013.
27. Craig Innes and Subramanian Ramamoorthy. Elaborating on learned demonstrations with temporal logic specifications. 2020.
28. Craig Innes and Subramanian Ramamoorthy. ProbRobScene: A probabilistic specification language for 3D robotic manipulation environments. In *International Conference on Robotics and Automation (ICRA)*, 2021.
29. Juha Itkonen and Kristian Rautiainen. Exploratory testing: a multiple case study. In *2005 International Symposium on Empirical Software Engineering, 2005.*, pages 10–pp. IEEE, 2005.

30. Nidhi Kalra and Susan M Paddock. Driving to safety: How many miles of driving would it take to demonstrate autonomous vehicle reliability? *Transportation Research Part A: Policy and Practice*, 94: 182–193, 2016.
31. Diederik P Kingma and Max Welling. Auto-encoding variational Bayes. *stat*, 1050: 1, 2014.
32. John E. Laird, Kevin Gluck, John Anderson, Kenneth D. Forbus, Odest Chadwicke Jenkins, Christian Lebiere, Dario Salvucci, Matthias Scheutz, Andrea Thomaz, Greg Trafton, Robert E. Wray, Shiwali Mohan, and James R. Kirk. Interactive Task Learning. *IEEE Intelligent Systems*, 32 (4): 6–21, 2017. ISSN 15411672. https://doi.org/10.1109/MIS.2017.3121552.
33. Sarah Maslen and Jan Hayes. Preventing black swans: incident reporting systems as collective knowledge management. *Journal of Risk Research*, 19 (10): 1246–1260, 2016.
34. Rok Pahic, Andrej Gams, Ales Ude, and Jun Morimoto. Deep Encoder-Decoder Networks for Mapping Raw Images to Dynamic Movement Primitives. In *Proceedings - IEEE International Conference on Robotics and Automation*. IEEE, 2018.
35. Svetlin Penkov, Alejandro Bordallo, and Subramanian Ramamoorthy. Physical symbol grounding and instance learning through demonstration and eye tracking. In *Robotics and Automation, 2017 IEEE International Conference on*, Singapore, June 2017.
36. Charles Perrow. *Normal accidents: Living with high risk technologies-Updated edition*. Princeton University Press, 2011.
37. Affan Pervez, Yuecheng Mao, and Dongheui Lee. Learning deep movement primitives using convolutional neural networks. In *IEEE-RAS International Conference on Humanoid Robots*, pages 191–197. IEEE, 2017.
38. Amir Pnueli. The temporal logic of programs. In *Proceedings - Annual IEEE Symposium on Foundations of Computer Science, FOCS*, pages 46–57, 1977.
39. Diligent Robotics. Home Page. https://diligentrobots.com/, 2021.
40. Eric Rohmer, Surya PN Singh, and Marc Freese. Coppeliasim (formerly V-REP): a versatile and scalable robot simulation framework. In *Proc. of The International Conference on Intelligent Robots and Systems (IROS)*, 2013.
41. Johan M Sanne. Incident reporting or storytelling? Competing schemes in a safety-critical and hazardous work setting. *Safety Science*, 46 (8): 1205–1222, 2008.
42. Patrice Y Simard, Saleema Amershi, David M Chickering, Alicia Edelman Pelton, Soroush Ghorashi, Christopher Meek, Gonzalo Ramos, Jina Suh, Johan Verwey, Mo Wang, et al. Machine teaching: A new paradigm for building machine learning systems. *arXiv preprint arXiv:1707.06742*, 2017.
43. Starship. Home Page. https://www.starship.xyz/, 2021.
44. Waymo. Home Page. https://waymo.com/, 2021.
45. Jingyi Xu, Zilu Zhang, Tal Friedman, Yitao Liang, and Guy Van Den Broeck. A semantic loss function for deep learning with symbolic knowledge. In *35th International Conference on Machine Learning*, volume 12, pages 8752–8760, 2018.
46. Zipline. Home Page. https://flyzipline.com/, 2021.

Part III
Practical Considerations

Chapter 7
Verification Test Methods and Metrics

Jeremy A. Marvel, Craig Schlenoff, Jennifer Case, and Anthony Downs

Chapter 7 Editorial Note

This chapter focuses on the development of metrics and measures for evaluating autonomous systems. Although it does not identify a solution (and in fact, asserts that a solution that works across the enormous diversity of autonomous applications is unlikely to exist), it does provide a set of tools and approaches from which metrics may be developed and test approaches may be selected. It specifically addresses issues related to the stakeholder questions of assessing both the usefulness and the trustworthiness of a given system. Technical challenge 16 (consistently and accurately describing the autonomous capability embodied by a robot) is still very much an unsolved problem, but this chapter provides a starting point for thinking about it. At the very least, it addresses challenge 21 (identifying needed tools and techniques) by offering a set of tools to enable operators to quantitatively evaluate the tasks they want the robot to do. Stakeholder questions are presented in the Chap. 1, Introduction. Technical challenges 16 and 21 are addressed in more detail in Chap. 2, Verification Challenges.

Abstract In this chapter, we discuss the tools and mechanisms leveraged to verify and validate robot autonomy. From the identification of stakeholder preference and key performance indicators to the design of experiments and evaluation of sampling costs, we highlight the considerations and common practices of the testing methodology. To complement these discussions, we also present metrology use cases in which the performance of robotic systems with varying degrees of autonomy is evaluated.

J. A. Marvel (✉) · C. Schlenoff · J. Case · A. Downs
NSWC PCD, Panama City, FL, USA
e-mail: jeremy.marvel@nist.gov; craig.schlenoff@nist.gov; jennifer.case@nist.gov; anthony.downs@nist.gov

S. Redfield et al. (eds.), *Verification of Autonomous Systems*,
https://doi.org/10.1007/978-3-031-88546-4_7

7.1 Introduction

As has been discussed earlier, there are a myriad of challenges faced throughout the research and development cycle of autonomous systems. From the initial efforts to identify application domains and key stakeholders, all the way through deployment in the field, these challenges must all be met with some means of ensuring the goals and performance requirements are met. Whether it is for the purpose of benchmarking system performance against known archetypes or assessing technology readiness levels (TRLs) for system development, consistent test methods and metrics are essential for testing and evaluation. However, the selection of measurement science[1] (i.e., "metrology") tools for this purpose is not always intuitive, as it is critical that they both accurately capture the performance of the system and actually test what they purport to test. In other words, both the systems under test and the test methods themselves must be verified and validated.

The effort to verify and validate robot autonomy is complicated by the inconsistent definitions of autonomy that exist in the literature. For example, some researchers define autonomy as the robotic system's ability to achieve a goal assigned to it with the level of autonomy depending upon the complexity of the task [1–3], while others have defined autonomy by robot capabilities that enable the previous definition of autonomy or used autonomy as a synonym for an automatic process. Still, others may present autonomy in terms of independence from a rigidly defined infrastructure for electrical power, processing, or communications. With so many definitions and approaches for characterizing autonomy, it is nearly impossible to define test methods and metrics for the verification and validation of autonomous robotic systems.

As such, in this chapter, we will use the definition of autonomy provided by the Autonomy Levels for Unmanned Systems (ALFUS) group. They defined autonomy as "an unmanned system's (UMS) own ability of integrated sensing, perceiving, analyzing, communicating, planning, decision-making, and acting/executing to achieve its goals as assigned by its human operator(s) through designed human-robot interaction or by another system that the UMS communicates with." Autonomy is then broken down further into root autonomous capabilities (i.e., the "sensing, perceiving, analyzing, communicating, planning, decision-making, and acting/executing capabilities") and contextual autonomous capabilities, which are defined by the assigned goal into three different aspects of mission complexity, environmental complexity, and human independence [1]. The ALFUS group proposed a guideline framework of measuring the level of autonomy using these three aspects [4]. It should be noted that many of the metrics proposed to test for "autonomy," which will be discussed later, are task-specific. Thus, it is more accurate as a measure of how well a robot can perform a specified task without human intervention rather than a measure of true autonomy.

---

[1] In the context of this chapter, we define measurement science as the collection of test methods, metrics, and measures used to evaluate and characterize the properties of objects or events.

Given this broad categorization of autonomy, where is the robot's autonomy injected into the robot, capabilities, or task? For example, consider the task of fetching and administering medicine for in-house patient care. For this example, the following information is necessary to complete the task:

- What medicine is required?
- When is the medicine required?
- Where is the patient located?
- Where is the medicine located?
- How does the robot get to the medicine's location?
- How does the robot fetch the medicine?
- How does the robot get to the patient?
- How does the robot give the patient the medicine?
- Does the patient know what to do with the medicine?
- Does the patient need assistance administering the medicine?
- Does the robot know whether the medicine has been administered correctly?

Autonomy can be injected into any or all of these sub-elements of the medicine administration task. Which elements, if any, are actually expected to be autonomous? Moreover, what if there is uncertainty in the process? What happens if the medicine is not where it is supposed to be? What happens if the medicine is empty? What if the patient cannot be found, is unwilling to take the medicine, or is unable to take it? Is the robot expected to have autonomy under such circumstances, too? Evaluating the correctness and appropriateness of the robots' actions and behaviors is, ultimately, part and partial to the verification and validation process. These evaluations are influenced by the task, the people involved, the expected outcomes, and the expectation of uncertainty in the system.

It is also important to take a moment to better explain the concepts of verification and validation. Though the general idea is understood, the processes and goals of each can be somewhat muddled. In some cases, verification and validation are confused with one another or even considered synonymous or interchangeable. Part of this stems from different domains using the terms in contradictory contexts. What is considered "verification" in industry may be termed "validation" in metrology and vice versa.

For the purpose of this chapter, we define verification and validation as:

- **Verification**—the means by which it is determined whether a system, process, or framework meets the needs or requirements as specified.
- **Validation** the means by which it is determined whether the system, process, or framework is suitable for the intended goals outlined by stakeholders .

From a metrology perspective, these concepts can be summarized as follows: Verification asks whether the measurements were done correctly, while validation asks whether the correct factors are being measured. These definitions align with the intended use as outlined by the Project Management Body of Knowledge[2] [5] but are opposite to the meanings used by the International Organization of Standardization (ISO, e.g., as used in ISO 9001:2015[3] [6]). Regardless, the processes of verification and validation are independent of each other.

For the purpose of discussion, we will present the concepts of verification and validation from two separate but related contexts: a systems engineering approach and a measurement science approach. The former is intended to evaluate the robotic system as a whole and measure aspects with regard to the robot's expected performance. The latter investigates the measurement methodology, itself, and is focused on the means by which information is gathered and assessed. In other words, the systems approach looks at the measurements, while the measurement science looks at the means by which measurements are taken.

7.2 What Is Verification?

Verification seeks to assess whether the specified requirements were met. From a robot systems perspective, this means that the robot(s) being evaluated (i.e., "systems under test") performs within some range of specifications. Robot specifications may be given in terms of positioning repeatability and accuracy, functional performance, strength, remaining useful life, return on investment (ROI) , or literally any number of other metrics. In contrast, from a measurement science standpoint, verification would fall into categories such as measurement consistency, statistical distributions, and so on.

In the end, however, the goal is ultimately the same: Does the system under test perform as we expect it to? There are, of course, a number of ways "expected performance" can be interpreted. Such values can be assessed along a continuum of values, a binary pass/fail, or even an expression of distributions. Any number of factors can be measured either independently or aggregated to form a singular score metric.

For example, let us assume the metrics we want to verify are focused on task completion time, t, and task quality, q, for an exploration operation. Given a nominal map for exploration (Fig. 7.1), we want to identify any possible changes of features (blue boxes). Here, we define the task quality as a percentage of map exploration, $q \in [0, 1]$, where 0 is 0% of the map explored, and 1 is 100% of the map explored. Exploration of the map is visually indicated by the map background becoming progressively more white, and the objects being remapped as red blocks. In this example, there is a clear discrepancy in the map that will be corrected as the robot explores the area.

[2] The Project Management Body of Knowledge is a collection of standards, terms, and guidelines used throughout the project management process.

[3] ISO 9001:2015 defines the requirements for quality management systems.

Fig. 7.1 Example map for a robot exploration task. The robot (circle in the lower left) is responsible for updating its internal map. The objects in the existing map (blue features) are being updated as the robot explores the gray regions (turning them white to indicate they have been explored). Object features are updated on the new map as red objects

To evaluate t, the time elapsed from the beginning of the trial is recorded in the range of $t \in [0, \infty]$. Obviously, one cannot wait indefinitely for the task to be completed, so a timeout, $t_{timeout}$, is selected to determine whether the completion time is reasonable for the application. This timeout may be on the order of seconds (e.g., for an assembly) or even months or years.

$$t = \begin{cases} t & \text{if } t < t_{timeout}, \\ t_{timeout} & \text{otherwise.} \end{cases} \tag{7.1}$$

For n number of trials, the average time performance is calculated simply as

$$t_{avg} = \frac{1}{n} \sum_{i=1}^{n} t_i. \tag{7.2}$$

From this, there are several different ways that we could verify the performance of t. For example, we could evaluate relative to a benchmark, $t_{benchmark}$, where the performance gain, $t_{gain,avg}$, is simply the difference between a target value and the average measured time:

$$t_{gain,avg} = t_{benchmark} - t_{avg}. \tag{7.3}$$

If t_{gain} is positive, the average performance is better than the benchmark. Otherwise there is no improvement as compared with the benchmark.

Of course, this only looks at the average case. For some verification trials, one may only want to look at the worst-case scenario. In which case, you simply replace t_{avg} with t_{worst}

$$t_{gain,worst} = t_{benchmark} - t_{worst}, \tag{7.4}$$

where

$$t_{worst} = \max(t_i), \forall t_i, 1 \leq i \leq n. \tag{7.5}$$

So the question remains: how do we interpret these numbers? From a statistical testing perspective, there are a variety of ways of looking at these numbers, each informing us different aspects of the trial data. For example, looking only at t_{avg} gives us insight into the *average performance* for t. For a sufficiently large enough n trials, this informs us about how we should expect the robot to perform *on average*. In general, this is actually quite useful, but it does not provide much insight into the *range* of values we should expect. For the average (mean) performance range, we recommend $\left(t_{avg} - kt_\sigma, t_{avg} + kt_\sigma\right)$, where k is some constant (usually 1 or 2), and

$$t_\sigma = \frac{1}{n} \sum_{i=1}^{n} \left(t_i - t_{avg}\right). \tag{7.6}$$

For example, in ISO 9283 [7], the accuracy (mean) and repeatability (standard deviation) of the robot's tool center point at an end point and along a path are leveraged as the de facto performance metrics, given as reportable specifications for industrial robots. In practice, however, typically only end-point repeatability is reported.

Alternatively, we could look at a *quantile performance*, where we investigate the quantity, $q_{benchmark}$, of performances that meet or exceed our expectations:

$$q_{benchmark} = \frac{1}{n} \sum_{i=1}^{n} q_i, \tag{7.7}$$

where

$$q_i = \begin{cases} 1 \text{ if } t \leq t_{benchmark}, \\ 0 \text{ otherwise.} \end{cases} \tag{7.8}$$

The *mode*, for example, is another representation of the quantile average in which the reported value is that which is most likely result based on the frequency of occurrence.

Of course, all of these tests evaluate performance along a scale. For simplicity, we could also report our verification testing as a binary, where we return *true* if the value we achieve meets or exceeds our expectations (e.g., if $t_{avg} < t_{benchmark}$ or if $q_{benchmark} > 0.5$). Otherwise we return *false*. Neither scalar nor binary values are necessarily better than the other, as they are both useful in their own right. Binary reports, for example, serve as a quick litmus test for expected performance. In contrast, scalar results can be leveraged to guide systematic improvements and identify biases in the robot's performance.

There are, of course, other reports that can be leveraged to gain useful insights into performance. For example, given a sufficiently large sample size, averages other than the mean (see Eq. 7.2, which is a weighted, distributed range of recorded values around a central point), such as the median (i.e., the unweighted center of an ordered set of values) and mode (i.e., the most common occurrence of the values measured), provide useful information based on the entire collection of the data rather than a conglomeration of measurements rolled into a single value.

7.3 What Is Validation?

While verification seeks to ensure that the robotic systems are performing within their specifications, validation is targeted at ensuring that the specifications, themselves, are correct. For example, if the specifications require that the operating temperatures of the robot never exceed 25 °C, ensuring that this metric is achieved across the entire line of robots is ineffective if the measurement is intended to assess robot position repeatability. More specifically, measuring the robot's temperature is an *invalid* metric for pose repeatability. This is not to say that operating temperature does not influence pose repeatability (it can!), but only that temperature itself is insufficient to meet the needs of the end user for assuring task performance.

Similarly, from a metrology perspective, validation is focused on determining whether a given methodology captures what is intended to be measured. Do the measurements make sense from a measurement science perspective? Does changing the parameters result in meaningful performance variations for the factors that are important to us? Or, more fundamentally, does the test methodology give us any *new* insight into the robot's performance?

As an example, let us say that the robot's pose repeatability in a given quadrant of the testing environment is particularly poor. We observe that the quadrant is close to where one student's desk is located and decide the robot just does not like that particular person. To test this theory, we devise a series of tests in which the student moves to different parts of the laboratory and remeasure the robot's repeatability for each new location for the student. The results, themselves, are independently verifiable, but one could argue the test methodology is not valid since it is based on a flawed assumption: that the robot does not like or is even aware of a particular person.

The validation process is necessarily different from verification, as its purpose is to ensure the quality and appropriateness of the testing methodology. The types of questions the validation is targeted at answering are:

- Do these measurements make sense, or have we provided sufficient justification for measuring specific factors?
- Will our stakeholders find this information useful?
- Is there another way someone could interpret these results, or are we misinterpreting the results? Is it possible there are other explanations for the performance we are seeing that is buried in the data?
- Do they provide useful information about the system under test?
- Are we measuring the correct factors/variables?
- Are we measuring everything we should be? Are we measuring things we should not?
- Is this a reasonable test to expect others to perform?
- Are the results repeatable (i.e., will we get similar results if we were to test the same factors all over again)?
- Are the tests reproducible (i.e., if someone were to try repeating our tests under identical conditions, would they get the same results)?
- Have we accounted for all sources of uncertainty, or will this methodology expose the sources of error?

Throughout the remainder of this chapter, we will focus on identifying the key performance indicators for robotic applications and outline the test methods and metrics necessary for evaluating robot performance and assessing validity of the test methodology and measurement results.

7.4 Identification of Performance Metrics

Arguably, the single most important step in the verification and validation processes is the selection of the metrics by which performance will be assessed. These metrics specify the elements to be measured, how they will be measured, and how they will be assessed and evaluated. Whether it is the robot's speed, its efficiency, the ROI, or even how people respond to the robot, the evaluation of the robotic system is dependent on the measurements gathered before, during, and after the robot's design, integration, and use. The selection of these metrics is critical for the purpose of ensuring progress of technology evolution (e.g., for assigning TRLs , [8]), establishing baseline performances for improvement, and determining whether certain changes push the system toward performance goals.

Unfortunately, there is no single "correct" way of performing verification and validation, given that the metrics for assessing robots are as vast and varied as the robots' applications. A given set of measures simply cannot be all things to all people and all applications, and any attempt to be exhaustive would be impossible to enact. There are some efforts to provide test methods and metrics for setting baseline specifications (e.g., ISO 9283 [7], which outlines specifications for measuring industrial robot accuracy and repeatability), yet these reflect only the bare minimum for measuring specific factors that may not always be applicable to every task or robotic system. Moreover, there are no industry-specified standard test methods or metrics that touch on autonomy, though it is expected regulatory bodies will develop guidelines for autonomous vehicles driving on public roadways.

As such, the metrics selected for analysis are different based on the robot's application, the equipment or sensors being used by the robot, and so on. Simultaneously, the selected metrics are expected to capture the factors that are the most important, often while being kept to as minimal as possible given that data collection may be expensive. At the heart of the identification and selection of metrics are the preferences of the stakeholders who have a vested interest in the performance of the robot.

In this section, we will discuss factors and considerations for the selection of metrics and the impacts these considerations have on the verification and validation processes from the vantage point of stakeholder preferences . Among these, we will discuss the key performance indicators, which are those elements that are critical to the performance of the system or process, as defined by the stakeholders.

7.4.1 Who Are the Stakeholders?

Every product, service, market, society, and population has a set of stakeholders who have a vested interest in how things operate, the tasks they perform, the needs and wants they fulfill, and the impacts they have on the world. For autonomous systems, these stakeholders fall into five broad categories: mechanic/developer, supervisor, operator, teammate, and bystander [1].

Here, *mechanics/developers* are the people and companies who are designing, programming, and producing the robots. These can include companies, universities, and researchers who produce robots from scratch or modify existing systems to be introduced to the market as new products. For autonomous systems, the mechanics/developers are active participants in the verification and validation processes since the code running the autonomous behavior will likely need to undergo many iterations before the system is performing satisfactorily. *Supervisors* oversee the autonomous systems performance with respect to accomplishing mission goals. The supervisor can assign mission-level goals to the systems or assign an operator to work with one of the systems. *Operators* work with autonomous systems on a closer level. This may mean controlling the system through teleoperation in case of an error or providing additional information to the system to enable it to run semiautonomously. If the error can be resolved, control of the robot is returned to the supervisor; otherwise the mechanic/developer may be required to step in and address the issue. *Teammates* are the people working alongside autonomous systems to complete a mission. The teammate does not necessarily have to be an active participant but is impacted directly by the robots' actions. For example, for healthcare robots, the patient is the teammate of the robot. And *bystanders* are people who exist in the environment that the autonomous system is operating in that may be directly or indirectly affected by the system. The bystander is distinguished from the teammate in that the bystander is not necessarily expected to interact with the robot. For example, a robot giving directions and information to people in a museum may be surrounded by bystanders who are largely moving around the environment, not interacting with the robot. Here, the bystanders could potentially become teammates if they interact with the robot to gain information, but such an interaction is initiated by the human and not the robot.

It is worth taking a moment to briefly discuss the *regulating* stakeholder . This was touched upon briefly at the beginning of this section. Regulatory bodies and government agencies (local, regional, national, and international) focused on establishing and enforcing requirements have an inherently vested interest in the specific performance characteristics of the goods and services they regulate. Many such regulations are focused on product and worker safety, while others may target specific characteristics such as material composition, sourcing of components, or signage. In terms of autonomy, the regulators would be most invested in the perspectives of liability. Specifically, if the autonomy of the robotic system being used results in injury or property damage, who will be held liable? Is it the manufacturer who designed and built the robot? Is it the integrator who decided to use the robot in a particular environment or application? Is it the government, itself, for allowing such autonomous systems to be used in public? Or is it the end user who purchased the system and actually owns and operates the robot? In many cases, it is all of the above, but to different degrees. Thankfully, these are the questions and concerns that fall outside of the scope of this book. The regulators responsible for addressing liability and culpability are also responsible for specifying the requirements that must be met, the means by which they will assess compliance, and the consequences for failure to meet conformity.

There are three driving factors that one must consider to understand the key factors relevant to each stakeholder :

1. Who are the stakeholders (i.e., the people or groups within each broad category that need to evaluate the robots)?
2. What are the responsibilities of each stakeholder?
3. What do the stakeholders need to know to be able to do their jobs?

Consider, for example, a line manager in a facility that uses robots to produce widgets, which are then sold to other manufacturers for use in their products. The line manager represents one stakeholder in the broad category of "end-users." The line manager's responsibilities include ensuring that the robots are available when they are needed, that the robots are performing their tasks at the rate that is needed to ensure production throughput, that whatever autonomy is employed does not introduce hazards to personnel or other equipment, and that downtime on the robot line is minimized. To accomplish these, the line manager must know the robot's task throughput, production error rates, robot availability and efficiency, and health monitoring data to track the expected remaining useful life of the robot and its tooling. In contrast, the company's president, who is responsible for ensuring profitability of the company's production, may be more concerned about ROI of the purchase of the robot workcells and the impacts that using robots will have on production costs, employee health, and product quality. And a line worker, who is responsible for day-to-day operation of the robots, is likely to be more concerned about the robot's task performance and operational safety. Within a single company, there may be any number of stakeholders whose preferences may be considered. Similarly, their perspectives on autonomy may be more or less pronounced than others.

Or, as another example, consider the director of an elderly care facility that is looking to use robots to assist with the daily distribution of medicines for the residents. The director's responsibilities are less hands-on than the nursing staff's in terms of interactions with the residents. However, their job is to ensure the facility is solvent and operating effectively and efficiently. So when considering the director as a stakeholder in the procurement and utility of robots, what are their responsibilities and what do they need to know when considering a robot purchase? From the vantage of ensuring solvency, factors such as purchase cost, maintenance and support fees, equipment reliability, and value depreciation may be common concerns. A robot that is relatively expensive to purchase and maintain may be seen as a source of diminishing returns and may be assessed differently from alternatives that are cheaper or require less maintenance.

Of course, it is not necessary to capture all of these stakeholders' preferences, and different individuals within a single stakeholder category may weight different factors more or less than others. Deciding which stakeholders to consider and how much to factor their preferences into the verification and validation methodology is, ultimately, the decision of the organization in question.

To bring this topic back to the focus of this book, we can consider the question of verification and validation for autonomy as a filter for identifying those stakeholders who are most invested in the autonomy performance of robotic systems. Who would be most concerned about the autonomy of a robotic system? Obviously, those stakeholders who are most directly impacted would be the people who move and work with or around the robots. These are the pedestrians, patients, nurses, doctors, workers, astronauts, families, students, teachers, chefs, retailers, and so on who are intended to benefit from (or potentially be injured by) autonomous robots.

7.4.2 What Are the Key Performance Indicators?

As name implies, key performance indicators (KPIs) are those factors that uniquely capture how the system as a whole is faring as it performs its task. These KPIs will vary from stakeholder to stakeholder , and some may be more important than others to each individual or groups. Moreover, some of these factors may be independent of one other another, while others are largely meaningless when taken alone and out of context. As such, the first step in identifying the KPIs is to define the context in which they will be assessed. Luckily, this is typically related direction to the identification of the stakeholders' roles and preferences determined in the previous step. Moreover, many of these indicators may be leveraged to reveal specific aspects pertinent to different stakeholders . For example, both maintenance personnel and quality assurance testers are concerned about the robot's mean time to failure , as the nature and rate of failures impacts both the robot's remaining useful life and the quality of the products being made by the robots.

Frequently, many of these KPIs will be task-specific and not universally applicable. For example, production throughput has little meaning to a laparoscopic surgeon using a robot to mend an ailing person, nor does the robot's ease of programming impact the forklift driver responsible for taking the pallets packed by the robot to the loading dock. Moreover, it is important to note that some applications may have different KPIs depending on where in the development lifecycle the robots actually are. For example, in the development of technologies for space exploration, government agencies may leverage different KPIs depending on the TRL [9], which is used to capture the maturity of the robotic technologies from initial conception to final deployment.

It is important to note that not all KPIs directly capture the autonomy of robotic systems but can be influenced by a system's autonomy. For example, fully autonomous systems may require less direct assistance from human operators, thus impacting labor costs and injury rates. Similarly, robots capable of autonomous navigation and planning require less front-loaded programming or testing, reducing the time necessary to commission a system.

7.4.2.1 Examples of KPIs Indirectly Impacted by Autonomy

For the purpose of illustration, let us discuss a number of common metrics that are often leveraged as KPIs and may be indirectly indicative of autonomy performance. In many cases, these KPIs are focused on capturing performance related to profit, production quality, and maintenance. These metrics include *return on investment*, *overall equipment effectiveness*, *failure rate*, and *efficacy*.

Return on Investment From a business perspective, one of the principal metrics for evaluating the effectiveness of any robotic system is the ROI . Frequently, ROI is used for evaluating material purchasing decisions as an assessment of the timeline to recuperate investment costs for purchase and integration (i.e., cost) as a function of expected financial gains (gain) over a given period of time (e.g., 1 year). ROI is calculated as

$$ROI = \frac{gain_{term} - cost}{cost}, \tag{7.9}$$

where $gain_{term}$ is the gross profit, and $cost$ is the combination of equipment cost, integration fees, and expected expenses (e.g., maintenance and training). An alternative perspective of ROI calculates profit margins over the foreseeable life of the equipment, $gain_{total}$:

$$ROI = gain_{total} - cost. \tag{7.10}$$

Neither metric is inherently superior to the other, and the selection of one or the other is ultimately at the discretion of the end user. It is also worth mentioning that certain costs may be assessed differently depending on the institution. For example, some companies may only consider charges associated with actively producing goods in the computation of ROI , whereas support costs (e.g., programming or maintenance) may be accrued as overhead.

How could ROI be impacted by robot autonomy? Largely, the factor of autonomy that impacts ROI most directly is its cost. Fully autonomous systems tend to be more expensive than more traditional, automated solutions. Much of this comes from the costs of development, but factors such as manufacturer support, integration costs (including those for required, ancillary equipment that supports the robots' autonomy), and product liability. Integrators generally cannot support autonomous functions of robots built into the system, which means the manufacturer must maintain staff, equipment, and facilities for product and service support. In some cases, end users may expect fully autonomous solutions to have lower costs long term, but only so long as the tasks and operations to which they are applied are supported by the robots' software.

Overall Equipment Effectiveness When making adjustments to a given robotic operation or process, it is frequently necessary to capture the impacts such changes have on the efficiency and effectiveness of the equipment in general. One such means of capturing this is to evaluate the compounded overall equipment effectiveness (OEE, [10]):

$$OEE = availability \times performance \times quality. \tag{7.11}$$

Here, $availability$ is the percentage of scheduled time the equipment is available to operate

$$availability = \frac{scheduled\ time\ -\ lost\ time}{scheduled\ time}, \tag{7.12}$$

and *performance* is the percentage of operational speed of the equipment

$$performance = \frac{actual\ machine\ speed}{design\ machine\ speed}. \tag{7.13}$$

Of course, other measures for evaluating performance may be substituted and leveraged in lieu of task speed, such as task accuracy (e.g., diagnosing symptoms in patients or finding and tracking targets of interest), accrued force, motion precision, and so on. Whatever metric is selected, the value is expected to be in the range $performance \in [0, 1]$.

Unlike availability and performance, *quality* is generally considered more task-specific. In manufacturing, $quality$ can be measured as the ratio of good to bad parts produced

$$quality = \frac{number\ of\ good\ parts}{total\ parts\ produced}. \tag{7.14}$$

In medicine, it could be measured as a normalized prognosis as a result of autonomy-applied patient care. In tactical situations it could be the effective provision of ground support or efficiency in protecting targets of interest. And in self-driving vehicles it could be comfort of the ride (e.g., as a function of acceleration profiles or on road roughness), routing efficiency, fuel efficiency, etc.

In total, the combined OEE is not only an effective tool for assessing the performance of the robot as a whole but also a derivative of other useful factors that may, themselves, be leveraged as KPIs. Virtually all elements may be impacted by levels of autonomy, particularly with regard to the robots' ability to compensate for uncertainty and maintain high levels of performance.

Failure Rate In terms of predicting when maintenance will be required or non-repairable assets replaced, it is necessary to calculate some form of failure metric. There are many such metrics to choose from, but we will highlight three here. The first, the *mean time to failure* (MTTF), is one of the more common metrics to use when exact failure rate information is not available. The MTTF is calculated as

$$MTTF = \frac{total\ operational\ time}{number\ of\ assets\ in\ use}. \tag{7.15}$$

The MTTF is leveraged to provide a rough estimate as to how often *something* in a robotic workcell will require maintenance or replacement. In contrast, if failure rate data for a given robot is available, one may instead provide a more accurate estimate of that robot's failure rate by calculating the *mean time between failures* (MTBFs):

$$MTBF = \frac{total\ operational\ time}{total\ number\ of\ failures}. \tag{7.16}$$

If, instead, you would like to know the *impacts* of failures on equipment availability, you can alternatively calculate the *mean time to repair* (MTTR), which can be leveraged as an estimate of equipment downtime:

$$MTTR = \frac{total\ maintenance\ time}{total\ number\ of\ repairs}. \tag{7.17}$$

In many ways, failure analysis can be viewed as a balance of the tenuous relationship between hardware and software. On the one hand, the longevity of the system may be determined by hardware degradation, which cannot be controlled by software. On the other hand, one could argue that self-monitoring, system diagnostics, and performance prognostics are reasonable functions of autonomous systems. Ultimately, the question is one of balancing the *task autonomy* with the *self-awareness* to provide an optimal level of performance while minimizing human intervention.

System Efficacy And, finally, a useful and meaningful metric for almost any robot application is to measure the robot's *efficacy* as it pertains to performing its task. Once again, this can be interpreted in a number of different ways. For example, *effectiveness*, or the quantification of how well the task is completed, may measure any number of different manufacturing factors but ultimately reduces down to two basic elements: (1) the number of completed operations (assembly, packaging, inspection, etc.), o_t, and (2) the amount of waste (e.g., the number of damaged or scrapped parts, the quantity of wasted food or ingredients, or the number of failed missions), o_w:

$$effectiveness = \frac{a\,(o_t - o_w) - (1 - a)\,o_w}{o_t}, \tag{7.18}$$

where $0 \leq a \leq 1$ is a user-defined weight specifying the company's preferences in balancing quantity versus quality.

Another mechanism for measuring efficacy is based on characterizing the robot's timing efficacy. Specifically, useful metrics include assessing the *average time to complete a task*, t_{avg} or its inverse, the task *throughput*, over a given period of time, t_{comp}:

$$t_{avg} = \frac{t_{comp}}{o_t}, \tag{7.19}$$

or its inverse, *throughput*:

$$throughput = \frac{o_t}{t_{comp}}. \tag{7.20}$$

Either of these timing measures may be sufficient for capturing efficiency in many application domains (e.g., response robotics [11]). However, in the case of manufacturing, many companies also take into account the minimization of waste as a factor for *efficiency*:

$$efficiency = \left(1 - \frac{o_w}{o_t}\right)\left(\frac{o_t}{t_{comp}}\right) = \frac{o_t - o_w}{t_{comp}}. \tag{7.21}$$

To this end, the measure of task efficacy is the closest means of capturing the impact of robot autonomy without actually explicitly measuring robot autonomy. Ultimately, the purpose of an autonomous robot system is to perform a task. And that robot's effectiveness at performing that task compared with an automated solution is the greatest testament to the effectiveness of applied autonomy.

7.4.2.2 Autonomy KPIs

Now that we know how autonomy can indirectly impact different metrics of robot system performance, how do we go about verifying and validating autonomy directly? This is significantly more challenging, as the interpretation and manifestation of autonomy takes many different forms. Looking at the corpus of works regarding robot autonomy, most approaches to verifying autonomy are indelibly tied to the task or application of the robotic system [3]. In other cases, autonomy is inferred from the independence from some otherwise limiting factor (e.g., the inclusion of onboard power or processing [12]).

So how is autonomy measured? There are, of course, a number of methods for approaching autonomy metrics. For example, Clough presented a series of characteristics of unmanned vehicle autonomy measured on a spectrum from 0 (full tele-operation) to 10 (human-like autonomy) [3]. These metrics include "perception/situation awareness" (the capacity to recognize and infer the intent of others operating in the same environment as the robot), "analysis/coordination" (being able to set and receive team-based goals and coordinate with others as necessary), "decision making" (independently making plans and executing strategies), and "capacity" (the ability to perform the job without guidance or assistance individually or as part of a group). Such subjectively qualitative assessments must be made by expert stakeholders and are based on observable actions and behaviors from which different capabilities are inferred.

For a more objectively quantitative metric, an alternative means of measuring autonomy evaluates communications and notifications from the robot and observing task interruptions from humans or other robots to correct for errors. From these, the combination of communications, specifically detailing service requests and the number of times humans, must intervene to correct for robot errors (also known as "utility of mixed initiatives" [11]) and is used to determine how dependent the robotic system is on external help.

The ALFUS group's [1]) framework proposed metrics of autonomy based upon three aspects: mission complexity, environmental complexity, and human intervention. The metrics for mission complexity are level of task decomposition, type of tasks, complexity of tasks, decision space structure, agent perspective, interface/data sharing perspective, and situational awareness. The metrics for environmental complexity are given as search space, variability of the environment, and objects observability. The metrics for human intervention are given as robotic initiation, ratio of human intervention time to mission time, ratio of human planning time to mission time, level of human interaction, and level of human workload. The metrics are defined and have brief descriptions of how to measure them in [4] and are not repeated here. Antsaklis proposed using measures of the set of goals and uncertainties to define autonomy [2].

For each set of metrics, the metrics are dependent upon the performance of the system for a specified task. In the short term, it is likely that developers of autonomous systems will need to declare the performance of their system against various tasks required/expected by end users. For example, self-driving vehicles may need to specify that they can drive in sunny weather but are unable to operate autonomously in heavy rain or snow. If there are certain tasks that a system is unable to handle, it is important that the system recognizes that it is unable to safely operate, there are procedures in place to hand control back to a human operator, and there are alternative procedures should a human operator not resume control when requested by the system. An additional concern with some of the metrics for autonomy is their subjective nature. There is currently no regulating body that defines or advises how tasks should be defined or tested.

7.4.3 Making Sense of Measurements

One approach to defining performance metrics is to consider the context and scope of each level within a robotic system. This methodology has been applied to many domains, including assembly [13, 14], safety [15, 16], driving [17], and urban search and rescue [18, 19]. The performance requirements for the overall system can be derived based on an understanding of the application-level domain and the context within which the robot(s) must operate and, in other words, using an application analysis to guide the metrology.

The process begins with an enumeration of the major tasks the system will be performing as defined by the stakeholders. For example, maintenance personnel may focus on health monitoring and failure analysis, while day-to-day operators are more interested in applied task performance and reliability. From this, we then derive the "skills" that are required for the stakeholders to perform these tasks and define, to the extent possible, the external variables (i.e., the elements or factors that can be adjusted or tuned) that affect performance. These variables may be indicative of the overall application performance of the robot or more indicative of lower level aspects from which higher level performance is derived.

From these variables, we then identify the means by which measurements will be taken. This includes identifying system reports, external sensors, and the means of sampling data. Some applications may necessitate continuous monitoring (i.e., an in-line process, where measurements are being taken throughout the trial) or batch sampling (where representative samples are taken to speak for all time-steps or products). These measurements can be assessed as binaries (usually as a basis of being compared to a threshold), binned according to ranges, or evaluated along a spectrum. Similarly, values may be evaluated singularly or as part of a collective where the relationships between measurements are considered metrics in and of themselves.

Similarly, as was seen previously, measures can be either objective (i.e., independent of personal perspective or interpretation, such as measuring time) or subjective (i.e., where assessments can vary depending on who is performing the test) in nature. These measures may also be quantitative (i.e., assigning numerical values) or qualitative (where information is derived from relationships or personal

Table 7.1 Examples of both objective and subjective metrics that are either quantitative or qualitative

	Objective	Subjective
Quantitative	The robot's task time is 15 s	The robot is 7 out of 10
Qualitative	Robot A is on top of robot B	This robot smells funny

assessments). Examples of qualitative versus quantitative and objective versus subjective measures are given in Table 7.1.

7.5 Design of Experiments

The verification of robot performance is wholly dependent on the ability to capture pertinent information about the robot that gives insights into how well it can execute its task. The means by which this information is gathered constitutes the data collection methodology and is determined by the *design of experiments* .

Simply put, the purpose of the design of experiments is to identify:

- The factors that are most important to capturing the performance of the robot (i.e., the metrics)
- The impacts each factor has on the system performance
- The means by which the metrics will be measured (i.e., the test methodology)
- The interactions that the various factors have when determining the performance of the robot system

The design of experiments achieves these goals by capturing performance data in a meaningful way. More importantly, the design of experiments organizes and plans the verification and validation processes of the robot to ensure an optimal methodology. The benefits of a formalized design of experiments include ensuring testing consistency and repeatability, providing traceability (i.e., identifying who designed the design of experiments, who approved it, who performed it, etc.) , and enabling testing and sampling efficiency and effectiveness.

7.5.1 Significance of Results

As mentioned briefly in Sect. 7.1, the design of experiments may have any number of intended purposes. It could be intended to report absolute performance (as measured by a reference system), establish relative performance against a benchmark, identify sources of error, validate the selection of KPIs , or even identify those variables that have the most impact on the robot's performance. This, therefore, begs the question: how do we know if the results are significant?

At the heart of a design of experiments is the assumption the results are, in fact, significant. To verify this, the test methodology must necessarily include a statistical analysis of the data such that *null hypothesis* (i.e., the assumption that there is no significant difference in the results and that any differences observed are simply due to sampling error or a flaw in the original experimental design) is false. The most common means of assessing this is to evaluate whether the variance in the data is meaningful. If all that is important is the average performance, the Student's t-test [20] is among the most common mechanisms for testing for variance in the statistician's toolbox. If, however, you wish to assess the variation over populations of data points, an analysis of variance (ANOVA, [21]) is a more appropriate tool.

The computation of ANOVA is fairly straightforward:

1. Calculate the mean value for each element for each group (i.e., the *group mean*).
2. Calculate the *total mean* or the mean value of all elements in the population.
3. Find the *within-group variation*, or the total deviation of each element from its respective group mean (i.e., calculate the standard deviation within each group).
4. Find the *between-group variation* or the standard deviation from each group mean from the total mean.
5. Compute the F statistic: $\frac{between\text{-}groupvariation}{within\text{-}groupvariation}$.

This F statistic is then used in conjunction with the P value to determine if the variance is a result of randomness in the system (i.e., the null hypothesis cannot be rejected) or is actually indicative of statistical significance. The P value is the probability that the results achieved are not random. For example, a P value of 0.95 means there is a 95% chance that the data is not the result of noise. The F statistic is an "in-the-weeds" view of the data, while the P value is a high-level perspective. The P value is given in the ANOVA table (often given relative to α, where $\alpha = 1 - P$), and the F statistic is compared with an *F critical* number. If the P value is greater than the target *alpha* (i.e., the probability of resulting from random chance is too great) or the F statistic is greater than the F critical value (i.e., the variance is too large), then the null hypothesis cannot be rejected.

7.5.2 Accommodating Noise in the System

The real world is nondeterministic, and there is always variation in how robots perform. There is noise in the environment, noise in the system, and noise in the sensors. There are several means by which these sources of noise can be accommodated. In a laboratory environment, for example, researchers may attempt to control for the noise to measure its impacts on the robot's performance. But how does one accommodate noise under real-world conditions?

The most rudimentary way of handling noise is to accept it and try to understand it, usually by increasing the number of samples. Under real-world conditions, the presence of noise often necessitates recording numerous iterations of a single variable value and then calculating a mean and standard deviation to determine the most likely "true" value. Determining how many such iterations to capture is not straightforward and is dependent on the assumptions regarding the distribution, nature, and magnitude of the noise. It also depends on how certain you want to be of the results. If all that is necessary is "rough ballpark," three or four samples (enough to be able to compute the standard deviation) may be all you need. The more precise you want to be, the more samples you need to take. As a rule of thumb, many statisticians will recommend no fewer than 30 samples to ensure statistical significance of the test results.

Another way to address noise is to filter it out. If you know something about the nature of the noise, then you can attempt to remove it. For example, if you know the frequency of the signal variation that you are looking for, you can use various pass-through filters to exclude the data that is outside of what you are looking for. For example, low-pass filters will remove high-frequency variation in the data, while high-pass filters remove low-frequency variation. Band-pass filters will remove variation outside of a specified frequency range. Still other methods attempt to characterize or model the noise and attempt to remove its influences on the data. A prime example of this is the linear Kalman filter[4] [22], which takes a noise model (most typically, assuming a normal distribution of known magnitude and variance) and, through a least-squares approximation, removes the noise from the system.

7.5.3 A Brief Note About Data Collection Costs

There are costs associated with all elements of the testing process (colloquially, "there is no such thing as a free lunch"), and some of these costs may be excessive. Prototypes and measurement systems can be expensive. Experiments may take hours, days, or months to complete, and time can be very expensive, especially as the number of people required for testing increases. And the more variables that are being tested, the more time will be required to complete the tests. For example, if it is known that all variables are independent of one another, then the total time necessary to complete the tests for m variables, with each variable, i having a per-trial time cost of t_i for k_i trials will require a total time of

$$t_{test} = \sum_{i=1}^{m} t_i k_i. \tag{7.22}$$

If we assume each trial takes 2 minutes to complete and each variable is sampled at regular intervals resulting in ten samples each, then for $m = 2$, the time required to complete the test will be 20 minutes and 30 minutes for $m3$, simple

[4] In contrast with the linear Kalman filter, the *extended Kalman filter* is leveraged to make sense of nonlinear data, where the first- or second-order derivatives are used to approximate the nonlinearity.

and straightforward. But how do you know if the variables are independent of one another? That, itself, will require an additional design of experiments to determine.

However, if the variables are dependent, tuning variables will require a factorial combination of factors to assess their impacts on performance. For large numbers of variables, testing along a continuum for each controllable factor is impractical. As such, reference points—most typically maximum and minimum values for variables—are selected as bases of analysis. This, of course, assumes robot performance varies linearly with the adjustment of the variables. Otherwise one or more midpoints for each variable may also be required. With these reference points, *full-factorial testing* will determine the full spectrum of variable impacts.

Given m variables, choosing a low and high value requires 2^m trials. For example, if $m = 2$, then a given test will require four trials. If $m = 5$, then a test will require 32 trials. If, however, a midpoint is required for each variable, the total number of trials will be 3^m (9 trials for $m = 2$ or 245 for $m = 5$). Recall that each trial takes a nonzero amount of time to complete, so there is additional incentive to be intentional about the design of experiments to keep testing to a minimum. Table 7.2 shows how the cost increases exponentially as a function of the number of variables being tested, assuming each trial takes 2 minutes to complete. For small numbers of variables, the testing times are reasonable but quickly become burdensome as the number of variables increases.

For most, the time and costs required to perform such tests are unreasonable or even impossible. A more conscious design of experiments must therefore be created to gather the most information in as few trials as possible. However, this is an extensive topic in and of itself, and we suggest the reader reference guides specifically for information regarding the design of experiments for small sample sizes (e.g., [23]).

Table 7.2 Full-factorial testing if each trial takes 2 minutes to complete

m	Low and high	Time	Low-mid-high	Time
1	2	4 minutes	3	6 minutes
2	4	8 minutes	9	18 minutes
3	8	16 minutes	27	54 minutes
4	16	32 minutes	81	2.7 hours
5	32	1.07 hours	243	8.1 hours
6	64	2.13 hours	729	24.3 hours
7	128	4.27 hours	2187	3.04 days
8	256	8.53 hours	6561	9.11 days
9	512	17.07 hours	19,683	27.34 days
10	1024	34.13 hours	59,049	82.01 days

7.6 Change Management and Regression Testing

Throughout the robot development life cycle, nearly every step of the process—from initial conception to implementation—involves some form of verification and validation to ensure changes do not negatively impact the broader operational specifications. Moreover, as changes to the robotic system are made, the potential for legacy components and subsystems to lose functionality or degrade in performance is of significant concern. Higher level decision functions may be impacted as lower level capabilities are altered and consume more computer processing power and memory. New sensors or motors can have negative consequences on battery life. And reading and writing software states and variables in new programs may alter behaviors resulting from the execution of other programs.

To this end, it is imperative that the robot systems be periodically reevaluated to verify there has not been any degradation in performance, that no unwanted or emergent behaviors are manifesting, and that the system as a whole remains both functional and safe. This is the cornerstone of *regression testing*.

Both regression testing and change management (i.e., the process by which changes to the system are planned, implemented, and tracked) are topics of study in their own right and cannot be fully addressed in this chapter. Instead we will briefly focus our attention on four basic levels of system testing that ensure changes to the system do not adversely impact system functionality or performance:

- Unit testing
- Integration testing
- System testing
- User requirements testing

From user requirements down to unit testing, adding or changing functionality at different levels of the robot development has the potential for impacting performance and functionality of the levels above them. At each level, specific tests are performed and compared with known or expected results.

7.6.1 Unit Tests

At the lowermost level, as individual components of the robot system are modified, each must be verified and validated. Whether addressing software or hardware, the *unit test* are intended to identify and eliminate issues at the unit level. Here, a *unit* is considered the smallest functional component that exists by itself, and unit testing ensures that said components still work as intended when isolated from the rest of the system. Unit tests also provide a convenient mechanism to quickly and reliably verify functionality over time and can be leveraged to perform root cause failure analysis when higher levels experience inconsistent or broken behavior.

This then begs the question: what is the smallest unit of autonomy? Like virtually all things robotics-related, there are different levels of functionality. Indeed, the question itself is somewhat of a red herring. Autonomy is the result of the sum of its parts working together. And, as will be discussed in subsequent sections, the granularity of autonomy is not always obvious. Even seemingly non-autonomy components such as power management or sensor noise filtering can have lasting impacts on the autonomous functionality of the robot. For example, sensing, navigation, collision avoidance, reactive balancing, actuation, and natural language processing are all building blocks of many social robots intended to function autonomously. Similarly, recall from earlier the example of the medicine-administration task. Each aspect of the medicine locating, fetching, and administering process consists of sub-steps, all of which may be subject to errors and uncertainty as the software, hardware, and operating parameters are modified. Ultimately, the components still need to function *as expected* if the code running the "autonomous" behavior is going to function correctly.

7.6.2 Integration Tests

At the architectural level, changes to individual units may impact the ability for two or more units to work together to achieve their intended tasks. Here, the *integration test* is intended to identify and resolve issues stemming from interfaces between units. At this level, tests target whether units intended to work together can do so to achieve their collective functions.

Moreover, integration tests are also used to ensure that units can exist and function independently without negatively impacting functions of unrelated units. For example, the autonomy of a service robot's navigation and locomotion, largely independent of its autonomy for washing dishes, may negatively impact the robot's efforts to scrub a pile of plates clean if the navigation functions are constantly pulling the robot in different directions if a family dog happens to enter the kitchen.

7.6.3 System Tests

At the application level, changes in the integration of units can negatively impact the overall intended functionality of components coordinating to complete systematic goals and functions. The *system test* is intended to ensure application-level goals and expectations are met. This is where the bulk of robot verification and validation takes place from a performance perspective and typically results in adjustments at the unit and integration levels for operational performance tuning. In many ways, system tests can be viewed as a higher level of abstraction of integration testing, as they are frequently leveraged to ensure that both coordinated and independent systematic functions are working correctly and do not interfere with each other (e.g., walking and chewing gum simultaneously).

7.6.4 User Requirement Tests

While system tests are targeted at ensuring robot applications are functioning as intended, the *user requirement tests* (or user acceptance tests) are used to ensure that the applications actually meet the end user's needs and expectations. This consists of both a verification that the application-level goals meet the requirements of the user and that the robot—as delivered—performs as expected in the user's environment. Ultimately, the user requirement test is intended to ensure the robots can do their tasks in real time, in the real world.

Most importantly, given the somewhat amorphous and subjective nature of autonomy, the user requirement tests are, ultimately, the deciding factor in measuring, verifying, and validating the autonomy of the system. As will be discussed shortly, the *stakeholders* are responsible for defining and characterizing the robot's autonomy from their own individual perspectives. And it is the stakeholders whose preferences will define the verification and validation of the robots.

7.7 Use Cases

In this section, we present two different use cases in which different levels of autonomy are illustrated. In the first (Sect. 7.7.1), robots with full autonomy are demonstrated performing material handling tasks. In the second (Sect. 7.7.2), robotic systems with shared autonomy are demonstrated using different interfaces in collaborative pick-and-place tasks.

7.7.1 Use Case: Robot Agility and the Agile Robotics for Industrial Automation Competition (ARIAC)

For the purpose of this case study, we define agility in terms of the robot's ability to move quickly and easily. In manufacturing terms, agility refers to the idea of responding effectively to changing customer needs in a volatile marketplace by handling product variety and by introducing new products quickly [24, 25]. In the context of this chapter, we define agility as "the ability of a robot system to succeed in an environment of continuous and unpredictable change by reacting efficiently and effectively to changing factors." While there is no agreed upon definition of robot agility in the literature, this definition is consistent with proposed definitions of agility which involves not only robot agility but also agility of the manufacturing process as a whole [26, 27].

Robotic systems need to be able to operate safely in collaboration with humans or other robots, be easily tasked and re-tasked, and be integrated into the rest of the enterprise seamlessly and quickly. These systems can greatly help small and medium manufacturers facing rising raw material and labor costs, stiff prices and offshore competition, quality concerns, skilled worker shortages, worker safety issues, and limited resources that hinder growth and profitability. Once the role and the definition of an agile robotic system are given, we need a way to measure the agility of such a system and verify that it is working properly. Defining and measuring agility will allow manufacturers to select the right system to address challenges such as (1) swapping robots in and out without introducing extended downtime or reprogramming, (2) fast replanning when a new order is provided to it, or (3) responding to changing environmental conditions (e.g., non-fixtured tray moves), due to new product designs.

A literature review for robot agility shows only a few examples that address the verification and validation of assembly-type manufacturing use cases and change cases. For the assembly-type use cases, Quinn et al. [28] describe an example assembly task with four plastic parts that get snapped and inserted together. This is described as a typical light assembly task for the workcell being tested, which includes two robots working together. Frei and Di Marzo Serugendo [29] describe an example assembly of an adhesive tape roll dispenser assembly, which is a slightly more complicated assembly than the first use case as it requires a screw for locking the pieces together. The authors also described a change case in the assembly of the adhesive tape roll dispenser assembly, where the environment gets changed to a different locking method for the assembly process, in this case, changing from a screw-lock assembly method to a snap-fit method of assembly. Another change use case was described by Gou et al. [30], wherein three cases are described. The first case is used as a baseline to compare performance for the other two cases. The second case is a new high priority order coming into the system to invoke a re-prioritization. The third case is a variation on the second, but the re-prioritization is caused in this case by a machine breakdown, causing the system to adjust to absorb the workload.

To advance the agility performance of manufacturing robotics assembly systems in unstructured and dynamic environments and verify their performance, the Agile Robotics for Industrial Automation Competition (ARIAC) was initiated in June 2017 by the National Institute of Standards and Technology (NIST) in collaboration with the Open-Source Robotics Foundation (OSRF). ARIAC is designed to be a simulation-based competition to allow competitors around the world to utilize latest advances in artificial intelligence and robot planning to address real-world industrial challenges pertaining to kitting (or kit building) applications. The latest iteration of the competition was held between April and May 2020. The purpose of the annual occurrence of ARIAC is twofold. First, NIST intends to use the results and knowledge gained from ARIAC to further its efforts to develop metrics and test methods to measure and validate robot agility, as well as tools for manufacturers to assess the agility of their robotic systems. Second, ARIAC aims to encourage competitors to develop the most effective solutions to address manufacturing processes while keeping the cost down.

The ARIAC organizers needed a way to be able to measure the effectiveness of how a robot system was implemented to handle manufacturers' agility challenges. Because of the variety of trade-offs and options that a robot system developer chooses during design, the organizers needed a set of metrics to compare the systems both against other systems and compare the system across different trials where different agility challenges are in play.

7.7.1.1 Current Metrics

There are currently three general metrics that can be used individually to compare systems in terms of their agility performance: Cost Factor, Completion Score, and Efficiency Factor. These three individual metrics are also combined along with some constant factors into the Ranking Score, which is the equation used for ranking the teams during the finals.

Cost Factor The first general metric is to compare how expensive the systems are based on the choices made. The main idea for this metric is that a lower cost system is better than a higher cost system, all other factors being equal. This cost would be a combination of the costs of the robot, the sensors, and the infrastructure. In the specific cases of ARIAC , the competitors are limited to only one robot, and thus the main cost comes from the choice of sensors. Each sensor is assigned a nominal cost value based on its usefulness in the scenarios and other factors by the competition organizers. The costs of the competitor's chosen sensors are summed up and represented as shown in Eq. 7.23. A baseline cost (BC) is determined by the organizers by using a representative, midrange number of sensors in the environment. This baseline cost and the total cost (TC) are used as shown in Eq. 7.24 to calculate a team's cost factor (CF).

$$TC = \Sigma_{i=1}^{n} Cost_i \tag{7.23}$$

$$CF = \left(\frac{BC}{TC}\right). \tag{7.24}$$

Completion Score The next general metric for comparison is the kit completion score (CS). For an individual kit order submission, S_j, with i parts in the order, the following points are available:

- One point (up to i points) for each part of the correct type being placed in the kit tray
- One point (up to i points) for each part being placed in the correct position ($\pm$ 3 cm) and orientation (± 0.1 rad)

- i points awarded if the above two cases are maxed out
- One point (up to i points) for each part of the correct color being placed in the kit tray

The completion score is calculated by adding up the four categories above, and with each category having a maximum of i points, there is a maximum possible completion score of $i \times 4$, as shown in Eq. 7.25. In the case where a trial contains multiple different orders, each order would have a separate completion score and will be used in different places in the overall scoring equation.

$$CS_{S_j} \leq i \times 4. \tag{7.25}$$

Efficiency Factor The last of the three main comparison metrics is the efficiency factor (EF). The main idea on this metric is that in general, a faster system is better than a slower system, which leads to greater throughput overall. When an order is sent to the competitor's system, a timer is started that begins counting up until that order is completed and delivered. For a given trial, j, all of the competing teams have their time, T_j, averaged together, represented as AT_j. As an edge case, if a team's system times out (taking longer than 500 simulation seconds), their efficiency factor is set to 0, and the trial's time is not included in the average. Otherwise, the efficiency factor is calculated as shown in Eq. 7.26.

$$EF_j = \left(\frac{AT_j}{T_j}\right). \tag{7.26}$$

7.7.1.2 The Future of ARIAC and the Associated Metrics

ARIAC remains an evolving entity with a close ear to both industry, robotics, and simulation software advancements. The ARIAC organizers continue to look to both industry and academia for relevant challenges that address the industrial challenges of today, as well as the verification and validation aspects that relate to them. Future ARIACs will include assembly at various levels of abstraction. Initially, this will include placing assembly pieces in three-dimensional orientations that do not lie on a single plane. Later iterations of ARIAC will also include how these parts are assembled, like screwing, placing, or pressing parts.

7.7.2 Use Case: Human-Robot Interaction

As the second use case, we investigate the design of experiments for measuring the ease of use of a novel interface in a robotic pick-and-place operation with shared autonomy. An emerging application domain of metrology regards the measurement of performance in human-robot teams. "Teaming" has many different meanings within the field of human-robot interaction (HRI) and has been characterized by any number of factors, including proximity, task goals, interactions, and physical contact. As robots become safer and the factors impacting humans more well known, the expectation is that robotic technologies will be more ubiquitous in industrial, commercial, residential, educational, and medical environments. To maximize their efficacy, however, means by which their performance is assessed and optimized are required.

As an example, let us take the simplified use case of an operator learning to use a robot using a novel, tablet-based interface. This is a hypothetical trial used as a test case for a future planned human subject study.

The application is to remotely adjust the motions of robot with a pneumatic, parallel gripper to arrange ten blocks in a simple pick-and-place operation. The operator is using a custom tablet-based interface to communicate with the robot and affect joint and/or Cartesian space motions, operate the gripper, and move to common poses (e.g., "home"). The developers of the system want to optimize the interface layout and functionality to accomplish the following:

- Minimize learning time
- Improve teaming performance
- Assess utility of the interface
- Track attention of operator (robot vs interface)
- Reduce the mental effort on the operators

The developers propose a design of experiments intended to capture metrics that support the above goals. In this set of experiments, a human operator is standing at a control station, where the tablet is statically mounted. The robot is mounted separate from the operator for safety purposes but is clearly visible from the control station.

7.7.2.1 HRI Metrics

Recall from Sect. 7.4.3 that metrics may be objective or subjective and quantitative or qualitative. In the previous examples, many of the metrics identified fell into the "objectively quantitative" designation. However, given the human element in HRI, it should necessarily be expected that many metrics will be qualitative and subjective in nature. For an overview of some of the more common metrics in HRI, see [31]. In this section we describe briefly the metrics leveraged for assessing each of the aforementioned KPIs and the means by which they are measured.

Learning Time Generally speaking, there is no practical way of capturing the time necessary to learn how to use a robotic system, as "learning" how to use a robot is a personal assessment that is difficult to pin down. One tool, however, proposed in [32], for quantitatively approximating learning time is to compare the amount of time necessary for a novice use to achieve the same performance as an expert user of a given system.

In this example, we look at the time necessary for each block placement $t_{block,i}$, which we can assume is bounded by the beginning of each block handling process and the point where it puts the block in place. Mathematically, this can be expressed as $t_{block,i} = t_{drop,i} - t_{init,i}$, where $t_{init,i} = t_{drop,i-1}$. This permits the measurement of each cycle time to be automated, so no human input is required to assess when the task is completed.

The "learning time" can be assessed as a linear regression of the number of steps or iterations necessary for a novice user to consistently approximate the performance of an expert user for each material handling task. In the event that such performance is not achieved, another way of assuming that, when the novice user achieves a steady state (i.e., performance improvements cease improving as a function of practicing the task), the user has "learned" the system. For the purpose of this use case, the developers were concerned more about the time necessary to complete the task and largely ignored the quality of the block arranging.

Interface Utility Measuring the interface utility is much more straightforward in its implementation. Each control interface is presented on a separate tab of the tablet software application. At a low level, the application generates a log of timestamped records for each button press of the digital interface. Each press and release action is recorded separately, recording the identity of the button being pressed, the timestamp of the action. This enables us to know exactly how many times the operator used the interface, what parts of the interface were being used, and for how long they were used. All data collection is performed automatically and in situ, as it is built into the interface itself.

Attention Tracking Operator attention is not always intuitive as to how it should be measured. While some studies elect to leverage human observers guessing what the person is looking at, the developers instead opted to leverage a motion capture system already installed in the lab. Each human operator was instructed to wear safety glasses that had passive motion capture markers attached, thus enabling us to identify where the glasses were and where the lenses of the glasses were pointed. The system then determined whether the glasses were pointed at the robot or the interface based on known locations of both in the motion capture system's measurement volume. Because both the tablet and the robot were statically mounted in the test configuration, their placement could be controlled to ensure marked head orientation toward either.

Therefore, for attention tracking, each "gaze" at the different components was classified as the focus of attention: at the robot, at the tablet, or something else. The number of occurrences for each was recorded, as well as the length of time for each occurrence.

Effort In contrast with the previous three metrics, capturing operator effort is difficult to achieve either automatically or objectively. While some researchers have suggested using electroencephalograms to estimate mental effort (e.g., [33]), for this study it was decided to use a survey tool for the operator to complete a self-assessment at key intervals. The tool the developers selected was the National

Table 7.3 Summary table of the elements of the NASA-TLX [34] survey tool used in this study. Subject responses are scored on a Likert scale of 0–10

Factor	0	10
Mental demand	Very low	Very high
Physical demand	Very low	Very high
Temporal demand	Very low	Very high
Performance	Perfect	Failure
Effort	Very low	Very high
Frustration	Very low	Very high

Aeronautics and Space Administration's Task Load Index (NASA-TLX, [34]). The NASA-TLX is an oft-used staple of HRI research and attempts to capture the mental, physical, and temporal demands and the operator's personal perspectives on their performance and mental state. A summary of the NASA-TLX is given in Table 7.3.

7.7.2.2 Test Methodology

As was discussed earlier, each human operator was positioned at a control station with a tablet interface. The operator and the robot were physically separated, but the robot's motions and workspace could be seen clearly from the control station. The purpose of this exercise was for each operator to remotely operate the robot to arrange a series of ten blocks, with each pick-and-place action consisting of one cycle. A feeder presented the operator with a single block, which specified its destination on a linear array on the table in front of the robot. The blocks and target destinations were clearly visible and identifiable to the operator. Each operator wore safety glasses that had motion capture markers to identify the direction (at the robot, at the tablet, or something else) in which the operator's head was oriented. The tablet interface had metrology tools integrated into its software to log button presses for later processing. The interface had separate tabs for Cartesian control, joint control, gripper actuation, and moving the robot to predefined poses.

Each operator was told to control the robot to take the presented blocks and place them at the specified locations. Each operator was told to do the entire operation once, and the order of block presentation was identical for each individual person. The trial would be paused at the 30%, 60%, and 100% completion points of the exercise, and the operator would fill out the NASA-TLX to capture their perception of effort. In total, 100 simulated subjects participated in the trial. Each simulated participant was given a unique, nominal performance, and effort profile, to which random noise was added. Of these simulated participants, five were designated "expert" users of the robot.

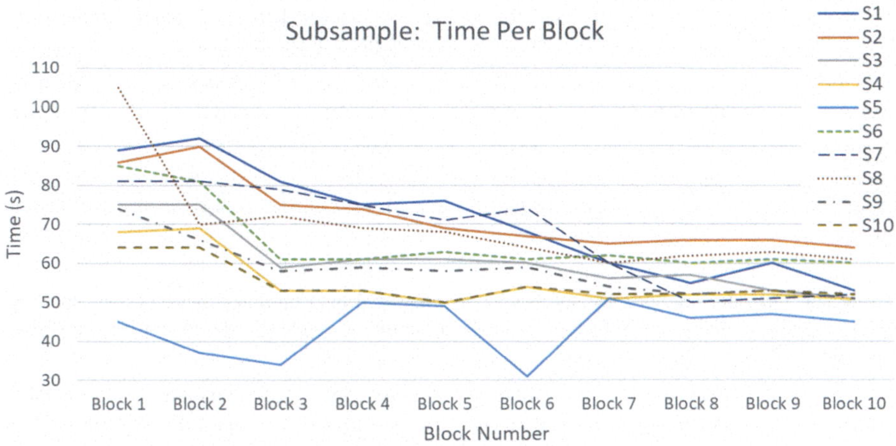

Fig. 7.2 Subsampling of the trial data, with ten subjects, selected at random completing a ten-block arrangement task

7.7.2.3 Example Results

For the purpose of reporting high-level conclusions, ten of the simulated participants were selected at random. These test subjects, labeled S1 through S10, include one expert user and nine novice users. The task timing performance is reported for these ten subjects below. One subject, S7, was randomly selected for individualized reporting and analysis.

Learning Time In broad terms, as each operator used the interface more, the time necessary to perform a pick-and-place operation decreased. Based on the results shown in Fig. 7.2, many of the operators achieved a steady state in task timing by the time they controlled the robot to pick-and-place block 8.

Subject S5 was designated an expert user of the system, as is evidenced given the relatively consistent performance and low required time for assembly. When compared to S5, subject S7 required a total of about 690 seconds (11.5 minutes) to complete the task, while S5 required a total of about 446 seconds (7.4 minutes) to do the same task. In other words, S7 took 55% longer than S5 to complete the pick-and-place operation for all ten blocks. However, worth noting is the improvement of performance of S7. For block 1, S7 required 89% more time to complete the block placement task. But, by block 8, S7 was able to complete the task requiring only 11% of the time taken by S5, implying that the operator's skills had increased as a function of repetition.

Interface Utility As seen in Table 7.4, subject S7 spent most of their time interacting with the Cartesian control tab but had more interactions with the joint control tab over the duration of the test. The average duration of each interaction with some element on the Cartesian control tab was about 9.6 second, while the each interaction

Table 7.4 Interface utility for subject S7, for the four control tabs of the tablet application. Interactions are characterized by pressing and releasing parts of the interface

Control tab	Number of interactions	Total time (s)	Average time (s)	Standard dev. (s)	Median time (s)
Cartesian	48	452.3	9.63	1.895	7.6
Joint	205	103.3	0.572	0.113	0.3
Tool	45	29.2	1.53	0.302	1.5
Preset	23	12.3	1.8	0.354	1.9

Table 7.5 Operator attention for subject S7, as measured by a motion capture system tracking head movement. Attention is factored based on vector of head orientation toward the robot, toward the tablet, or somewhere else

Target	Number of interactions	Total time (s)	Average time (s)	Standard dev. (s)	Median time (s)
Robot	25	207	8.28	0.272	9.1
Interface	109	455.4	4.42	0.137	0.8
Other	13	27.6	4.6	0.070	2.76

with an element on the joint control tab was only 0.6 seconds, implying the Cartesian tab was the principal mechanism for controlling the robot. The second most-used tab on the interface was the preset motion tab. This indicates that means by which the time necessary for operators access the controls on these tabs (e.g., by combining the functions onto a single tab) are likely to improve task performance.

Attention Tracking As seen in Table 7.5, a bulk of subject S7's time was spent looking at the interface. It is also worth noting that the total time looking at the interface is less than the time the operator actually spent *using* the interface, implying S7 was occasionally looking at something other than the interface while using it. Judging by the attention breakdown, the operator was either holding down buttons on the interface while watching the robot move or had their hand positioned such that button presses could be made from memory of their locations on the interface.

Effort Based on the simulated responses to the NASA-TLX from S7, Fig. 7.3, it can be observed that the subject's mental demand went down and their performance improved as they became familiar with the interface. However, there does not seem to be any significant impact on physical or temporal demand, and the overall assessments of effort and frustration remain relatively high. During the next iteration of interface development, it is likely that asking S7 targeted questioned to identify areas of improvement that could possibly result in lower frustration and effort.

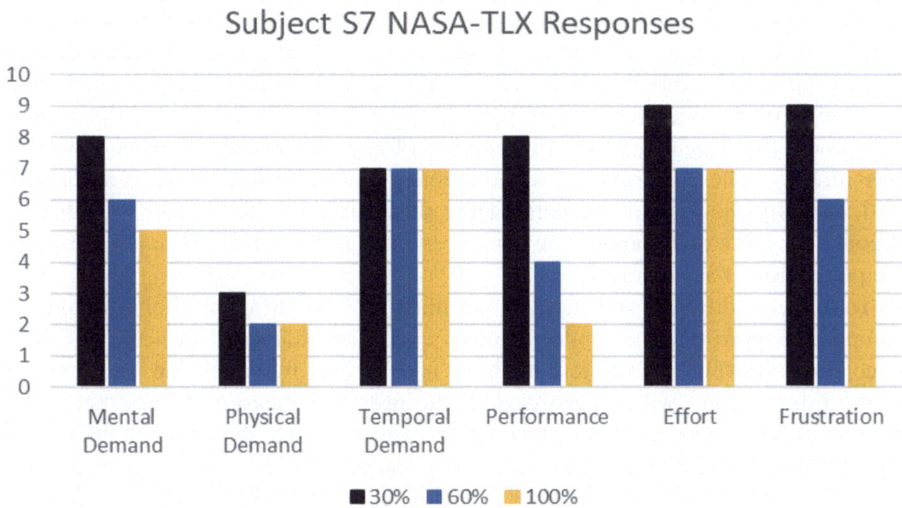

Fig. 7.3 Responses of subject S7 to the NASA-TLX questionnaire at 30% (black), 60% (blue), and 100% (orange) through the exercise

7.7.2.4 Interpretation of Results

This case study simulates a single step in the interface design process for a new robot controller with a specific application in mind and may not reflect the user experience for a similar interface for other applications. While these initial results imply that the tablet interface is learnable, with task performance improving over time, the results from subject S7 imply some improvement could be needed to reduce frustration and effort, possibly by such a small change as moving the tablet where both the interface and the robot can be seen with minimal movements of the head. However, such results may be an outlier, and a full evaluation of all 100 subjects would be required.

7.8 Conclusion

The autonomy of a robotic system is strictly bounded by the application for which it was built or task it was programmed to perform. These applications and tasks define the rules by which the robot's autonomy is assessed. A robot designed and programmed to autonomously map the ocean floor can be evaluated for its ability to navigate, perform its mapping task, and correct for positioning and measurement uncertainty while operating independently of human inputs or external power. This same robot, however, cannot be assessed for its autonomy in cleaning a child's bedroom because it was not meant to perform this task.

The juxtaposition of multiple functional processes, external factors, inputs, outputs, internal states, and corrective actions are used to capture and characterize the autonomy of a robotic system. Evaluating the performance, interactions, and integration of the many composite parts is central to the validation of the system as a whole. Specifically, while any number of components may impact—directly or indirectly—the performance of the autonomous system, the robot's autonomy is defined by the sum of its parts. How these different parts work together to accomplish tasks and compensate for uncertainty in the world ultimately impacts the perception of autonomy.

To this end, different stakeholders are likely to have different interpretations of autonomous performance. Understanding what these stakeholders need and want is central to developing the test methods and metrics by which the performance of the robot system will be assessed. If the tests and measure do not match the key performance indicators the stakeholders are expecting, these stakeholders will have little basis to feel the robots' autonomy has been verified. Similarly, if the test methods have not been validated to actually capture the aspects the stakeholders are expecting, evaluations may result in inconsistent or potentially meaningless conclusions.

Along these same lines, it is also worthwhile recognizing that different people have different interpretations of the term *autonomy*. While it would be preferable that everyone works from the same common definition, being aware of and accommodating alternative expectations for autonomy is critical. Similarly, knowing what aspects of a robotic system are expected to be autonomous will ultimately define the means by which said systems are measured and assessed.

Therefore, as was presented in this chapter, the processes verifying and validating autonomous robotic systems must necessarily begin with defining the metrics by which said systems will be evaluated. These metrics will cement the definition of autonomy and guide the design of experiments for testing and evaluation. Selecting these metrics is accomplished by means of capturing the preferences of the stakeholders , who must ultimately be satisfied to declare the sufficient degree of autonomy achieved.

References

1. H.M. Huang, Autonomy Levels for Unmanned Systems (ALFUS) Framework. Volume I: Terminology. Tech. Rep. 1011-I-2.0, National Institute of Standards and Technology (2008)
2. P. Antsaklis, Annual Reviews in Control (2020)
3. B.T. Clough, Metrics, schmetrics! how the heck do you determine a UAV's autonomy anyway. Tech. rep., Air Force Research Lab Wright-Patterson AFB OH (2002)
4. H.M. Huang, E. Messina, J. Albus, Autonomy Levels for Unmanned Systems (ALFUS) Framework. Volume II: Framework Models. Tech. Rep. 1011-II-1.0, National Institute of Standards and Technology (2007)
5. The Standard for Project Management and A Guide to the Project Management Body of Knowledge. Standard, Project Management Institute (2017)

6. 9001: Quality management systems — Requirements. Standard, ISO (2015)
7. ISO 9283 - Manipulating industrial robots - Performance criteria and related test methods. Standard, ISO (1998)
8. J.A. Marvel, K. Saidi, R. Eastman, T. Hong, G. Cheok, E. Messina, in *Proceedings of the Workshop on Performance Metrics for Intelligent Systems* (2012), pp. 109–113
9. G.F. Dubos, J.H. Saleh, R. Braun, Journal of Spacecraft and Rockets **45**(4), 836 (2008)
10. P. Godfrey, Manuf. Eng. **81**(3), 109 (2002)
11. A. Steinfeld, T. Fong, D. Kaber, M. Lewis, J. Scholtz, A. Schultz, M. Goodrich, Proc. 1st ACM SIGCHI/SIGART Conf. on Human-Rob. Interact. pp. 33–40 (2006)
12. R.S. Pierre, S. Bergbreiter, Annual Review of Control, Robotics, and Autonomous Systems (2019)
13. M. Shneier, E. Messina, C. Schlenoff, F. Proctor, T. Kramer, J. Falco, Measuring and Representing the Performance of Manufacturing Assembly Robots. Tech. Rep. 8090, National Institute of Standards and Technology (2015)
14. J.A. Marvel, R. Bostelman, J. Falco, ACM Computing Surveys (CSUR) **51**(1), 1 (2018)
15. J. Falco, J.A. Marvel, R. Norcross, Proc. Int. Conf. Safety Indust. Autom. Syst. (2012)
16. J.A. Marvel, J. Falco, I. Marstio, IEEE Trans. Syst., Man, and Cybern.: Syst. **45**(2), 260 (2015)
17. C.I. Schlenoff, J.M. Evans, A.J. Barbera, J.S. Albus, E.R. Messina, S.B. Balakirsky, in *Mobile Robots XVII*, vol. 5609 (International Society for Optics and Photonics, 2004), vol. 5609, pp. 116–127
18. E. Messina, A. Jacoff, J. Scholtz, C. Schlenoff, H.M. Huang, A. Lytle, J. Blitch, NIST Draft Report (2005). URL https://www.nist.gov/document-12931
19. A. Jacoff, H.M. Huang, E. Messina, A. Virts, A. Downs, in *Proceedings of the 10th Performance Metrics for Intelligent Systems Workshop* (ACM, New York, NY, USA, 2010), PerMIS '10, pp. 161–168. DOI https://doi.org/10.1145/2377576.2377606. URL http://doi.acm.org/10.1145/2377576.2377606
20. A. Al-Achi, Research & Reviews: Journal of Hospital and Clinical Pharmacy **5**(1), 1 (2019)
21. A. Gelman, et al., Annals of statistics **33**(1), 1 (2005)
22. G. Welch, G. Bishop, et al., An introduction to the Kalman filter. Tech. rep., University of North Carolina, Dept. of Computer Science (1995)
23. J.K. Telford, Johns Hopkins APL technical digest **27**(3), 224 (2007)
24. P. Lindbergh, International Journal of Operations and Production Management **10**(2), 94 (1990)
25. H. Sharafi, Z. Zhang, International Journal of Production Economics **62**(1–2), 7 (1999)
26. L. Jin-Hai, A. Anderson, R. Harrison, Business Process Management Journal **9**, 170 (2003). https://doi.org/10.1108/14637150310468380
27. N.R. Bagul, M.P. Ahire, International Journal of Advanced research in Science and Engineering **5**(2), 282 (2016)
28. R.D. Quinn, G.C. Causey, F.L. Merat, D.M. Sargent, N.A. Barendt, W.S. Newman, V.B. Velasco Jr, A. Podgurski, J.Y. Jo, L.S. Sterling, Y. Kim, IEEE Transactions **29**(10), 901 (1997)
29. R. Frei, G. Di Marzo Serugendo, in *IEEE International Conference Self-Adaptive Self-Organizing Systems* (Venice, Italy, 2008), pp. 97–106
30. L. Gou, T. Hasegawa, P.B. Luh, S. Tamura, J.M. Oblak, in *Proceedings of the Fourth International Conference on Computer Integrated Manufacturing and Automation Technology* (1994), pp. 142–149. https://doi.org/10.1109/CIMAT.1994.389081
31. J.A. Marvel, S. Bagchi, M. Zimmerman, B. Antonishek, ACM Transactions on Human-Robot Interaction (THRI) **9**(4), 1 (2020)
32. T.L. Roberts, T.P. Moran, Commun. ACM **26**(4), 265 (1983)
33. N.H. Liu, C.Y. Chiang, H.C. Chu, sensors **13**(8), 10273 (2013)
34. S.G. Hart, L.E. Staveland, Adv. Psychol. **52**, 139 (1988)

Chapter 8
Review on Verification for Marine Autonomous Systems

Mae Seto and Ali Bassam

Chapter 8 Editorial Note

The chapter reviews the limitations and constraints imposed by the maritime environment and maritime applications on current verification processes, tools, and techniques, engaging with four of our stakeholder questions. It addresses how the marine environment impacts the practice of verifying a system and its functionality, and how one determines whether a system is useful in the context of marine applications. Its focus on understanding how the marine environment impacts the autonomous systems technology design provides improved understanding of how operating in a marine environment drives the most complex verification challenges. Defining the tasks that marine robots undertake is complex, but this chapter addresses technical challenge 16 (consistently and accurately describing the autonomous capability embodied by a robot) by reviewing the verification of marine system functionality across abstraction levels. Existing tools as they are applied in a variety of tasks are reviewed, addressing challenge 22 (verifying system fitness) by providing a clear picture of how we can currently verify the fitness of a marine system for the task it is being asked to accomplish. Stakeholder questions are defined and discussed in the Chap. 1, Introduction. Technical challenges 16 and 22 are discussed in more detail in Chap. 2, Verification Challenges.

Abstract Marine autonomous systems require autonomy to operate within an environment that is dynamic and harsh for navigation, sensing, and communications. This chapter presents the marine autonomous system challenges and how the autonomy enables operations despite them, and then reviews what the methods the community has used for their verification. It concludes with suggestions on future research work.

M. Seto (✉) · A. Bassam
Mechanical Engineering Department, Dalhousie University, Halifax, NS, Canada
e-mail: mae.seto@dal.ca; al903300@dal.ca

S. Redfield et al. (eds.), *Verification of Autonomous Systems*,
https://doi.org/10.1007/978-3-031-88546-4_8

8.1 Introduction

This chapter addresses the verification or provable correctness, performance, and reliability of marine autonomous systems (MAS) and their autonomy. Marine includes both autonomous vehicles that operate underwater (autonomous underwater vehicle, AUV) and those that operate on the water surface (autonomous surface vehicle, ASV) to achieve goals or objectives through their *missions*. This review of MAS verification developments within the last decade or so provides a broad view of the topic, including design, operational, and autonomy verification, as well as presenting promising areas for research aimed at autonomy developers and testers.

First, a review of important definitions and terminology is in order.

8.1.1 Terminology and Definitions

Vehicle is used interchangeably with robot, autonomous system or marine autonomous system (MAS).

Oceans are distinct bodies of water that are far from land, whereas *seas* are surrounded by, or touch on, land to some extent. *Marine* is used as an adjective for ocean or sea. The MAS referred to operate in oceans or seas. It is implied that they could also be deployed to smaller inland bodies of water like lakes, rivers, and ponds. The marine environment refers to above and under water.

The MAS operational environment is described as dynamic or unstructured which means it changes rapidly in space and time and is difficult to predict. *Harsh* refers to extreme forces in the robot's operational environment due to winds, sea state, pressure, temperatures, corrosion, etc.

The *mission* is the instantiation of the steps by the MAS' sensing, actuation, deliberation, and processing to realize MAS goals. The mission could be adapted or changed to achieve a MAS' objective or goal. Mission is referred to here as a payload or autonomy capability for goals, e.g. a survey mission to detect specific targets.

System can refer to entire MAS or a component, subsystem, or autonomy software module which is a part of it.

Autonomous is distinct from *automatic* and are both used here as adjectives. Automatic control refers to a system following closed-loop control laws or a script. Autonomous implies some level of local decision-making or adaptability in the control, however small.

A truly autonomous system governs itself with its ability to make decisions, act on the decision, evaluate the efficacy of its actions, and adapt as needed to achieve its goals. The autonomous system tries to accomplish its goals without human intervention even given uncertainty or unanticipated events.

Autonomy, used as a noun or adjective, refers to advanced abilities of an autonomous system that previously required a human operator be in-the-loop.

Autonomous capabilities include re-planning missions, optimizing resource use (e.g. sensors and energy), analyzing sensor measurements in situ to select MAS actions, and on-going evaluation of whether decisions and subsequent actions maximize the MAS's successful achievement of its goals. Autonomy varies from the limiting case of minimal to no autonomy (i.e. direct operator remote control, automatic control) to complete autonomy where an operator is not required and the system is autonomous. Autonomy, here, is instantiated as software.

Self-adaptive is another descriptor for systems with autonomy.

MAS can combine some onboard autonomous abilities (e.g. autonomously create plans or navigate a survey around an exclusion zone) and others that are more automatic (depth-keeping or waypoint following). Most MAS have some combination of automatic and autonomous abilities with the more capable possessing more autonomous abilities.

Requirements are the necessary and correct performance, behaviours, or services of/for a system. *Specifications* are a precise technical description of how a requirement might be met.

Validation ensures a system's capabilities fulfill specific requirements. *Verification* addresses whether the MAS' capabilities are correctly designed to fulfill the specific requirements, that is, its provable correctness, performance, and reliability. This chapter is concerned with verification more than validation.

When a MAS subsystem, process, or service operates *correctly,* it does not exceed desired thresholds (speed, depth, attitudes, and pose uncertainty) and collide (unintentionally) with another object, etc. Verification ensures that the MAS (hardware and software) do not violate this.

Reliability is the likelihood that a system, process, or service performs its correct function over a specified period of usage and under specified operating conditions, in a manner that meets or exceeds requirements. Reliability verification, also called reliability testing, uses models, statistics, and other verification methods to evaluate the reliability of systems, processes, or services like cars, watches, cyber-physical systems, etc. One can also consider reliability to capture a system's capabilities. Reliability is achieved through verification. A system that is reliable is a system that has been verified.

Certification refers to a proof or process that shows a system has reached a defined performance standard. Certifications often fulfill a legal rather than engineering or scientific requirements.

Operator and *user* are used interchangeably here. They are distinct from *developers, designers*, and *testers* who create and test the system.

8.1.2 Autonomy and Its Verification

The goal of deploying MAS (i.e. the mission) may be to make a measurement, survey to find a target (e.g. a mine), or monitor a water column substance (e.g. phosphorus)—to name a few. Basic MAS operational capability includes achievable speed, depth, stability, maneuvering, waypoint-following, basic navigation, and localization. Mission autonomy assumes that the MAS' operational capabilities are adequate to support the mission and covers the autonomous capabilities relevant to mission requirements. This separation enables a distinction to be made between onboard operational (or vehicle) and mission (or payload) sensing, driven by the mission. For example, conductivity, temperature, and depth sensors aid MAS operations but they can also be mission sensors if the goal is to map the temperature of a region.

This review is based on behaviour-based autonomy [1] which focuses more on adapting its responses (behaviours) based on information, in-water sensor measurements, and onboard deliberation than on modelling events.

Some operators are reluctant to integrate autonomy for mission and safety-critical operations due to lack of familiarity and the perception of risk associated with autonomy. Verification of the autonomy algorithm and its software instantiation address this to some extent, although verification of autonomous systems can also be a barrier to rapid adoption of these systems for routine operational use. In systems with static and predictable conditions, a constant level of trust can be assigned, but in self-adaptive systems that operate in unstructured environments, it is challenging to quantify trust. A verification framework to evaluate autonomous vehicle missions can provide a foundation for proper and trustable operation of autonomy-enabled MAS, thus increasing trust.

Typically, verification is performed to ensure the system-under-test supports its operator correctly. Autonomy verification is different because it focuses on verification of the system's ability to perform its tasks independent of its operator. Ceding control of the MAS to its autonomy means that it is not directly guided by human judgement and may be perceived as incurring more risk. Human judgement may make it better or worse as human error is linked to human judgement. While human operators have checks and balances to minimize the risk of error, the same could be applied to MAS autonomy. The concern is valid, but the situation is not inherently more serious without direct human judgement.

Therefore, the MAS autonomy is intended to adapt to uncertainty with little or no human intervention. This runs counter to systems engineering approaches that mitigate ambiguity and add precision at design time. One must be vigilant that MAS design time specifications do not constrain their ability to adapt to evolutions in their state, environment, interactions with the environment, and mission. Test and verification must occur at all stages of autonomy development and not just late in the process where it may be difficult to detect and correct errors from earlier stages. Verification should not end after an autonomy capability has been delivered. Since post-delivery verification tends to focus on formal methods, and pre-delivery verification may incorporate both formal and informal methods, this chapter will review MAS verification using both formal and informal methods and from requirements definition to the operational system.

After a brief introduction to autonomous underwater and surface vehicles, the environment in which they operate, and requirements for MAS missions, existing verification practices are discussed. For non-autonomous systems, roboticists currently use formal methods and informal methods like extensive simulations with tools like Gazebo [2] followed by in-water scenario-based testing. This is insufficient for complex autonomous systems, where these practices have failed to manifest errors from software bugs despite extensive testing and operational hours [3, 4].

Finally, tools and techniques that have been, or could be, used to address the MAS challenges are reviewed, other topics like test scenario generation and metrics, are discussed and the lack of examples for MAS are highlighted, and promising research areas are presented, culminating in a summary of the state of MAS autonomy verification.

8.2 Introduction to Marine Autonomous Systems (MAS)

As robots, marine autonomous systems are complex concurrent systems, made more complex by interactions with their harsh environments. Like space robotics, verification of MAS requires insight into how their performance is inextricably linked to their environment.

As with their verification, marine autonomous systems' development, operations, and autonomy have lagged their counterparts in the ground, aerial, and especially space domains.

Initially, naval organizations drove AUV development with the first AUV fielded during Operation Iraqi Freedom (2003) for NMCM [5]. They were driven by security and safety requirements given the strategic and tactical advantages AUVs conferred. AUVs operate in contexts where errors have large consequences such as determining if an in-water contact is friendly or not in anti-submarine warfare. This situation is further compounded by the greater environmental uncertainty (poor underwater acoustic signal-to-noise ratio), impacting underwater communications, localization, navigation, and sensing. If the autonomy includes operator decision aids to recommend a decision based on in-water measurements, how would the verification for this be accomplished given these large uncertainties (high false positives and negatives)?

Given their challenging environment, MAS autonomy is a considerable strength, but it presents challenges in its verification for provable correctness, performance, and reliability. The autonomy [6, 7]:

- is an NP-complete system with characteristic discontinuities where small changes in the input can lead to significant changes in the algorithm accuracy and performance
- has an input–output behaviour that is difficult to state subsequently and verify (i.e. it is non-deterministic)
- responds to a wide variety of inputs and not just a nominal scenario
- has a declarative model-based architecture so it is difficult to predict its execution path

While MAS verification has been applied to deterministic faults, controllers [8, 9], navigation [10], path-planning [11], and communications [9], this chapter focuses on the verification in those areas where it is non-deterministic (e.g. faults).

8.2.1 Autonomous Underwater Vehicles (AUV)

Autonomous underwater vehicles (AUV, Fig. 8.1) and autonomous surface vehicles (ASV, Fig. 8.2) are used for a wide range of military, scientific, and commercial missions including marine surveys, inspections, and payload deliveries [14, 15]. They vary in size from vehicles small enough to manually deploy with two operators to those that are 10 m long and require a crew to deploy over-the-side of ships or jetties.

AUVs are free-swimming torpedo-shaped robotic vehicles powered by onboard medium energy density sources.

AUVs move in three-dimensions and can dive to depths of more than 3 km [8]. They have a front seat vehicle control computer (VCC) that provides basic vehicle control. The above-water global navigation satellite system (GNSS) is not available to submerged AUVs for navigation or positioning. Their onboard navigation and localization solution is based on a fusion of sensors the AUV uses for dead-reckoning (e.g. INS/IMU, DVL, compass, ultra-short baseline, etc.) aided by other methods [16]. Additionally, once submerged and underway, above-water radio frequency (RF) communication is not available and underwater acoustic

Fig. 8.1 REMUS 100 AUV used by USN [12]

Fig. 8.2 SeaRobotics USV-2600 in the Canadian Arctic [13]

methods are used for communication with the operator/support ship. These are low bandwidth and short range compared to above-water RF.

The AUV's inability to access above-water localization (and navigation) and communication methods makes verification challenging. For example, when submerged, there are no universal positioning systems like GNSS to provide some ground truth for robot pose. This makes it difficult to verify a localization algorithm. The underwater acoustic communications used by AUVs are challenging compared to RF as the underwater communications channel is sensitive to in-water conditions like bathymetry, underwater ambient acoustic conditions, proximity to surface, and sound velocity profile. Lack of ground truth for the underwater communications channel makes it difficult to verify communication methods like modulation schemes. AUVs are ideally suited for deep sea missions and often work collaboratively with ASVs.

8.2.2 Autonomous Surface Vehicles (ASV)

The important difference between ASVs and AUVs is that ASVs have part of their hull above the water surface. This enables ASVs to use RF/cellular for above-water communications and GNSS for navigation and localization (although most ASVs have a navigation and localization solution that does not rely on GNSS as this is not always available). Given their unique place in the water column ASV is frequently used to relay between AUVs and above-water points. An additional advantage of an ASV's above-water surface expression is its ability to use air-breathing engines for forward propulsion, since liquid fuel is 60+ times higher in energy density than the best current lithium-ion batteries.

While RF is used for communications between a support ship and an ASV, some applications control the ASV through a satellite link like Iridium [17]. This makes it possible to control the ASV from a much greater distance, but its expense can also limit the communications bandwidth. Also, given the ASV is at sea level, it is not always possible to get a strong RF or satellite link, especially in high sea states.

8.2.3 MAS Control and Autonomy

Modern MAS are cyber-physical systems (CPS): embedded systems with networked computing nodes that communicate to achieve their purpose and interact with their environment through sensors and actuators. They have discrete and continuous dynamic states (i.e. hybrid systems), with software and hardware components that support communication between subsystems of sensors and actuators and closed-loop control systems integrated through an onboard VCC (vehicle control computer or front seat driver) as an embedded processor. It is worth noting that MAS functions like sensing, decision-making and planning, actuation, and control run concurrently. Simultaneous control of multiple concurrent tasks means multiple processes that run in parallel. These multiple intertwined and interacting processes add to the complexity of the system, increasing the difficulty reproducing and capturing the range, volume, and level of interactions between multiple processes in verification.

MAS autonomy, as opposed to scripted missions and responses, can adapt to a wide range of conditions that may not have been considered at the design stage or even at mission runtime. This is especially relevant to marine systems where the operator cannot communicate at high bandwidths, if at all, with the MAS. Unanticipated or anomalous issues or events may arise in its state, its environment, or interactions between the environment and mission. As MAS cannot halt for an operator to re-configure it once deployed, the autonomy must find a solution without operator support. This autonomy enables MAS deployment in dynamic environments where it is not possible to have an operator close-by, but these types of requirements drive significant verification challenges. For example, faults triggered by environmental harshness that require a mission re-plan are not always anticipated, so a remedy may not be available. Even when identified in advance, these conditions are difficult to create to verify an autonomous remedy. While all sensors related to vehicle state are logged, it is not possible to log all quantities in the dynamic environment or its interaction with the vehicle.

Mission-specific or payload sensors can be integrated on a separate embedded processor called the 'back seat driver' [18, 19]. As shown in Fig. 8.3, the front seat VCC and the back seat processor divide up the responsibilities of controlling the vehicle. The mission or payload sensor (e.g. synthetic aperture sonar) would be controlled by the payload sensing autonomy algorithms hosted on the back seat processor which interacts with the VCC.

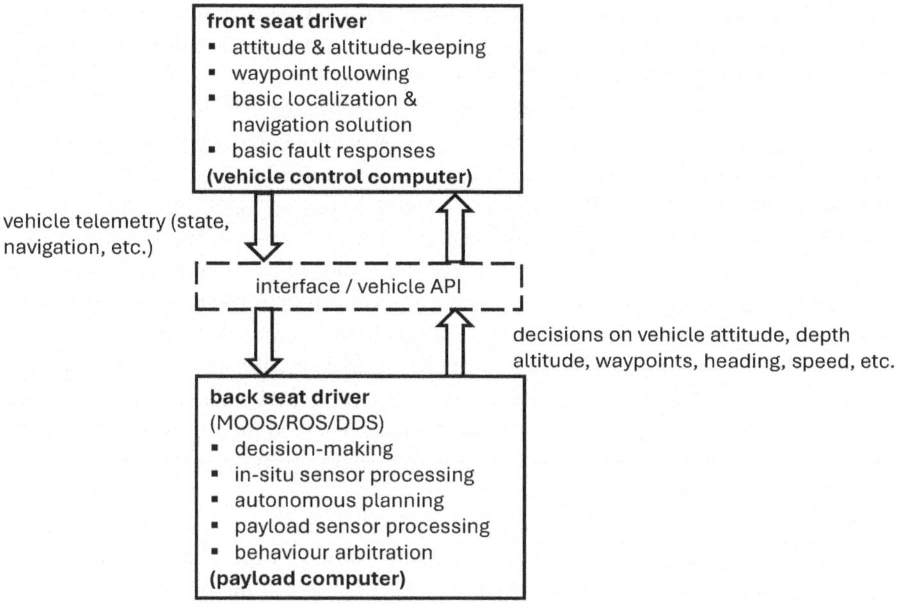

Fig. 8.3 Front seat/back seat architecture for autonomous underwater vehicles. The front seat provides the basic vehicle control, and the deliberative autonomy is on the back seat

This abstraction cleanly decouples MAS basic operations from the mission/autonomy and makes it possible for MAS to support other payload sensors, processing, and missions. Instead of designing a unique platform for each task, the user selects an AUV/ASV platform upon which to build their mission/autonomy. The MAS' inherent operational capabilities (endurance, speed, energy, etc.) limit the missions it can support so this selection must be made with these considerations in mind. While this decoupling simplifies operational autonomy development for developers, it introduces complexity into the verification process. The front and back seat drivers interact across an interface. Verification of the autonomy must take this into consideration and be specific to the shared interface. The VCC, back seat driver, and shared interface must be verified independently and as an integrated whole. Access to VCC information is needed (but not always available) to verify the back seat. The front and back seat may use this information differently, adding to the complexity of the verification.

The naval mine countermeasures (NMCM) use case provides an example of how the operational side interacts with the mission side. For a survey mission, the AUV autonomy software hosted on a payload embedded processor processes the side-scan sonar data stream to detect specific targets with its automated target recognition. For a given detected target, the autonomy opportunistically re-plans a mission to have the side-scan sonar re-scan the target from a different aspect to verify the detection. Here, the autonomy would (1) process the side-scan sonar images to extract a thumbnail image of the target based on a fusion of multiple detectors (e.g. matched filter); (2) decide whether the target is interesting enough for a multi-aspect imaging; and (3) if so it may interrupt the current mission with its re-planned mission and work with the VCC to implement the new mission. In this example, there is in situ processing of an in-water sensor, decisions made, and then actions from these decisions. This mission could also have been performed with an operator who decides whether the target was sufficiently interesting and whether the underway mission should be interrupted for an opportunistic re-imaging of the target. The ability to make decisions and arbitrate conflicting actions are the hallmarks of autonomy, but the lack of an a priori correct answer causes significant problems with verifying the safety and functionality of these systems.

Controllers can range from basic vehicle control to fully autonomous control. In the case of advanced controllers for a complex autonomous mission, their verification challenges are driven by the reduced (or completely absent) operator presence to make decisions or intervene. Controllers are both safety and mission-critical systems.

While the sequence of the nominally required actions is clear, the challenge in the NMCM missions' AUV verification is compounded by conflicting requirements that arise at the decision stage. It amounts to verification of the behaviour arbitrator which is difficult given the range and types of decisions it arbitrates.

In Fig. 8.4, the MAS capabilities (in red) for a mission are ordered in increasing autonomy from left to right along the x-axis. These capabilities could be implemented by (1) the operator given sufficient communication bandwidth with the MAS (far left); (2) a combination of operator and onboard autonomy (centre section), or

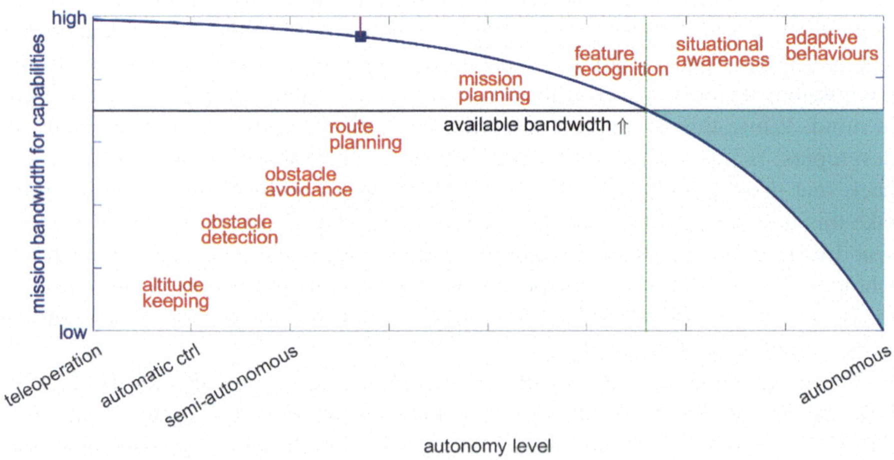

Fig. 8.4 Impact of operator communications bandwidth on MAS autonomy—less available bandwidth implies more autonomy required. The mission bandwidth curve (blue) shows the bandwidth needed to incrementally support capabilities with increasing required autonomy capability from left to right

(3) the onboard autonomy (far right). It is assumed that all capabilities up to the selected point on the x-axis are simultaneously engaged for the MAS mission, so the use of mission planning assumes the use of altitude keeping, obstacle detection, obstacle avoidance, and route planning.

The y-axis is normalized so that the maximum operator communications bandwidth to support any given point on the x-axis lies at the high end. The blue curve represents the degree to which autonomy reduces the required operator communications bandwidth. For example, at the blue square on the curve, there is sufficient bandwidth to combine altitude-following, obstacle detection, obstacle avoidance, and route planning. The required operator bandwidth, to support these four, is reduced by the amount marked by the magenta line due to the increased autonomy at that level. The magenta line (the distance between high bandwidth and the blue bandwidth-with-autonomy line) illustrates the autonomy's contribution to reduce the operator's bandwidth at that level.

The black horizontal line is the available bandwidth for a given environment and communication system combination, while the dashed green vertical line shows the autonomy level where the blue mission bandwidth and the black available bandwidth intersect. At this point, all capabilities to the left can be supported solely through autonomy. All capabilities to the right of the dashed green line require operator control. As the available bandwidth decreases, the operator is limited to higher level control and the MAS must provide more autonomy.

Every point along the autonomy level axis has different verification requirements. To the far left, the limiting case is no autonomy—the operator has direct control of the vehicle and completely drives it, and existing verification techniques suffice. Automatic control can be as basic as closed-loop control about the optimal set point value (e.g. altitude-following). Where defining the set point value is the autonomy's responsibility, the verification ensures the optimal set point is selected. Its verification could be driven by why the set point needs to change (e.g. area with potential seabed obstacles).

Verifying route planning (i.e. autonomously re-plan a mission based on in situ sensor measurements, available maps, and sound velocity profiles) addresses different requirements and different techniques, potentially including provable correctness, route safety subject to the availability and reliability (correctness) of environmental information such as bathymetry, underwater weather, and sound velocity profile (if communications is required). The verification must model the very large span of permutations for a given bathymetry map, seasonal sound velocity profile, and supplementary in situ measurements.

Beyond route planning, verification of the arbitration of multiple behaviours running concurrently as well as the adaptivity of the behaviours in response to vehicle or environment changes poses a new challenge. Adaptive behaviours require exhaustive verification as the operator is not involved at all. For systems with many behaviours, the verification must verify each behaviour independently, all permutations of multiple incrementally integrated behaviours and finally, verification of all behaviours as a whole and any interactions with the operator, such as defining the contexts within which the operator deploys the behaviour.

Given the indeterminate and variable nature of MAS autonomy, the outcome spans a multi-variate space, and the verification must reflect this. The MAS autonomy verification methods must be aligned with the autonomy level or expected operator interaction/decision-making. The MAS architectures that could support this autonomy spectrum are discussed next as the architecture also drives the verification requirements.

8.2.4 MAS Software Architectures

Modern real-time robotic autonomy architectures are built around middleware like MOOS-IvP (Mission Oriented Operating Suite—Interval Programming, MIT) [20], ROS (Robot Operating System) [21], and DDS (Data Distribution Service, Object Management Group) [22]. With MAS, the middleware integrates the hardware and software system components to provide low-latency data connectivity, extreme reliability, and a scalable architecture. For distributed systems like MAS, the middleware is the software layer between the operating system and applications. The system components easily communicate and share data. This simplifies distributed system development by providing a framework so developers can focus on applications rather than passing information between applications and systems. The details of the operating system are thus abstracted from the application.

The advantage of these types of middleware is that prior to in-water verification, the autonomy control software developed in MOOS-IvP, ROS, or DDS frameworks for the mission embedded processor is ported near-seamlessly to the robot with the physical robot replacing the simulator accessed by the middleware. Because the simulated behaviour and the behaviour implemented on the robot are identical, the behaviour's design and function only need to be verified once, and only its interaction with the platform needs to be verified after it transitions to the hardware. Consequently, this accelerates the development and verification spiral of the autonomy and its software implementation.

But this simplification does not affect the difficulty of verifying the platform in its operational environment. The complexity of a robot's environment drives the autonomy required for its interaction and operation within the environment, driving increased verification complexity.

8.3 Marine Autonomous System Operational Environments

The marine environment is amongst the harshest on Earth for robots and consists of two very different environments. The underwater (submerged) environment is closer to aerial environments, supporting three-dimensional motion, ground topography, and properties analogous to wind and weather, while the on-the-water (surface) environment is a dynamically changing plane. Each has different autonomy requirements and consequently, verification challenges.

8.3.1 Underwater

The underwater marine environment's dynamic and spatio-temporal variations are not easily predicted. This means that MAS have extraordinary constraints as autonomous real-time embedded systems compared to those in other environments [23]. The spatio-temporal variability of the underwater environment changes on the order of hours (solar heating), days (weather systems), and weeks and seasons (currents and migration patterns). The environment is dependent not only on time of day or day of the year, but also on where it is on Earth. Most models are applicable to only specific environments (e.g. shallow water near-shore). This lack of accurate models manifests as difficulties with:

Communications: Underway AUV communications utilize acoustic propagation and are characterized by low bandwidth, short range and frequency, and range-dependent attenuation, resulting in poor communication between the vehicle and the operator. The autonomy's verification must therefore guarantee some level of connectivity or quality-of-service supporting safety- and mission-critical scenarios. This is difficult given the large spatio-temporal variability which directly impacts the underwater acoustic communications channel's link quality and packet loss.

Navigation and localization: Since underwater lacks absolute positioning references like the GNSS available above-water, AUVs generally use dead-reckoning methods. These are generally verified based on establishing either an error accumulation rate based on the distance travelled or specific displacement error at specific locations. This challenge is to verify localization and navigation algorithms without a ground truth measure to verify against.

Sensing: The underwater environment suffers from poor exteroceptive sensing, that is, low signal-to-noise ratio, high variability, limited sensing modalities, and short ranges. Most underwater sensors are active, with transmitters that emit acoustic pulses and receivers that detect the transmitted pulse 'returns' which have been distorted and delayed by the environment. From these returns' two-way travel time and knowledge of the local sound velocity profile, an object's range, velocity, and direction could be inferred. The verification challenges lie in the large uncertainty due to the low signal-to-noise ratio and uncertainty about the impact of complex environmental conditions on the local sound velocity profile.

Fault-prone: A physically dynamic environment (extreme motions, temperatures, and pressures) increases the likelihood of faults and failures. This requires not only autonomy but also a fault detection, identification and remedy/recovery (FDIR) system. The verification of such FDIR systems is challenging as the same fault can manifest for different reasons which makes its identification doubly difficult. Similarly, the remedy may need to be specific to environmental conditions or system behavioural state as opposed to a generic remedy for say, a loss of power.

8.3.2 *On the Water Surface*

ASVs on the water surface can be subjected to pounding sea states or wave motions that can be as high 19 m [24], powerful forcing wave periods that can be as low as almost 30s, high winds, and extremely variable temperatures, from almost freezing to 100 °F. The surface environment is different from the submerged environment but no more easy to operate in.

The marine ASV occupies a unique position at the top of the water column and the bottom of the atmosphere. To some extent, its autonomy verification methods can be borrowed from uncrewed ground vehicles for which there is work to date [25]. The communication, localization, and navigation and sensing challenges above-water are not as severe as underwater, and the challenges of verifying these systems are reduced simply because it is easier to observe and interact with them.

8.4 MAS Use Cases

Naval mine countermeasures (NMCM) and arctic survey missions are chosen to represent the state-of-the-art in AUV verification as they are the most mature. The NMCM mission requires mission re-planning, in situ sensor processing, and decisions to be made—all critical actions that require verification if performed autonomously. The NMCM mission is relatively short as the vehicle is recovered daily and the targets (mines) are static. In an Arctic under-ice survey mission, the vehicle is deployed for an extended duration, to the point where both the AUV and the environment have changed since deployment.

Autonomous shipping and anti-submarine warfare missions represent the state of the art in ASV verification.

8.4.1 *AUV Use Case: Naval Mine Countermeasures*

The steps of the NMCM mission with AUVs are shown in Table 8.1. Previously, this was performed by NMCM ships towing sidescan sonar with the neutralization performed by ship-based divers. AUV autonomy has advanced to address (1)–(4) since their initial deployments in the 1990s Persian Gulf conflict. In place of the diver, the operator oversees MLO imagery acquisition and analysis to decide on

Table 8.1 Added value of using autonomy-enabled AUVs for the NMCM mission

Mission steps with MLO		Legacy	With autonomy-enabled AUVs
1	Search (detect)	Ship-towed side-scan sonar	Re-plan missions that adapt to currents, exclusion zones, shallow water, etc.
2	Classify (mine? type?)	Operators analyze side-scan data	Detect MLO in side-scan sonar imagery with automated target recognition
3	Map (localize)		Autonomous mapping with simultaneous localization and mapping option
4	Reacquire and identify	Ship-towed side-scan sonar	Re-plan a mission to reacquire target; refined localization and navigation to increase likelihood of reacquires
5	Neutralize—if mine	Divers	Remotely operated vehicle

whether step 5. (neutralization) is required. Doctrine in some nations requires this so the operator intercedes between steps (4) and (5). With the appearance of AUVs for NMCM, the NMCM ship deploys AUV(s) from a safe stand-off and the AUVs enter the potentially mined waters to perform steps (1)–(4) to gather/transmit back MLO imagery for operators to base decisions on for step (5).

Verification is particularly important for the localization and navigation autonomy and of the identification algorithms for mine-like objects. Localization and navigation autonomy is critical to this mission. If a map of targets is the outcome of the search survey, with the intent that another AUV goes back to reacquire the targets, sufficiently accurate localization and navigation is critical to ensure that the targets are correctly geo-referenced, or the later stages of the mission cannot proceed. The identification of mine-like objects can be performed using matched filtering or machine learning, and its accuracy is affected by sensor type, AUV pose within the environment, bottom cover, etc.

The relatively short duration of the AUV NMCM missions mean that while the environment may change dramatically from one mission to the next, the environment is expected to be relatively stable within the context of a single mission, limiting the complexity of testing and evaluation at the individual mission level.

8.4.2 AUV Use Case: Arctic Surveys

Over long duration missions lasting weeks or more, both the AUV and the environment evolve. The AUV may present faults that evolve to failures, so long-endurance AUV autonomy requires a more complex architecture that includes mitigation and remedies for onboard faults and failures as outlined in the section on MAS control and autonomy. The challenges with localization and navigation and with communication are intensified by the under-ice environment, since the vehicle cannot surface to access GNSS or above-water communications infrastructure and must travel much longer distances. At these distances, 100s of kilometres, underwater acoustic telemetry is on the order of bytes, all the way to the right end of the x-axis in Fig. 8.4—a heartbeat to confirm the AUV is still alive.

The Autosub 2 AUV loss and its aftermath was a pivotal event in AUV operations and verification. It underscored the real, not just assumed, risk in deploying AUVs in dynamic underwater environments.

The National Oceanography Centre (NOC) in Southampton, UK is a world leader in the development of working class (7 m length, 400 km endurance, 1600 m depth rating, 0.1% navigation accuracy) AUVs for ocean observations. NOC's contributions in ocean science include topics that study the correlation between ocean-ice and climate. Autosub 2 was designed to operate in dynamic underwater under-ice environments and had successfully performed under-ice deployments for 2 years. The NOC AUV development team used sound engineering practices, and the system's prior reliability had informed redesign decisions. In February 2005, this system was deployed under the Fimbul ice sheet in the Antarctic and ended up stranded 15 km under the ice sheet. Given the remote location Autosub 2 was never recovered [26].

In the AUV community, vehicle loss is an acknowledged risk. In this case, an independent board of enquiry was struck to comprehensively investigate the loss and reduce the risk of similar losses. As the AUV was never recovered for direct examination, their findings were based on NOC's 14-years (240 missions) of experience [27] with the Autosub AUV class.

They recommended more attention to faults from AUV manufacturing and assembly, as well as improved maintenance and reliability or technical risk assessment processes to support design decision-making. These recommendations motivated research and development in AUV fault tolerance, diagnostics, reliability, and/or its verification [28–35]. Faults and failures may occur not only from defects in manufactured or assembled components, but from MAS interactions with their environment [34].

The environment is a critical factor in verification of under-ice MAS. They are deployed from a surface that floats on the ocean and can change position and state while the MAS are deployed. The MAS must incorporate homing mechanisms to find its retrieval point, even when that point is the same place on the ice it was launched from. This adds an extra layer of difficulty to the verification, both by increasing the complexity of the system and by increasing the complexity of the required environmental modelling.

However, under-ice and NMCM AUV autonomy verification is consistent in assuming that the operator is attempting to support the system with local assets. This is not always the case for surface vehicles.

8.4.3 ASV Use Case: Autonomous Ships

Autonomous ships may be controlled remotely over RF and satellite communications networks. In addition to the onboard decision-making autonomy that must be verified for AUVs, verification for surface vessels must include communication with land-based crew as well as avoiding collisions with other surface vessels.

World economies are connected through global trade with a predicted 30% increase in maritime trade by 2030. Such an increase in marine traffic means an increased risk of marine incidents and accidents at sea. To date 90% of marine casualties and incidents are attributed to human error. Autonomous ships (a special case of ASV) are one way to mitigate this. These ASVs are emerging as a transformation for maritime shipping. With no onboard crew, they are remotely monitored by land-based crew over satellite and radio communications and operate at the other end of the scale on Fig. 8.4 from AUVs. This reliance on high bandwidth and frequent communications with their remote operators to perform autonomous behaviours with oversight from their operators means that verification must assess whether the system has sufficient communication range and bandwidth to monitor sensors and allow remote control when human intervention is needed. This link between the ASV and its virtual captain is vital as there are consequences if the operator does not intervene in a timely manner for some scenarios. While these ships are on the water surface (i.e. with antennae at very low altitudes), the weather and satellite position can strongly affect performance. This complex interaction is very difficult to reproduce for verification as it cannot be controlled, but it is a priority as it is a safety-critical system even for ships with high levels of autonomy.

Traditional maritime shipping practices were not designed for autonomous operations, but the community is slowly adopting autonomy. For autonomous ships, their autonomy levels [36] span 0 (traditional ships) to 6 (supervised operations where decisions are made and actioned by the system during the mission). At autonomy level 6, the ASV determines the consequences and risks autonomously without human intervention, meaning that verification of the autonomy itself is also essential.

Beyond the usual ASV operational requirements and challenges, autonomous ships must also interact with crewed and uncrewed vessels and buoys, detect other vessels, and adapt to avoid collisions by observing maritime traffic rules. Given all this, the system must be verified for safe operation to avoid damage to property and people.

On the water surface, navigation protocols are clearly defined by the International Maritime Organization which provides the Convention on the International Regulations or COLREGS [37] (rules that ships, and other vessels must observe). ASVs and autonomous ships fall under this as they must actively work to avoid collisions.

8.5 Marine Autonomous Systems Verification

8.5.1 MAS Models for Verification

There are two components in a verification analysis—the system's performance or behaviour and the properties or requirements it is to be verified against. Models are required to represent them in a form that supports the verification. A system model captures the behaviours *or* performance that must be verified against *requirements, specifications, or properties* which must also be modelled.

A model is an abstract description of the MAS, its environment, and its interactions with the environment which describes correct operations in the context of missions like those described above regardless of the implementation. The specific model depends on the verification method and its formality.

8.5.1.1 System Models for MAS

Formal models of state machines and transition systems are used to model system behaviours. Mathematical methods like model checking and theorem provers require rigorous mathematical or logical language.

These models can be used to generate the MAS design, or the MAS could be developed some other way, and the formal model used to verify the MAS' behaviours. The model is a description of the correct MAS behaviour in its operational environment and forms the basis for the MAS' verification.

Many tools exist to create informal or semi-formal models of autonomous systems and their behaviour. Model-based systems engineering uses languages with semi-formal semantics like Unified Modelling Language (UML) [38] based tools to model the components, classes, and interactions (sequence diagrams) in a system [39]. Languages like SysML [40] are commonly used to design and implement software. SysML is especially appropriate for use in modelling autonomous systems as it describes the system hardware and software (unlike methods that are too abstract to capture instantiations) using a finite set of diagrams and symbols to create state machine diagrams for example. UML tools capture the system structure and behaviours as well as data structures for hardware and software, but UML-based methods are not appropriate for algorithms or systems analysis.

Tools like MATLAB/Simulink [41] are used at the design stage to create a model for systems analysis. This model may later be discretized to be a verification model [30]. Once the system model is abstracted and created in this way, there are tools that automatically generate code which satisfies the designed properties and constraints [8]—the model itself is not formal, but guarantees can be made that the system itself matches its model. However, this can be difficult for more complex algorithms as they are more difficult to abstract well. This is usually applied only to a MAS subsystem, for example, guidance and navigation as it is intractable to create a model that can describe the whole MAS. With such a model, it is possible to apply mathematical analyses of correctness, completeness, and optimality.

But creating the system model alone is insufficient. The requirements must also be expressed, preferably in a semi-automated manner so verification can be efficiently performed.

8.5.1.2 Requirements Models

As with other complex systems, verification methodologies change as a MAS component, subsystem, or autonomy software module progresses in its development and increases in maturity. Engineering design review addresses requirements capture, review, testing, and verification to ensure efficiency and compliance in creating the system [42]. This process formalizes and manages the process to verify a MAS component, subsystem, or autonomy software module. But autonomous behaviour requirements are defined in a context of deliberate uncertainty, driving non-determinism related to the environment in which the system will be operating and the specific states the system moves through as it operates.

Currently, representing autonomous system specifications in a formal language [43] is a barrier. Specifications are usually expressed in natural language which can be ambiguous and less amenable to formal analysis. To use formal methods to verify the system, informal requirements must be formulated in a formal specification language. Such languages have clearly defined syntax and semantics, that is, strict rules on valid statements that can be clearly and unambiguously defined and interpreted. In this form, requirements can be subject to mathematical analysis, are machine readable, and can be used for automated reasoning.

In contrast to the formal models of the system (rather than its behaviour), formal models of autonomous system requirements and properties often utilize variants of temporal logics. These specify temporal behaviours like ordering of events, timing, and avoiding deadlocks (where two processes share the same resources and each prevents the other from accessing the resource, both processes stop). The most basic temporal logic is linear temporal logic (LTL) which operates on Boolean signals in discrete time. Spatial Temporal Logic (STL) is used to specify real-valued signals over continuous time including time constraints. STL uses robustness semantics which give a quantitative number on how robustly a formula satisfies a temporal logic formula rather than a binary true/false or pass/fail evaluation. Therefore, STL has a syntax to specify behaviours and a metric to evaluate its compliance which increases its applicability to cyber-physical systems.

The Formal Requirement Elicitation Tool (FRET, [44]) is open source and developed by NASA Ames to write, understand, and formalize requirements. Requirements are written in FRET in a restricted natural language called FRETISH [45]. This takes the articulated natural language expressed requirement and decomposes it into fields (e.g. scope, condition, component shall, timing, etc.). FRET also has a simulator, as well as textual and diagrammatic semantic explanations.

ARSENAL [46] converts natural language requirements into formal models using SAL (Symbolic Analysis Laboratory) [47] which is a formal language to specify transition systems in a compositional way and in linear-time temporal logic. An additional benefit is the ability to automatically generate monitors on AUVs and ASVs to perform self-verification (discussed in Self-Verification subsection).

With a system model and formally expressed requirements in place, verification methods may now be addressed.

8.5.2 *MAS Verification Methodologies*

Verification methodologies as applied to safety and mission-critical AUV and ASV systems vary in the level of formality required for their systems and requirements models. With greater formality for both, the verification is more efficient and complete (exhaustive).

Given the complexity of MAS and the unstructured environments they interact with, instead of being a separate engineering requirement after the fact, system verification should be required at all stages of MAS development and operations: (1) initial design; (2) detailed design; (3) software implementation; (4) in-water; and (5) on-going through its operational life. Verification beyond the initial stages is required given that the MAS' hardware and software (vehicle and autonomy) evolve over their operational life from software/firmware, actuator, and sensor upgrades.

The appropriate verification method depends on the MAS (subsystem or autonomy) development stage and consequently, the acceptable abstraction level. This makes a case to create compositional verification frameworks for not just an AUV/ASV class but for individual vehicles.

Functional components (sensors, actuators) are verified with informal models and frameworks (e.g. ROS) given the parameters that must be modelled. Decisional components (planning, learning) may be verified with formal methods as they can be expressed with formal unambiguous semantics. Even a basic MAS mission will involve multiple functional components that interact with deliberative components.

8.5.2.1 Formal Methods

Formal methods like model checkers and theorem provers provide exhaustive verification at the cost of the greatest abstraction. Greater abstraction means more assumptions about the modelled system, its environment and its interaction with the environment. This is most useful in the preliminary phase, assisting with system concept development and requirements definition, and is often used for mission and safety-critical algorithms (e.g. obstacle avoidance).

As a system matures beyond a concept and system details must be modelled and verified to a finer granularity, model checkers and theorem provers become challenging due to the system and requirements complexity. This leads to a *state explosion* where the complexity is too great for the computing resources available.

Model Checking

In model checking, systems are modelled as state-transition systems and properties or requirements with a logical language. Model checking (MC) is a principled approach to system development that applies formal methods to verify the correctness of a system design based on models. After an appropriate model for the system is developed, the model must be adapted to the model checker's formalism so requirements can be verified against the model to detect violations and provide insight into its cause. MC supports prediction of potential violations by reasoning on potential behaviours. When the model violates a requirement, the MC presents at least one counterexample to base further analysis upon. Formal verification methods will find an erroneous state in the system if it exists, and if it exists it can be shown that this is an issue in the system model.

Additionally, MC supports analyses to reason on competing adaptation strategies which are relevant for MAS autonomy. The strength of MC, with the verification problem as posed, is that it is possible in principle to achieve verification goals through exhaustive scenario-based testing of all possible system responses to an event. As shown in the earlier sections, there can be near-infinite combinations of responses, so a scenario-based approach is not tractable. Completeness (NP-complete), probability, and optimality are satisfied. This approach is often applied in safety-critical applications like MAS and autonomous systems in general.

Model checking tools available for industry scale systems include: PRISM [48], MCMAS (Model Checker for Multi-Agent Systems) [49], NuSMV v2 (FBK-irst, CMU, University of Genoa, University of Trento) [50], Alloy (MIT) [51], BLAST [52], UPPAAL [53], and UPPAAL STRATEGO [54] and have been applied to fault management, multi-agent communications, and above-water operation verification for MAS.

Faults

Model checkers can be applied at design time (Table 8.2, step 2) [8, 9], so design errors are detected in the early development stages to ensure higher reliability. They can also be applied in diagnostics and refinement of mature systems [29, 30] and have been applied at runtime both to support fault management and to assess operational decision-making.

The work of [28, 29] created an Integrity and Fault Assessment System (IFAS) for the Autosub 2 AUV to verify controllers (e.g. collision and obstacle avoidance). Natural language programming with 'system English' or *sEnglish* [55] abstracted continuous AUV states to discrete ones. This formulated AUV functionality at different abstraction levels and linked the labelled transition system's discrete abstractions both to Stateflow/Simulink and to the input language representation

Table 8.2 MAS verification stages and methodologies used. The methodologies are ordered in decreasing formality and abstraction from top to bottom

System development stage		Activity	Verification methodology	
1	Requirements capture	Develop concepts; capture requirements	• Model checking • Theorem provers • Runtime verification	Formal
2	Design	More granularity in model	• Model checking [8, 9] • Software-in-the-loop simulators	
3	Implementation	Real-time embedded software	• Hardware-in-the-loop simulators	Informal
4	Verification	In-water deployment	• Runtime monitoring	
5	Deployment	In-water deployment	• Self-verification • Compositional verification	
6	Refinement/more maturation	MAS subsystems and autonomy upgrades	• Self-verification • Runtime verification	

of the MC. In [29] the authors modelled standard AUV sensors like acoustic Doppler current profilers (ADCP) and forward-look echosounders as sensors for AUV collision and obstacle avoidance. In this approach, component failures can be captured as discrete events. Reliability can be verified under assumed component reliability and within the accuracy bounds of the discrete abstraction for the environment and controller feedback. While there were no illustrative examples, their verification system can be applied to network communications between the AUV subsystems and behaviour of human controllers.

The work of [30] built on the IFAS methodology of [28, 29], adding fault injection, which enabled exploration of fault tolerance verification and self-diagnosability for the Autosub 6000 AUV (a later variant of Autosub 2). Deliberate faults were designed and injected into the correct AUV model to create a mutated one. Then, the MC was used to analyze the mutated AUV's performance and compare it against the correct model's performance. This made it possible to infer system properties for correct execution and when faults arose and enabled them to evaluate the AUV's robustness to faults, ability to diagnose them and to recover from them.

The state space explosion problem limits the usefulness of MC methods in analyzing MAS that require complex logical processes to perform decision-making tasks. Model checkers are limited in their expression due to the restrictive logics they use, requiring stringent assumptions about physical components and environments. This problem can be partially addressed by decreasing abstraction by decomposing a system with an intractably large state space into multiple concurrent subsystems with tractable state spaces [56, 57] (also referred to as compositional verification). This suggests autonomy software should be designed so that the state space is easily decomposed into a tractable number of subsystems/subspaces. The outcome, though, is a heuristic (non-optimal, incomplete) solution.

Communications

The communication and control system of collaborating AUVs used in environmental monitoring was analyzed using UPPAAL [9]. Their AUV control architecture is based on the open-source MOOS-IvP (Mission Oriented Operating Suite—Interval Programming) middleware [20]. MOOS-IvP's star-shaped control topology makes it relatively flat (i.e. non-hierarchical) and MOOS vehicle capabilities are modular in structure. Both considerations make it relatively straight-forward to map the model to the UPPAAL automata, as UPPAAL does not support hierarchical structures. Models of the AUVs and their sensors were created for high bandwidth (surfaced, communicate with ground station) and low bandwidth (submerged, communicate with collaborating underway AUVs) communications. Their methodology injects faulted states that occur probabilistically to verify that the correct re-allocation of tasks occurs as brokered by the ground station occurred when a surfaced or submerged AUV stopped communicating with the AUV ground station.

Above-Water Navigation

Given the complex and populated environments ASVs operate within, they must have provably correct navigation algorithms and control strategies. Shokri-Manninen et al. [58] developed a navigation algorithm which complies with COLREGS to avoid collision with another ship and minimize fuel consumption. Their approach models the navigation problem as a stochastic priced timed game (from game theory) with UPPAAL STRATEGO which uses reinforcement learning and model checking. Notably, their navigation algorithm integrates wind and currents. Their navigation strategy was verified and optimized to safely reach the goal with minimal cost (fuel).

Theorem Provers

Theorem provers rely on specifications written as mathematical statements, where the tools apply proofs to these statements. Tools like KeYmaera X [59] and Isabelle [60, 61] perform the mathematical proofs.

These methods have been applied to AUV control architectures and underwater communications, using time domain multiple access (TDMA) where each AUV has a time slot to transmit in a cycle. The objective was to design distributed connectivity-preserving motion constraints for AUVs in a network that only knows the position of the other AUVs at the beginning [62]. The authors used the KeYmaera X theorem prover to develop formal proofs to verify their algorithm. While earlier motion constraints were built around, for example, flocking behaviours which is specific to an application, their formal verification method can interface with multi-agent path-planners, multi-agent searches, or formation control. With this method, a planner devising a search with guaranteed connectivity can create the plan so that algorithmic constraints are satisfied to prove desired safety properties. They can model packet drop-out (common in underwater communications), noisy measurements, and AUV tracking errors. In the process, they discovered edge cases where AUV agents transmit out-of-order with the consequence that they are inadvertently removed from the network. Such errors would have been difficult to detect without a formal approach.

MAS with hierarchical control architectures are difficult for verification tools like UPPAAL as there is a necessary step in the workflow to adapt the model so that it can be compatible with UPPAAL. However, hierarchical architectures are needed to manage the complexity of MAS controllers. Given the importance of verification, semi-automatic verification should be built-in to all phases of the controller development and operations. Tangirala et al. [8] developed their AUV controller with safety verification from the start using a commercial product, Teja NP. They developed a converter for their hybrid state machine representation into a form that was compatible with UPPAAL. They do not describe how their converter is designed or implemented, but they used UPPAAL for formal verification of AUV safety and performance.

Runtime Verification

Runtime verification (RV) is a formal method that observes system execution traces against formal specifications. RV is a practical application of formal verification, providing a less ad-hoc approach to complement conventional testing and debugging. Compared to formal verification, RV gains applicability by sacrificing completeness, as not all execution traces of a system can be observed.

Runtime verification inputs are the system model of its performance or behaviours to be checked and the properties, specifications, or requirements to be checked against during system execution.

The expression of the system properties or requirements can be in a formal specification language as discussed earlier. The advantage of verifying the system at runtime is that the system is able to take corrective action if a property or requirement was violated. Runtime verification consists of the following steps:

1. Monitor synthesis: creates a decision procedure for the property or requirement; this monitor consumes events from the running system and gives verdicts on the satisfaction of the requirement based on the history of received events.
2. System instrumentation: generates the events, requirements, or properties that are to be monitored.
3. Execution analysis: the monitor analyzes the execution; this can occur synchronously during execution or logged for examination after the run.

The advantage of monitors is that they are easy to implement and agnostic to the MAS design procedures used. While incorrect assumptions may lead to a faulty design, the monitor is decoupled from the design decisions and may more easily detect a subsystem failure. Onboard frameworks like JavaMOP [63] are designed to administer the runtime verification and can use various formalisms or logics for representation.

The research of [64–66] studies runtime *quantitative* verification (RQV) for AUV autonomy. RQV is a formal method to establish correctness, performance, and reliability (verification) for systems with *stochastic* behaviours. It has been successfully applied to dynamic power and cloud infrastructure management The interaction of AUVs with their dynamic and uncertain marine environment is stochastic and well-captured with stochastic means [16, 67, 68]. Using the PRISM probabilistic model checker to verify the approach, Calinescu et al. [64] proposed the integration of formal model checking with RQV at runtime to predict and identify requirement violations, plan strategies to prevent or remedy violations, and prove that the adapted software complies completely with requirements. In this way, RQV continually adapts and verifies its stochastic models in response to environmental (or other) changes and does so in a way that guarantees requirements are completely met. The authors' Markovian stochastic models had carefully selected abstraction levels to prevent state explosion so an initial proof-of-principle could be established for their proposed technique.

In [65], researchers directly addressed MC state explosion and general algorithmic efficiency to their RQV methodology by applying caching, lookahead, and nearly optimal reconfiguration. These are standard ways to increase computational efficiency in software. They integrated their runtime model checking to an AUV modelled with the MOOS-IvP middleware to create RQV-MOOS. Their results indicate that caching had the most impact in increasing efficiency when AUV state changes were small. When they became larger, lookahead in combination with caching, yielded the best efficiency gains at the cost of a larger cache. The combination of nearly optimal reconfiguration and caching achieved the all-round best results.

Calinescu et al. [66] focused on the RQV performance given changes in the AUV system requirement, the environment and the mission. This means the system model was verified on an on-going basis as triggered by changes it adapted to and was irrefutably proven as correct.

ROSRV [69] is a lightweight runtime verification that can be integrated without altering the core ROS (which, due to its deliberately open and dynamic framework, has no security protection). ROSRV addresses safety and security issues with a transparent monitoring infrastructure which intercepts and monitors commands and messages being published and subscribed. Operators provide formal safety and security specifications, and monitors are automatically generated and integrated into the system to ensure the robot's safety and security. Safety properties are defined in a formal specification language and ensured by automatically generated monitors.

Recently, Perez et al. [70] built on Huang et al. [69] to develop a formal approach to generate runtime monitors for autonomous robots with a more expressive specification language and guarantees on generating the monitors in real time. They use FRET with Copilot (a runtime verification framework). FRET specifies requirements with unambiguous semantics which are automatically translated into temporal logic formula. This was implemented on the Robonaut2 humanoid robot which collaboratively performs tasks on the International Space Station with the astronauts. With the R2U2 verification framework on the Robonaut2, faults were efficiently identified in real time.

To date, there are no reported MAS using runtime verification in ROS. This is an area that could be explored to verify MAS performance at specific sea states or novel navigation algorithms based on environmental features. While formal specifications for these autonomy capabilities could be expressed, the complex interaction of the MAS with its environment could not be verified in much detail using formal methods.

After an algorithm has been verified through formal methods, it is implemented as embedded real-time software on the AUV and that implementation's correctness is verified. It is impractical to use formal methods to verify the correctness of an in-water implemented system due to the sheer volume of variables and constraints that must be modelled. At this point, simulators and emulators are used. Simulators are less abstract than formal methods and emulators are even less abstract.

8.5.2.2 Informal Methods

Simulators include software- and hardware-in-the-loop (SITL and HITL) approaches. The verification evidence from simulators is not complete or exhaustive so they cannot provide adequate mission-critical or safety guarantees. Simulators complement the limitations of formal methods.

A simulator creates an artificial environment with the mission (autonomy and task), vehicle, and environment variables in the deployed MAS. An emulator mimics the hardware and software features of the deployed MAS. Simulators do not emulate the actual hardware that hosts the software. Since simulators create only software environments, they can be implemented with high-level programming languages. They only mimic system features that can be configured or definable using software, whereas emulators mimic hardware and software. Emulators occupy the middle ground between simulators and the actual MAS itself. A well-designed emulator is very close to the actual MAS with the correct responses achieved at the interfaces between the vehicle and autonomy control nodes, sensors, actuators, and environment.

SITL and HITL verification of vehicle and autonomy algorithms *mitigate the risk of experimental or prototype software that controls high-value MAS' deployed to dynamic underwater environments.* For modern MAS development, this is a necessary step prior to in-water verification. The instantiated algorithms and code must have as much risk mitigated as possible before the system goes into the water. The only uncertainty that should be left for in-water verification is the interaction with the environment. Verification at this stage accelerates development and reduces verification costs.

Software Tools

With software simulations, a software-defined model of the MAS (or its subsystems) operates in a virtual environment as a function of time, and the output is used to verify whether the system functioned correctly for a given scenario. The purpose of these simulators is to capture relevant aspects of the MAS, like its response to controllers (vehicle or mission/payload) interacting with the environment or the ability to easily design (plan) test scenarios and environments (e.g. currents, bathymetry) towards verifying the AUV/ASV and its mission.

Simulation of controllers is like testing in that it is limited to showing defects since it can only consider a small fraction of the MAS state space. Widely used robotic software simulation tools include programming platforms to design and analyze systems like MATLAB and its Simulink Toolbox [41] and GNU Octave [71] (open source) and the 3D physics-based Gazebo [2, 72] (open source).

Robotic middleware like MOOS-IvP and ROS have embedded simulators in their framework which model aspects like vehicle kinematics and controllers and may support verification of MAS dynamics in marine (under- or above-water) environments. Integrating simulators as plug-ins through interfaces for MOOS-IvP or ROS enable the MOOS-IvP or ROS framework (software) to be ported almost directly to the robot's embedded mission processor with the physical robot replacing the simulator across the interface for in-water verification.

Recently, there has been growth in marine robot simulators for controller development and verification. MATLAB/Simulink has the marine systems simulator (a third-party developed library for marine systems like AUVs, ships, and floating structures with guidance, navigation, and control algorithms for real-time simulations) [73, 74] as well as a well-developed third-party model developed by their student competition team [75, 76]. MATLAB and Simulink can generate C-code for implementation but it is still challenging to integrate it into a node-based publisher-subscriber architecture like ROS or MOOS-IvP.

UWSim [77] is an underwater simulator that runs missions for multiple marine robots operating underwater or on the surface. It has a visualizer so the robot(s) can be seen performing their mission in a marine environment and simulates the sensors, and is used to test control and perception algorithms prior to an in-water deployment. UWSim is available [78] but no longer supported.

The UUV simulator is a Gazebo plugin [79, 80] with basic hydrodynamics but it does not account for buoyancy or true underwater sensors. It was designed to support development of the ECA Group A9 AUV.

The work of [81] created an open-source underwater robot SITL simulator by extending simulators from ground and aerial robots to underwater ones. The authors captured the hydrodynamics of the underwater robots and simulated underwater sensors (scanning sonar and echosounder altimeter) and thrusters down to the level of sensor bias and noise. This SITL simulator can also be used to simulate path-planning and risk management. Their intent was to develop and verify guidance, navigation, and control algorithms for underwater robots and to do so with the exact software that would be running on the robot.

The USV physics-based simulation model of [82] accelerates the autonomy design, development, and validation for the ASV/Thales Halcyon implemented in MOOS-IvP. Uniquely, their environment has a sea surface where wind and wave-induced forces are modelled so the simulation can incorporate more realistic sea state conditions. The USV dynamics are modelled through the usual six-degree-of-freedom equations of motion (hydrodynamic derivatives from THARBM [83]). The hydrodynamic damping and added mass terms are also captured in the USV dynamics. To model the USV interaction with their sea surface environment, speed-dependent response amplitude operators were derived from the geometry of the wetted USV hull. Spectral wave models and a current influence spectrum were used to generate omnidirectional surface waves which include ocean swell, local wind, surface current, and shallow water depths. These interactions are especially relevant for USVs as they are smaller and thus more susceptible to significant excitations from a sea state compared to a surface ship.

Stonefish [84] is an open-source marine (underwater, surface and aerial) robotics simulator that captures hydrodynamics to target control and underwater vision applications with manipulators. It notably supports robot hydrodynamics, which Gazebo does not, and, uniquely, incorporates marine robot dynamics, vision, and underwater sensors. Their visual rendering of the robot in its environment is reported to be better than Gazebo's.

SITL simulations have been applied to create AUV models to efficiently develop both vehicle and autonomy controllers [85] for AUV formation control using potential field methods planning methods. As the AUVs were submerged, the authors also applied communications constraints through a potential field model. SITL results were consistent with the later stage in-water testing results.

SITL simulation has also been applied to verify real-time underwater sensor networks [86] deployed to monitor sea water quality, which enables improved AUV underway communications. The objective was a sensitivity analysis to identify the important parameters for their requirements related to duplicated packets, delays, and power consumption so that they could select parameters that should be studied during the in-water stage. Their SITL results helped guarantee the network's expected behaviour and reduce the amount of in-water verification necessary.

At this point, verification with model checking, runtime verification, and software simulators do not address the implementation of the real-time embedded autonomy software on the actual MAS.

8.5.2.3 Hardware-in-the-Loop (HITL)

HITL simulation is a half-way measure which allows some of the system hardware and all the software to be integrated through a dynamics simulator to capture aspects of testing in the environment that has been used for verification of embedded real-time software for over 30 years. The MAS controllers on the vehicle and payload processors are fundamentally embedded real-time software. However, the application of HITL to MAS verification is more recent.

This approach reduces the abstraction of software-only simulation as actual system hardware and software are incorporated to produce more realistic system responses and mitigate the risk of costly at-sea failures. For an AUV, this consists of integrating the physical AUV processor (with the actual vehicle software controllers installed), state estimation algorithms, and attitude and heading reference system.

HITL simulators verify that an autonomy module is correctly installed on the mission or payload computer and correctly integrated with the vehicle computer, as well as enabling scenario-based testing to verify the correctness of the detailed algorithm. Integration with the vehicle computer includes communications between the vehicle and mission computers that all mission nodes are correctly networked, and the mission decisions and commands to the vehicle elicit the correct vehicle response. The implemented mission/payload algorithms can be 'rehearsed' and verified in a way that is only one step removed from deploying the vehicle in-water. This mitigates operational risk prior to in-water deployment.

Rather than being integrated with the actual vehicle, the control system is integrated with the HITL simulator. The feedback from in-water sensors like altimeters, depth sensors, and the actual AUV motions is simulated by high-fidelity AUV dynamic models with sensors and actuators. This type of HITL simulation verifies control or mission algorithms. Comprehensive testing is possible—especially of failure scenarios to mitigate risk. It performs the simulation in real time.

HITL simulators have been applied to development of AUV vehicle and mission (autonomy) controllers [87, 88].

MARS [89] is a HITL simulator which models AUV swarms applied to environmental monitoring for fish swarms and vegetation. This helped to design and verify efficient collaborative behaviours to achieve the desired coverage rates. The cameras were developed with filters to simulate perception through suspensions in the water column. Their environment model has vegetation and fish swarms as targets as well as a temperature profile. What is noteworthy about this simulator is the complex environment and targets they were able to capture. This simulator was integrated with ROS through ROSJava and a generic TCP/IP interface.

HITL simulation was used to design and implement the embedded software control architecture for an autonomous underwater robot [90] with onboard autonomy and intelligence. The safety, high performance, and robustness verification was performed in real time. Their system uses the actual VCC, real actuators, and emulated sensors and a dynamic simulator of AUV forces and motions.

Vehicle control computer, actuators, real communications, inertial sensor simulators, and depth sensors simulators are used to simulate the real condition. Their HITL emulator was used to develop and verify the controllers for the autonomous underwater robot.

HITL has also been used to verify AUV obstacle avoidance algorithms [91]. The authors use the MATLAB real-time interface to integrate their real-time dSPACE DS1104 R&D controller board with Simulink for AUV dynamics, obstacle avoidance, trajectory planner, controller, and actuators dynamics. This hardware-software integrated simulator was applied to verify the performance of their multi-point potential field obstacle avoidance algorithm, realized in software on their vehicle controller. In this way, the vehicle and environment are abstracted, and the emphasis is on the AUV guidance, navigation, and control. Other groups verified underwater glider guidance navigation and control [92].

Underwater acoustic communication, an enabling capability for AUVs, has also been verified using HITL methods [93]. They emulated an underwater communications network by integrating an underwater acoustic transducer with Labview for the data acquisition and MATLAB to realize the signal transmission and reception. They were able to verify the efficacy of their novel physical layer modulation and de-modulation methods and verified their HITL results with in-water testing.

A novel OFDM/CDMA communications network that ran on full-duplex software-defined underwater acoustic modems detects, localizes, and tracked mobile underwater targets with multiple communicating AUVs [94] was verified through UnetStack's HITL simulations (UnetSim) and multiple in-water trials. Messages on AUV/modem location, speed, heating, etc. were securely and efficiently exchanged amongst the AUV collaborative.

8.5.2.4 Self-Verification

The multiple subsystems, sensors, actuators, and algorithms/software of AUVs and ASVs are often individually and incrementally developed where refined versions build on earlier ones. Over the course of a MAS' lifetime, its vehicle controllers, mission sensor, navigation solution, or mission autonomy algorithm will be upgraded as these evolve at a rapid pace. This means its systems' verification must be extended beyond its initial production and deployment and be on-going with a system model that is constantly evolving.

Self-verification can be thought of as a form of data-driven verification. It is similar to the runtime verification discussed earlier, but does not examine the intractably large state spaces of model checking. Instead, it directly analyzes select execution traces running on the MAS. While runtime monitoring is not as extensive as model checking, it scales relatively well to complex systems and improves the confidence of the resulting system analysis as it avoids the challenges of developing accurate system mathematical models.

Prior to initial deployment, self-verification provides developers with an extended runtime verification stage with more time and in-water opportunities for a more complete verification [95]. When the MAS are operational, self-verification should keep pace with the MAS' upgrades and changes in basic vehicle and mission systems as well as their ever-evolving autonomy.

Self-verification requires a lightweight onboard verification system that may require a separate embedded processor. Methods that reduce the state space or perform automatic lightweight proofs are beneficial. A promising workflow to generate monitors for ROS-based systems has been reported [70]. It uses a formal approach to generate onboard runtime monitors from requirements written in FRETISH. They use the lightweight runtime verification framework, Copilot [96] to generate the monitors, and use Ogma to integrate FRET's output with Copilot.

8.5.2.5 Compositional

As mentioned earlier, MAS are hybrid systems with continuous and discrete states. They are the sum of many modules (or subsystems, nodes, processes, etc.) executing concurrently—which adds to the complexity and state space explosion problem. Each module type can be at a different autonomy level (see Fig. 8.4), potentially requiring different verification methods. Compositional verification acknowledges this and aims to create frameworks that integrate or unify different verification methods for a system.

Verification methods can be classed as either static or dynamic. Static methods like model checking and theorem provers verify the system behaviour prior to its execution. They are applied to a system's model or software. Dynamic methods examine single executions of the system as with software-in-the-loop, hardware-in-the-loop, and self-verification.

The sheer complexity of real systems precludes using formal analysis to verify the entire system. Complexity also creates challenges for software testing, as generating sufficient test cases to span the full operational space is intractable. MAS are designed and built with modules that include but are not limited to navigation/localization, ballasting, propulsion, attitude and altitude control, and communications. Within the MAS, each module runs concurrently as a process. Running multiple modules concurrently exacerbates the state space explosion problem. In these conditions, compositional verification [97, 98] using a *divide-and-conquer* approach to decompose a system into multiple smaller concurrent processes to exploit their locality properties can be effective.

The system properties come from the properties of its components and their composition. Then a detailed verification method is chosen so that the properties of the constituent systems/processes are representative of whole system. Distinct modules often need different verification methods. For example, waypoint following requires software testing, whereas adaptive behaviours are more appropriately addressed through formal methods. With autonomous systems, the verification solution is an integration of the appropriate methods [99]. The state-of-the-art does not yet have many frameworks to link disparate verification methods within a complex system. Early work towards this [99] proposes a high-level first-order logic as a unifying language to express assumptions and guarantees. Contracts expressed in first-order logic guide the verification method for each node. A contract for an individual module describes its assumptions and the guarantees it provides [100]. The contract can also verify system global properties through compositional reasoning methods.

Deductive (Vice Algorithmic) Verification

A formal verification approach developed in [101] integrates simulation-based tests and formal verification. Numerical simulations performed in GNU/Octave were integrated with theorem provers (Isabelle). The robot dynamics are modelled as a system of the usual six ordinary differential equations for each degree-of-freedom, and their obstacle avoidance algorithm is automatically verified by the Hoare-auto tactic [102]. The target platform for this work is the ASV C-Worker 5 autonomous surface vehicle.

The difficulty of selecting methods is exacerbated both by the range of methods available and by external constraints driving which methods are practice to use in specific applications.

8.5.2.6 In-Water Verification

Model checkers and theorem provers can be expensive if open-source versions are not adequate. Similarly, software simulators can be less costly if open-source ones are adequate. The cost for these items would be similar for most autonomous systems regardless of domain.

High-fidelity software simulators reduce the needed in-water test cases by providing insight into the detailed performance space. If the boundaries of the performance space can be identified in software, then in-water tests can focus on gathering evidence there. Because in-water test weather is not controllable, these tests cannot capture all weather conditions but they can validate the simulation models to increase their verification range.

MAS are especially resource intensive to develop, usually requiring HITL simulators specific to a particular MAS. In-water testing happens after the algorithms and the software are verified. Initially, the system is tested in a structured environment in a sufficiently large indoor water tank to verify communications algorithms between nodes (operator and MAS or between MAS) or limited motion. Next, MAS are deployed for limited more costly and logistically complicated tests in a basin or lake to verify performance and behaviours on a smaller scale implementing runtime monitoring and self-verification. This requires the use of jetty-side infrastructure like small boats and cranes. Finally, the MAS are ready for large scale verification with all its vehicle systems and autonomy engaged in a sea or ocean environment, enabling more thorough assessment via self-verification and compositional verification. This requires an ocean-going ship and crew to test in deeper waters offshore.

Given the considerable expense for in-water tests, MAS autonomy verification should be dominated by evidence gathered prior to going in the water. In order to apply these techniques appropriately, both the test case selection and generation and the metrics to be used must be clearly defined.

8.6 Other Topics

Because exhaustive testing is not feasible, test case generation and verification performance metrics become critical. Responses to known edge cases in safety-critical behaviours that must observe constraints should be verified, so identifying those test cases is important. Because it is not possible to verify the MAS across the entire operational envelope of potential environments, the performance of the verification process itself should be quantified in terms of efficiency, effectiveness, and coverage.

8.6.1 Test Case Generation

The interaction between multiple autonomous behaviours or vehicle priorities also drives verification difficulty, as it is not obvious how these can be predicted or tested. Methods are needed to generate scenarios for these complex interacting controllers at critical performance points.

Mullins et al. proposed a method to generate test scenarios with emphasis on critical transitions in performance modes (events like obstacle collision and successful mission completion). A scenario is defined by the input parameters for the mission and environment and is a point in the test space. Parameters include mobile and stationary obstacles, environmental factors, time constraints, and mission types at a minimum. For a realistic scenario, the number of input parameters becomes large and cannot be tested. Performance boundaries, where small scenario changes create transitions between performance modes, are a good choice as these regions easily trigger decision-making by the autonomy. Their method adaptively searches for the autonomy's performance boundaries with the minimum number of samples. Supervised clustering was applied to group test scenarios by the subsequent performance modes. Their framework was able to test the autonomy realistically with generated scenarios that challenged the autonomy based on discovery and identification of performance boundaries.

In an AUV case study, the vehicle performed multiple missions and complied with all safety criteria. The mission objectives were to follow waypoints, station-keep in three prioritized mission areas for a given duration, and then reach a transmission area by a certain time. All three missions needed to be completed within a time window. Its safety criteria were to avoid all obstacles, remain within the operational area, outside of an exclusion zone, and reach a recovery point. Obstacle location provided the environmental uncertainty. The system's performance was assessed against high-level success criteria that included total failure (fail any safety criteria and any mission objective), safety success (pass all safety criteria regardless of mission objective performance), mission success (completes all mission objectives but fails any safety criteria), and finally total success (completes all mission objectives and all safety criteria).

This outcome provided the required metrics to quantify the verification performance.

8.6.2 Verification Performance Metrics

Because verification of MAS is necessarily limited, the value or quality of the verification must be assessed. Verification performance metrics include measures of efficiency, effectiveness, and coverage. Verification efficiency is related to the time and resources required for the verification and testing to be performed and is used most often. Verification effectiveness is quantified through the type and number of faults detected and is used second most often. While efficiency and effectiveness are related to performance, verification coverage addresses the comprehensiveness of the verification and its adequacy. Exhaustive coverage catches unforeseen edge cases and errors in the software design and/or implementation.

While only one group directly addressed MAS [103], the ground and aerial examples provided in the thorough recent review by [104] also provides useful insights.

Verification efficiency metrics can be based on

- Precision [103]
- Time to generate test cases or run the tests (e.g. execution time for model checking)
- In-water testing cost ($$/km)
- Test execution and simulation time
- Search time
- State space size
- Reduced test case execution time while verification effectiveness metrics can include
- The number of test cases, failures, and counterexamples
- Accuracy of simulations
- Reachability
- The number and probability of faulty scenarios generated

Diversity in test scenarios is a particularly appropriate verification coverage metric choice for MAS, where tests are costly [103]. Efforts to date have focussed on component-level assurances and have not considered the system level. The approach of [105] uses graph-based tools to achieve system-level testing with coverage guarantees, which is relevant to the later stages of MAS verification. While it is applied to autonomous cars, it could just as easily be applied to MAS. Coverage may also be assessed through requirement violations [106] or its compliance [107].

8.7 Research Opportunities

Autonomy and its verification for MAS lag its counterparts in ground, aerial, and space domains. Research opportunities specifically related to MAS verification include performing system modelling, program verification, and algorithm design with guaranteed efficiency and correctness, determining what should be verified at runtime, and how to design self-verifying systems. In the area of combinatorial methods, modularly constructed MAS would be well suited to compositional verification. Verification of complex decisions and arbitration within multi-component decision-making processes, defining rigorous measures of performance and adequacy, and providing coverage guarantees with system-level testing are all particularly challenging in MAS.

There is an emerging work for reliable integration of autonomous components [108] and IEEE standards that address this [109].

Formal methods for safety verification of autonomous surface ships [110] are still immature and not as widely used by the maritime robotics community as it is by the general robotics community. Rule 15 from COLREGS was verified in [58] for two vessels on a collision course, but other potentially relevant rules for an ASV could include Rule 2 (sense of responsibility), Rule 5 (situational awareness capabilities), and Rules 6, 8, 13, 14, 16–19 (collision avoidance action plan). Research is needed in verification for classes of collision avoidance action plans and in defining the requirements, specifications, or properties of a MAS' different behaviours given its highly variable communications bandwidth with the operator or other nodes.

Degraded performances during adverse weather conditions like fog, glare off the water, severe winds, and waves are especially challenging for AUVs and ASVs. Their situational awareness sensors may suffer errors, and their onboard radios may have poorer connectivity in a high sea state. While verification of other communication networks is amenable to formal methods, RF connectivity in realistic sea states is both vital to ASV operation and not amenable to formal methods, so research in verification under these circumstances must be prioritized.

Verification of systems with bounded behaviours can be achieved with standard methods like model checking, theorem provers, and runtime verification. But for non-deterministic systems verifying that it attempts to make the best decisions is difficult [111]. As discussed in [112], there are few approaches for such autonomy.

With model checkers, the correctness of a system with respect to a desired behaviour is verified by checking whether its model satisfies a specification as expressed in temporal logic. The advantage of formal methods is that the verification act is automatic, and a negative answer to a query is accompanied by a counterexample to the satisfaction of the specification. However, there is no additional feedback with a positive answer to the query. The positive answer might also be from an error in the system model of the behaviour or incompleteness in the specification. Since incompleteness of a specification is typical for non-deterministic systems, increased research by the formal methods community is needed.

8.8 Conclusions

The harsh MAS environment has a unique impact on its communications, navigation, sensing, actuation and control and by extension, and its verification. This complexity is manifested early in its development which puts more emphasis on informal verification methods, despite their incompleteness.

Verification must be built into marine autonomous systems from the early stages of development and continue through the operational life of a MAS, not as a separate engineering requirement prior to system delivery.

The verification of MAS operations, autonomy, and autonomy development lags those in other domains. Most recent reviews can only cite a few examples. While the ground, aerial, and to a lesser extent, space domains can leverage advancements from one another, this is possible less frequently in the maritime domain, and there are research opportunities that are unique to MAS verification.

When the consequences of decisions and actions for marine autonomous systems are high, as they may be for military, commercial, or scientific applications, they present a barrier to adoption of these systems. Verification of MAS operations and autonomy will help with such trust issues.

References

1. R. Arkin, Behavior-Based Robotics, The MIT Press, 1998.
2. "GAZEBO—Simulate before you b uild," Open Robotics, [Online]. Available: https://gazebosim.org/home. [Accessed February 2021].
3. C. Pecheur, "Verificataion and Validation of Autonomy Software at NASA," NASA Ames Research Center, Moffett Field, CA, USA, 2000.
4. H. Kress-Gazit, T. Wongpiromsarn and U. Topcu, "Correct, Reactive High-Level Robot Control," *IEEE Robotics and Automation Magazine,* vol. 18, no. 3, pp. 65–74, 2011.
5. D. P. M. Clegg, "User Operational Evaluation System of Unmanned Underwater Vehicles for Very Shallow WAter mine Countermeasures," in *OCEANS 2003 MTS/IEEE Conf.*, San Diego, 2003.
6. S. Redfield and M. L. Seto, "Verification Challenges for Autonomous Systems," in *Autonomy and Artificial Intelligence*, W. Lawless, R. Mittu, D. Sofge and S. Russell, Eds., Springer, 2017, p. 32.
7. G. Brat and A. Jonsson, "Challenges in verification and validation of autonomous systems for space exploration," in *Proc. 2005 IEEE Int. Joint Conf. Neural Networks*, 2005.
8. S. Tangirala, S. Kumar, S. Bhattacharyya, M. O'Connor and L. E. Holloway, "Hybrid-Model based Hierarchical Mission Control Architecture for Autonomous Underwater Vehicles," 2005.
9. H. Liu, R. Cheng, T. Yang and J. Wang, "Modeling and Verifying the Communication and Control of a Fleet of Collaborative Autonomous Underwater Vehicles," in *43rd Annual Conf. IEEE industrial Electronics Society*, 2017.
10. I. Porres, S. Azimi, S. Lafond, J. Lilius, J. Salokannel and M. Salokorpi, "On the Verification and Validation of AI Navigation Algorithms," in *Proc. Global Oceans 202*, Singapore—U.S. Gulf Coast, 2021.
11. J. Villa, G. Valicrosa, J. Aaltonen, P. Ridao and K. T. Koskinen, "Model-Validation and Implementation of a Path-Following Algorithm in an Autonomous Underwater Vehicle," *Appl. Sci.,* vol. 11, pp. 1–17, 2021.
12. United States Navy, "The Navy Unmanned Undersea Vehicle (UUV) Master Plan," United States Navy, 2004.

13. M. Seto and A. Crawford, "Autonomous shallow water bathymetric measurements for environmental assessment and safe navigation using USVs," MTS/IEEE, Washington, DC, 2015.
14. R. W. Button, J. Kamp, T. B. Curtin and J. Dryden, "A Survey of Missions for Unmanned Undersea Vehicles," Rand Corporation, 2009.
15. United States Navy, "The Navy Unmanned Surface Vehicle (USV) Master Plan," 2007.
16. L. Paull, M. Seto, J. Leonard and H. Li, "Probabilistic Cooperative Mobile Robot Area Coverage and its Application to Autonomous Seabed Mapping," *Int. J. Robotics Research,* vol. 37, no. 1, pp. 21–45, January 2018.
17. Z. Kassas, S. Kozhaya, H. Kanj, J. Saroufim, S. W. Hayek, M. Neinavale, N. Khairallah and J. Khalife, "Navigation with Multi-Constellation LEO Satellite Signals of Opportunity: Starlink, OneWeb, Orbcomm, and Iridium," in *IEEE/ION Position, Location and Navigation Symposium*, Monterrey, CA, 2023.
18. D. P. Eickstedt and S. Sideleau, "The Backseat Control Architecture for Autonomous Robotic Vehicles: A Case Study with the Iver2 AUV," *Marine Technology Society Journal,* vol. 44, no. 4, pp. 42–54, July/August 2010.
19. M. R. Benjamin, H. Schmidt, P. M. Newman and J. J. Leonard, "Nested Autonomy for Unmanned Marine Vehicles with MOOS-IvP," *J. Field Robotics,* vol. 27, no. 6, pp. 834–875, 2010.
20. MIT Ocean AI Group, "MOOS-IvP," [Online]. Available: https://oceanai.mit.edu/moos-ivp/pmwiki/pmwiki.php. [Accessed November 2021].
21. "ROS: Robot Operating System," [Online]. Available: https://www.ros.org/. [Accessed September 2022].
22. "What is DDS?," Object Management Group, July 2021. [Online]. Available: https://www.dds-foundation.org/what-is-dds-3/.
23. M. Seto, "Introduction to Marine Robot Autonomy," in *Marine Robot Autonomy*, Springer, 2012, pp. 1–46.
24. W. M. Organization, "19-meter wave sets new record," [Online]. Available: https://public.wmo.int/en/media/press-release/19-meter-wave-sets-new-record-highest-significant-wave-height-measured-buoy. [Accessed 2021].
25. N. Rajabli, F. Flammini, R. Nardone and V. Vittorini, "Software Verification and Validation of Safe Autonomous Cars: A Systematic Literature REview," *IEEE Access,* pp. 4797–4819, 2021.
26. J. Strutt, "Report of the inquiry into the loss of Autosub2 under he Fimbulisen," National Oceanography Centre Southampton, 2006.
27. G. Griffiths, N. W. Millard, S. D. McPhail, P. Stevenson and P. G. Challenor, "On the Reliability of the Autosub Autonomous Underwater Vehicle," *J. Underwater Technology,* vol. 25, no. 4, December 2003.
28. L. Molnar and S. M. Veres, "Verification of Autonomous Underwater Vehicles using Formal Logic," in *Proc. European Control Conf.*, Budapest, Hungary, 2009.
29. L. Molnar and S. M. Veres, "System Verification of Autonomous Underwater Vehicles by Model Checking," in *Proc. OCEANS 2009-Europe*, 2009.
30. J. Ezekiel, A. Lumuscio, L. Molnar, S. Veres and M. Peabody, "Verifying Fault Tolerance and Self-Diagnosability of an Autonomous Underwater Vehicle," in *Proc. Twenty-Second Int. Joint Conf. Artificial Intelligence*, 2011.
31. J. Bellingham, "Fault Detection, Diagnosis and Mitigation for Long-Duration AUV Missions with Minimal Human Intervention," Defense Technical Information Center, 2014.
32. M. Seto and A. Bashir, "Fault Tolerance Consideratioins for Long Endurance AUVs," in *Proc. 2017 Annual Reliability and maintainability Symposium*, Orlando, Florida, USA, 2017.
33. B. Raanan, J. Bellingham, Y. Zhang, M. Kemp, B. Kieft, H. Singh and Y. Girdhar, "Detection of unanticipated faults for autonomous underwater vehicles using online topic models," *J. Field Robotics,* vol. 35, pp. 705–716, 2017.
34. M. Seto and K. Svendsen, "Advanced AUV Fault Management," in *Autonomous Underwater Vehicles: Design and Practice*, F. Ehlers, Ed., IET Publication, 2020, pp. 419–445.

35. X. Chen, N. Bose, M. Brito, F. Khan, B. Thanyamanta and T. Zou, "A Review of Risk Analysis Research for the Operations of Autonomous Underwater Vehicles," *J. Reliability Engineering & System Safety,* vol. 216, 2021.
36. L. Register, "LR Code for Unmanned Marine Systems," 2017.
37. "Convention on the International Regulations for Preventing Collissions at Sea, 1972 (COLREGS)," https://www.imo.org/en/About/Conventions/Pages/COLREG.aspx, 1972.
38. Object Management Group, "OMG Unified Modeling Language," Object Management Group, 2017.
39. G. Callow, G. Watson and R. Kalawky, "System Modelling of Run-Time Verification and Validation of Autonomous Systems," in *Proc. 2010 th Int. Conf. System of Systems Engineering*, 2010.
40. Object Management Group, "OMG Systems Modeling Language," Object Management Group, 2022.
41. MathWorks, "Simulink is for Model-Based Design," [Online]. Available: https://www.mathworks.com/products/simulink.html. [Accessed September 2021].
42. "Systems Engineering Guidebook," United States Department Department of Defense, Washington, D.C., 2022.
43. K. Y. Rozier, "Specification: The bigget bottleneck in formal methods and autonomy," in *Proc. 8th Working Conf. Verified Software: Theoreis, Tools and Experiments*, 2016.
44. D. Giannakopoulou, T. Pressburger, A. Mavridou, J. Rhein, J. Schumann and N. Shi, "Formal Requirements Elicitation with FRET," in *Joint Proc. REFSQ-2020 Workshops*, Pisa, Italy, 2020.
45. D. Giannakopoulou, T. Pressburger, A. Mavridou and J. Schumann, "Automated Formalization of structured natural language requirements.," *J. Information and Software Technology,* vol. 137, no. Article 106590, 2021.
46. S. Ghosh, D. Elenius, W. Li, P. Lincoln, N. Shankar and W. Steiner, "ARSENA: automatic requirements specification extraction from natural language," in *NASA Formal Methods—8th Int. Symposium*, Minneapolis, MN, USA, 2016.
47. S. Bensalem, V. Ganesh, Y. Lakhnech, C. Munoz, S. Owre, H. RueB, J. Rushby, V. Rusu, H. Saidi, N. Shankar, E. Singerman and A. Tiwari, "An overview of SAL," in *Proc. 5th NASA Langley Formal Methods Workshop*, Williamsburg, VA, USA, 2000.
48. A. Hinton, M. Kwiatkowska, G. Norman and D. Parker, "PRISM: A tool for automatic verification of probabilistic systems," in *Proc. 12th Int. Conf. Tools and Algorithms for the Construction and Analysis of Systems*, 2006.
49. A. Lumuscio, H. Qu and F. Raimondi, "MCMAS: an open-source model checker for the verifiation of multi-agent systems," *Int. J. Software Tools for Technology Transfer,* vol. 19, pp. 9–30, 2017.
50. "NuSMV: a new symbolic model checker," [Online]. Available: https://nusmv.fbk.eu/NuSMV/papers/cav99/html/index.html also https://nusmv.fbk.eu/index.html. [Accessed January 2022].
51. D. Jackson, "Alloy: A Language and Tool for Exploring Software Designs," *Communications of the ACM,* vol. 62, no. 9, pp. 66–76, 2019.
52. T. Henzinger, R. Jhala, R. Majumdar and G. Sutre, "Software Verification with BLAST," in *Int. SPIN Workshop on Model Checking of Software*, 2003.
53. UPPAAL, "UPPAAL—Design Verification for Embedded Systems," UP4ALL, [Online]. Available: https://www.uppaal.com/. [Accessed November 2021].
54. "UPPAAL Stratego," [Online]. Available: https://uppaal.org/features/#uppaal-stratego. [Accessed March 2022].
55. S. M. Veres, Natural language programming of agents and robotic devices, vol. Series on Intelligent Agents, London, UK: SysBrain Ltd., 2008, p. 181.
56. E. Gat, "Autonomy Software Verification and Validation Might Not be as Hard as It Seems," in *2004 IEEE Aerospace Conference Proc.*, Big Sky, MT, USA, 2004.
57. S. Chung and B. Williams, "Significance of Decomposition in Spacecraft Autonomy," in *Infotech@Aerospace*, Atlanta, Georgia, USA, 2010.

58. F. Shokri-Manninen, J. Vain and M. Walden, "Formal Verificaiton of COLREG-Based Navigation of Maritime Autonomous Systems," pp. 41–59, 2020.
59. A. Platzer, "KeYmaera X: An aXiomatic Tactical Theorem Prover for Hybrid Systems," [Online]. Available: https://keymaerax.org/index.html. [Accessed October 2022].
60. L. Paulson, Ed., Isabelle: A Generic Theorem Prover, vol. Lecture Notes in Computer Science volume 828, Springer Berlin, Heidelberg, 1994.
61. "Isabelle," [Online]. Available: https://isabelle.in.tum.de/. [Accessed August 2022].
62. D. Muniraj, H. He, M. Farhood and D. J. Stilwell, "A Distributed Connectivity Maintenance Algorithm with Formal Gurantees for a Communication-Constrained Network of Unmanned Underwater Vehicles," *IEEE Systems Journal,* vol. 16, no. 2, pp. 1774–1785, June 2022.
63. D. Jin, P. O. Meredith, C. Lee and G. Rosu, "JavaMOP: Efficient Parametric Runtime Monitoring Framework," in *34th Int. Conf. Software Engineering*, Zurich, Switzerland, 2012.
64. R. Calinescu, C. Ghezzi, M. Kwiatkowska and R. Mirandola, "Self-Adaptive Software needs Quantitative Verification at Runtime," *Communications of the ACM,* vol. 55, no. 9, pp. 69–77, 2012.
65. S. Gerasimou, R. Calinescu and A. Banks, "Efficient Runtime Quantitative Verification using Caching, Lookahead and nearly-Optimal Reconfiguration," in *SEAMS*, Hyderabad, 2014.
66. R. Calinescu, S. Gerasimou, K. Johnson and C. Paterson, "Using Runtime Quantitative Verification to Provide Assurance Evidence for Self-Adaptive Software: Advances, Applications and Research Challenges," in *Software Engineering for Self-Adaptive Systems III. Assurances. Lecture Notes in Computer Science*, vol. 9640, R. de Lemos, D. Garlan, C. Ghezzi and H. Giese, Eds., Springer, 2017, pp. 223–248.
67. S. Thrun, W. Burgard and D. Fox, Probabilistic Robotics, The MIT Press, 2006.
68. S. Saeedi, L. Paull, M. Trentini, M. Seto and H. Li, "Efficient map Merging using a Probabilistic Generalized Voronoi Diagram," in *Proc. IEEE/RSJ Int. Conf. Intelligent Robots and Systems*, 2012.
69. J. Huang, C. Erdogan, Y. Zhang, B. Moore and Q. Luo, "ROSRV: Runtime Verification for Robots," in *Int. Conf. Runtime Verification*, 2014.
70. I. Perez, Mavridou, A., T. Pressbergurger, A. Will and P. J. Martin, "Monitoring ROS2: from Requirements to Autonomous Robots," in *Proc. Fourth Int. Workshop on Formal Methods for Autonomous Systems and Fourth Int. Workshop on Automated and verifiable Software sYstem DEvelopment (ASYDE)*, Berlin.
71. "GNU Octave—Scientific Programming Language," GNU, [Online]. Available: https://octave.org/. [Accessed June 2022].
72. N. Koenig and A. Howard, "Design and Use Paradigms for Gazebo, An Open-Source Multi-Robot Simulator," in *IEEE/RSJ Int. Conf. Intelligent Robots and Systems*, 2004.
73. T. I. Fossen and T. Perez, "Marine Systems Simulator (MSS)," 2004. [Online]. Available: https://github.com/cybergalactic/MSS.
74. T. I. Fossen, Handbook of Marine Craft Hydrodyanics and Control Motion, 2nd ed., Wiley, 2021.
75. MathWorks Student Competitions Team, "Model-based Design for Embedded Control Systems," 2023. [Online]. Available: https://www.mathworks.com/campaigns/offers/model-based-design-embedded-control-systems.html. [Accessed September 2021].
76. "Modeling and Simulation of an Autonomous Underwater Vehicle," Mathworks Student Competition Team, 2023. [Online]. Available: MathWorks Student Competitions Team (2023). Modeling and Shttps://github.com/mathworks/AUV-modeling-and-sim. [Accessed January 2023].
77. M. Prats, J. Perez, J. Fernandez and P. J. Sanz, "An open source tool for simulation and supervision of underwater intervention missions.," in *IEEE/RSJ Int. Conf. Intelligent Robots and Systems*, 2012.
78. "UWSim The UnderWater Simulator," 2012. [Online]. Available: http://www.irs.uji.es/uwsim/.

79. M. M. M. Manhaes, S. A. Scherer, L. R. Douat and T. Rauschenbach, "UUV simulator: A gazebo-baed package for underwater intervention and multi-robot simulation.," in *MTS/IEEE OCEANS 2017*, Monterey, California, USA, 2016.
80. "Unmanned Underwater Vehicle Simulator (UUV Simulator)," The UUV Simulator Authors, [Online]. Available: https://github.com/uuvsimulator/uuv_simulator. [Accessed January 2021].
81. E. H. Henriksen, I. Schjolberg and B. Gjersvik, "UW MORSE: The Underwater Modular Open Robot Simulation Engine," in *2016 IEEE/OES Autonomous Underwater Vehicles*, Tokyo, Japan, 2016.
82. P. H. Heins and B. L. Jones, "Design and validation of an unmanned surface vehicle simulation model," *Applied Mathematical Modelling,* vol. 48, pp. 749–774, 2017.
83. D. Hudson, "User Manual for Program Suite THARBM, Three-Dimensional Analysis of Rigid Body Motions," University of Southampton, 2000.
84. P. Cieslak, "Stonefish: An Advanced Open-Source Simulation Tool Designed for Marine Robotics, with a ROS Interface," in *MTS/IEEE OCEANS 2019*, Marseille, France, 2019.
85. Z. Zhang, W. Mi, J. Du, Z. Wang, W. Wei, Y. Zhang, Y. Yang and Y. Ren, "Design and Implementation of a Modular UUV Simulation Platform," *Sensors,* vol. 22, pp. 1–20, 2022.
86. A. Clavijo-Rodriguez, V. Alonso-Eugenio, S. Zazo and I. Perez-Alvarez, "Software-in-Loop Simulation of an Underwater Wireless Sensor Network for Monitoring Seawater Quality: Parameter Selection and Performance Validation," *Sensors,* vol. 21, pp. 1–25, 2021.
87. V. Kaliappan, A. Budiyono, D. Min, K. Muljowidodo and S. Nugroho, "Hardware-in-the-Loop Simulation Platform for the Design, Testing and Validation of Autonomous Control System for Unmanned Underwater Vehicle," *Indian J. Geo-Mar. Sci.,* vol. 41, pp. 575–580, 2012.
88. U. Chou, H. H. Chen, C. C. Wang and W. H. Chen, "A Hardware-in-the-Loop Simulation Platform for Development of AUV Control Systems," in *2019 IEEE Underwater Technology (UT)*, Taiwan, China, 2019.
89. T. Tosik, J. Schwinghammer, M. J. FeldvoB, J. P. Jonte, A. Brech and E. Maehle, "MARS: A Simulation Environment for Marine Swarm Robotics and Environmental Monitoring," in *MTS/IEEE OCEANS 2017*, 2016.
90. P. Sarhadi, R. N. A. Niachari, M. P. Rad and J. Enayati, "Hardware-in-the-loop simulatio for real-time software verification of an autonomous underwater robot," *Int. J. Intlligent Unmanned Systems,* vol. 4, no. 3, pp. 163–181, 2016.
91. S. Subramanian, T. George and A. Thondiyath, "Hardware-in-the-Loop Verification for 3D Obstacle Avoidance Algorithm of an Underactiated Flat-Fish Type AUV," in *Proc. 2012 IEEE Int. Conf. Roboticcs and Biomimetics*, Guangzhou, China, 2012.
92. T. W. O. Putri, M. F. Ginting, B. R. Trilaksono, E. M. I. Hidayat and M. F. Sagala, "Hardware in the Loop Simulation Development of Guidance System for Autonomous Underwater Glider," in *2017 6th Int. Conf. Electrical Engineering and Informatics*, Langkawi, Malaysia, 2017.
93. Y. Li, Z. Zhou and Y. Su, "Design of a General Hardware in the Loop Underwater Communication Emulation System," *IEEE Access,* vol. 9, pp. 37685–37696, 2021.
94. J. Patel, A. Bassam and M. Seto, "Multi-UUV object detection, localization and tracking with secure, full-duplex communications networks," in *Proc. Meetings of the Acoustical Society of America*, 2021.
95. C. Luth, M. Ring and R. Drechsler, "Towards a Methodology for Self-Verification," in *6th Int. Conf. Reliability, Infocom Technologies and Optimization (Trends and Future Directions)*, Noida, India, 2017.
96. L. Pike, S. Wegmann, S. Niller and A. Goodloe, "Copilot: Monitoring Embedded Systems," NASA, 2012.
97. H. Garavel, F. Lang and L. Mounier, "Compositional Verification in Action," in *Formal methods for industrial critical systems*, 2018.

98. R. C. Cardoso, L. A. Dennis, M. Farrell, M. Fisher and M. Luckcuck, "Towards compositional verification for modular robotic systems," in *Proc. 2nd Workshop on Formal methods for Autonomous Systms*, virtual, 2020.
99. M. Farrell, M. Luckcuck and M. Fisher, "Robotics and Integrated Formal Methods: Necessity meets Opportunity," in *Integrated Formal Methods*, 2018.
100. A. Champion, A. Garfinkel, T. Kahsai and C. Tinelli, "CoCoSpec: A Mode-Aware Contract Language for Reactive Systems," in *Int. Conf. Software Engineering and Formal Methods*, 2016.
101. S. Foster, M. Gleirscher and R. Calinescu, "Towards Formal Verification of Control Algorithms for Autonomous Marine Vehicles," in *Proc. 25th Int. Conf. Engineering of Complex Computer Systems*, 2020.
102. C. Hoare and J. He, Unifying Theories of Programming, Prentice-Hall, 1998.
103. G. Mullins, P. G. Stankiewicz, R. C. Hawthorne and S. K. Gupta, "Adaptive generation of Challenging scenarios for testing and evaluation of autonomous vehicles," *J. Systems and Software,* vol. 137, pp. 197–215, 2018.
104. H. Araujo, M. R. Mousavi and M. Varshosaz, "Testing, Validation, and Verification of Robotics and Autonomous Systems: A Systematic Review," *ACM Transactions on Software Engineering and Methodology,* pp. 1–60, 2022.
105. I. Majzik, O. Semerath, C. Hajdu, K. Marussy, Z. Szatmari, Z. Micskei, A. Voros, A. Babikian and D. Varro, "Towards System-Level Testing with Coverage Guarantees for Autonomous Vehicles," in *ACM/IEEE 2nd Int. Conf. Model Driven Engineering Language and System*, 2019.
106. M. Tatar, "Enhancing ADAS Test and Validation with Automated SEarch for Critical Situations," in *Driving Simulation Conf.*, Paris, France, Europe, 2015.
107. Y. Selvaraj, W. Ahrendt and M. Fabian, "Verification of Decision Making Software in an Autonomous Vehicle: An Industrial Case Study," in *Int. Workshop on Formal Methods for Industrial Critical Systems*, 2019.
108. M. Fisher, V. Mascardi, K. Y. Rozier, B. Schlingloff, M. Winikoff and N. Yorke-Smith, "Towards a framework for certification of reliable aonomous systems," *Autonomous Agents and Multi-Agent Systems,* vol. 35, no. 8, pp. 1–65, 2021.
109. IEEE, "Standards for Fail-Safe Design of Autonomous and Semi-Autonomous Systems," IEEE Reliability Society, [Online]. Available: https://standards.ieee.org/ieee/7009/7096/. [Accessed August 2022].
110. T. R. Torben, O. Smogeli, I. B. Utne and A. J. Sorensen, "On Formal Methods for Design and Verification of Maritime Autonomous Surface Ships," in *World Maritime Technology Conference*, Copenhagen, Denmark, 2022.
111. L. A. Dennis, M. Fisher, N. K. Lincoln, A. Lisitsa and S. M. Veres, "Practical verification of decision-making in agent-based autonomous systems," *Autonomous Software Engineering,* vol. 23, no. 3, pp. 305–359, 2016.
112. M. Luckcuck, M. Farrell, L. A. Dennis, C. Dixon and M. Fisher, "Formal Specification and Verification of Autonomous Robotic Systems: A Survey," *ACM Computing Surveys,* vol. 52, no. 5, pp. 1–41, September 2019.

Chapter 9
Performance Boundary Identification and Automated Test Plan Development

Galen Mullins

Chapter 9 Editorial Note

This chapter documents a simulation environment that automatically identifies boundaries in system performance. These boundaries can then be used to define test sets for characterizing system behavior in the physical world or to demonstrate the degree to which the system can generalize from one environmental configuration to another. This descriptive capability addresses stakeholder questions like our need to determine whether the product is reliable, its capabilities and limitations, and its trustworthiness. Performance boundaries can be defined on the basis of safety and security. Further, this tool enables improved understanding of autonomous behavior by enabling accurate descriptions of its functionality and limitations. Technical challenge 14 (test set development) explicitly identifies the need to establish performance boundaries and fault locations in the larger state space of potential task and system configurations, which is enabled by the technology described in this chapter. This chapter does not address the development of test sets and data. This approach also addresses challenge 12 (test design for general capability assessment). It does not provide general principles for test design that would lead to tests that indicate general capability, but it does define the boundaries within which general capability can be expected to exist—if the robot succeeds at one task configuration within the boundary, we can expect the robot to succeed at the task in other configurations within the boundary. It defines the details of tests in which this general capability would be expected to occur. Because these performance boundaries are defined at the system level within the task and system state space, this approach also addresses technical challenge 13 (design reference mission selection). The boundaries enable the selection of design reference missions

G. Mullins (✉)
The Johns Hopkins University Applied Physics Laboratory, Laurel, Maryland
e-mail: gmullin3@jh.edu
https://www.jhuapl.edu/ISC

S. Redfield et al. (eds.), *Verification of Autonomous Systems*,
https://doi.org/10.1007/978-3-031-88546-4_9

that indicate general capability for the system as a whole, rather than at the component level. By enabling analysis of a given capability and separating out the impact of environmental parameters and system parameters, this chapter addresses technical challenge 25 (disambiguation between specification and environmental assumption failures and anticipation of unknowns) both by identifying classes of safe and unsafe situations and by enabling disambiguation between cases where the system fails due to incorrect environmental assumptions and incorrect assumptions about the system itself. Stakeholder questions are presented in the Chap. 1, Introduction. Technical challenges are defined and discussed in more detail in Chap. 2, Verification Challenges.

Abstract Developing and executing field tests for autonomous vehicles is a time-consuming and expensive process, where only a limited number of tests can be performed. Therefore, it is important that each test provides the maximum amount of information to the system evaluator. We propose creating these testing suites by using a simulation-based framework to generate adversarial scenarios which are tailored to the autonomy under test. The challenge in the past is that test case generation methods have been limited to a low number of dimensions and simple scenarios. Additionally, they are typically only verified in a simulation environment, and the difficulties of transfer from simulation to reality have not been explored. We address these issues by applying adaptive sampling to efficiently explore a large scenario parameter space with multiple competing scoring criteria that represent a realistic unmanned underwater vehicle mission. The result is a diverse set of test cases that represent the performance boundaries of the autonomy software. Finally, we present the results of executing simulation generated tests in a live test range, which at the time of execution was the first field test of its kind.

9.1 Introduction

There are few portions of any autonomous vehicle program that are as expensive as designing and executing field tests. Both the time and expense involved means the number of tests that can be performed is extremely limited. The current process is to have subject matter experts hand design vignettes meant to test each behavior of the system individually [1–3] or a single scenario meant to represent a nominal mission [4]. However, these isolated scenarios cannot provide adequate cover of all the types of situations that autonomous vehicles may face during a long duration mission [5] work that subject matter experts can often miss critical failure modes of the system by focusing too much on regions where the behavior is well defined [6].

Currently, the primary way of testing an autonomous vehicle's performance is through the use of realistic simulations environments [7–9]. Even with simulations it is not possible to exhaustively sample all possible scenarios, particularly as the missions become more complex and require greater numbers of parameters to describe. One solution to this problem is search-based testing procedures, which use optimization techniques to find tests which minimize or maximize the performance of the autonomous vehicle. Search-based techniques have been applied to autonomous cars [10], unmanned aerial vehicles [11], and unmanned underwater vehicles [12]. These techniques are capable of finding relevant scenarios in a fraction of the time that it would take exhaustive techniques which randomly sample the state space. The challenge to this point has been demonstrating that the failures seen in the simulation can be replicated in reality.

In our previous work [12, 13] we introduced a method for generating informative test scenarios where the autonomy software was running in simulation. The ultimate goal for the Range Adversarial Planning Tool (RAPT) program was to generate scenarios which could be executed on hardware at a testing range. The effective transfer of simulation generated results to the real world is one of the biggest hurdles facing the robotics community. There are many ways in which errors can be introduced which cause the simulated system to diverge from the real world. These include effects such as modeling errors, unaccounted sources of uncertainty, simulated versus real sensing data, or differences in the way the hardware platform communicates with autonomy software versus the simulated platform.

In this chapter, we will discuss our approach for addressing the problem of generating scenarios from a nondeterministic simulation and the results of running tests designed by RAPT in the field. In Sect. 9.2, we review similar technologies in the published literature. In Sect. 9.3 we provide an overview of the RAPT software. In Sect. 9.4, we will provide an overview of the mission and our experimental setup. In Sect. 9.5, we will discuss the test design process that was utilized to build the test plan for the field. In Sect. 9.6, we will discuss the results of the field tests. Finally, in Sect. 9.7, we present our conclusions for the final demonstrated performance of the RAPT software.

9.2 Related Work

Despite the wealth of research available for testing black-box software [14], developing tests for autonomous vehicles is still an open problem [5, 15]. Approaches have been extensively explored as a method for generating test scenarios and test plans [11, 16–19]; however they fall primarily into two categories. The first is methods for optimizing priority and coverage [20–22]. These attempt to characterize the entire test space but require known models of the system which means they can only be applied for white-box testing. The second is global optimization methods which utilize objective functions that describe the quality of the mission [11, 23, 24]. These are more appropriate for characterizing black-box systems but have several drawbacks. The first is that these objective functions require extensive knowledge of the mission and the software to design. The second is that global optimization techniques will provide highly refined solutions in a

Fig. 9.1 IVER UUV used for the on-water tests of the autonomy software. These photos were taken during the hydrodynamic tests conducted in the Chesapeake Bay

single region but will not explore enough of the space effectively characterize the test system. Our approach solves both these problems by utilizing adaptive sampling to preferentially explore the space. Instead of requiring an objective function, it derives one from high-level mission performance metrics.

After tests have been discovered the next issue becomes transferring those results to the hardware platform (Fig. 9.1). Generating tests that can be replicated in controlled indoor testing environments has been successfully demonstrated [25], but these results are not always guaranteed. This is sometimes referred to as the "reality gap" [26] and has always been an issue when developing robotic systems. The problem has received additional interest with the rise of deep reinforcement learning and perception systems. The current model-free approaches require substantial amounts of data, far more than can practically be acquired through hardware trials. One approach is to utilize data augmentation techniques that introduce noise into the data or simulation [27]. The result is ideally a learned policy or a classifier which is robust against uncertainty and thus robust against the delta between simulation and reality. This is also known as domain randomization and has been successfully applied to autonomous drones [28]. In developing our we also incorporate data augmentation into our process via a dithering study, where noise is added to the scenario parameters to capture both inaccuracy in our simulation and uncertainty in scenario execution.

9.3 Range Adversarial Planning Tool (RAPT)

This section provides a high-level overview of the methodology used by RAPT to generate salient test scenarios for autonomous systems. For a full discussion of this framework and the underlying algorithms, the reader is directed to [13]. An architectural overview of this tool is illustrated in Fig. 9.2. The process begins with the test engineer specifying two configurations for the autonomy under test (AUT): the testing state space and the scoring space. The testing state space dictates the elements of the simulation that will be varied between scenarios including obstacle positions, ocean current magnitudes, and any other factors that could influence the autonomy decision-making. The scoring space then controls how the AUT evaluated for each completed scenario. Once the testing state space and scoring space are defined, the scenario generation process is broken into two phases: adaptive sampling followed by boundary identification.

9.3.1 Adaptive Sampling

In the first step, adaptive sampling is utilized to generate test scenarios that are predicted to represent the performance boundaries of the system. This is an iterative process which builds a surrogate model of the AUT's performance based on the results of completed simulations, where the inputs of the surrogate model are the parameters defined in the testing state space and the outputs are the evaluation criteria defined in the scoring space. This surrogate model is then used to generate subsequent scenarios in regions of the testing state space where performance boundaries are expected to occur, e.g., regions characterized by high variance in the

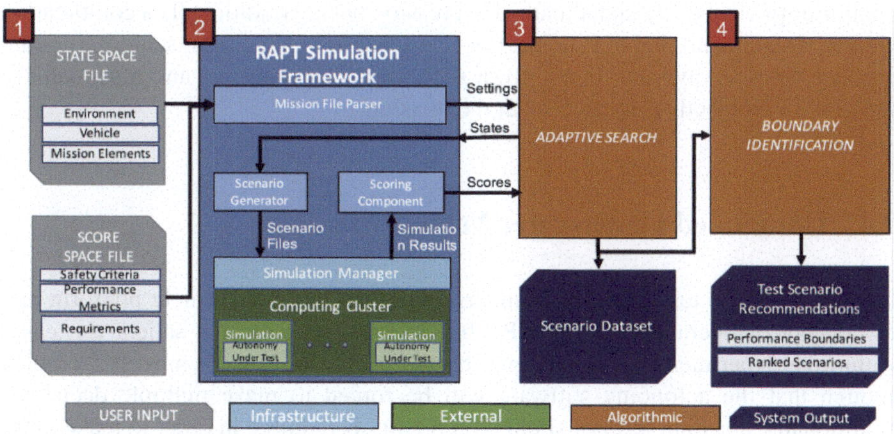

Fig. 9.2 Overview of the RAPT framework

AUT performance. A key feature of the adaptive sampling algorithm is the ability to balance the trade-off between fully exploring the testing state space while also preferentially focusing samples on regions of predicted performance boundaries.

9.3.2 *Boundary Identification*

After all the simulations have been executed, the second step is to characterize the performance boundaries of the system using the raw simulation data. However, because these performance boundaries are highly nonlinear and exist in high-dimensional spaces, they are not easily characterized analytically. What is possible automatically generating sets of adjacent scenario pairings that exhibit distinct differences in performance, where groupings of performance classes are identified using unsupervised clustering methods. These scenario pairings along the system's performance boundaries are then returned to the user. Ultimately, scenarios that straddle performance boundaries aid in diagnosing the AUT decision-making by attributing small changes in scenario parameters to large transitions in performance.

9.3.3 *Scoring Methodology*

The RAPT framework is domain agnostic and can accept any parameterized simulation and set of scoring criteria. It gives the user the capability to design studies based upon a combination of a testing space and performance metrics that define a mission. The user defines a score tree structure that describes the relationship between continuous metrics extracted from the simulation (e.g., distance from an obstacle) and the resulting scores (e.g., whether a collision occurred). These relationships can be any mathematical expression but are traditionally a combination of logical statements and thresholds. For example, the vehicle must achieve a certain distance from a waypoint to achieve a success at that objective and must achieve success at all objectives to succeed at the mission.

9.4 Unmanned Underwater Vehicle Mission

Designing test suites requires an understanding of both the vehicle platform and its concept of operations. The RAPT framework will only find salient test cases within the experimental space defined by the user. The mission must be complex enough that the autonomy software will be forced to make multiple decisions. Additionally, a certain level of simulation fidelity is required in order to make those results transferable to reality. In this section we outline the platform, simulation, and mission we utilized for this project.

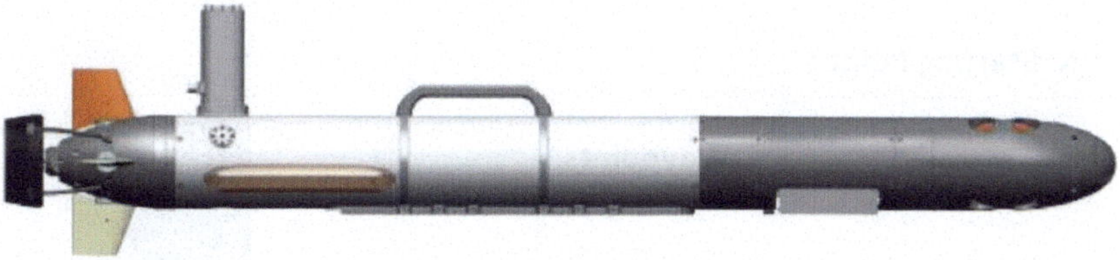

Fig. 9.3 A 3D rendering of the OceanServer Iver2 platform

9.4.1 Platform Description

The unmanned underwater vehicle (UUV) used was an OceanServer Iver2 unmanned underwater vehicle (Fig. 9.3). The vehicle is approximately 60 inches in length, has a diameter of 6 inches, and weighs approximately 60 pounds. It is controlled via a rear thruster and four external fins. The navigation sensors include a GPS, a Doppler velocity logger (DVL), a compass, and an inertial measurement unit (IMU).

9.4.2 Simulation and Autonomy

The simulation environment and autonomy were developed in-house using the Johns Hopkins APL Autonomy Toolkit (ATK) [29, 30]. The simulated system had a maximum speed of 3 m/s and used a 6 degree-of-freedom transit model for underwater vehicles model derived from [31]. The environment possessed a tidal current that varied based on the time of day between 3 m/s due northeast and 3 m/s due southwest with peaks occurring at 2 am and 12 pm, respectively.

When performing simulation-based testing, it is neither necessary nor desirable to utilize high-fidelity models for all aspects of the simulation. Increasing model fidelity incurs a cost in both development time and speed of the simulator, reducing the amount of data that can be generated during an experiment. Users should instead prioritize the properties that are most likely to impact the decision-making of the autonomy software and the performance of the mission objectives. The primary purpose of our field test (Fig. 9.4) was to examine how the autonomy software handles time and energy management and selects objectives. As such we are concerned with factors that impact energy consumption, vehicle speed, and navigational accuracy. We performed extensive hydrodynamic testing of the vehicle in the Chesapeake Bay to properly model the hydrodynamic properties of the platform, as buoyancy, drag, and surge speed. During this testing we also measured both current draw and battery drain under different motor loads which were fit to the model defined by Pals and Newman [32]. We incorporated a tidal current model which utilized NOAA tidal records [33]. This data was gathered

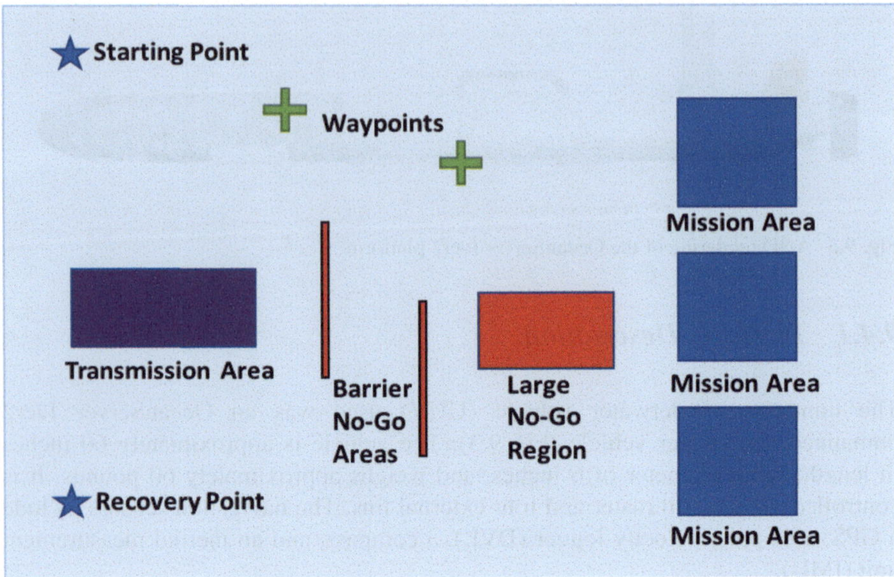

Fig. 9.4 Diagram of the Phase 3 mission

from oceanographic sensors managed by NOAA near the Keyport test range using averaged data from the months of September and November when the field tests were expected to occur. Finally, we updated the GPS/INS models to utilize incorporate realistic drift in the state-estimate of the UUV.

9.4.3 Mission Description

For this demonstration, the state space consisted of 13 variable parameters: (1) the start time, (2,3) the transmission window start time and duration, (4,5) the latitude and longitude of the transmission area, (6,7) the latitude and longitude of the no-go region center, (8–11) the latitude and longitude of the two barrier no-go areas, (12) the priority order for the mission areas, and (13) the starting battery capacity for the IVER. These provided representative cases of input states that have a strong effect on the autonomy (e.g., the tidal force imposed by the start time) and input states that have a weak effect (e.g., the priority order of the mission areas). The start time was varied in a 12-hour period between midnight and noon, and the transmission window was set to open between 30 minutes after mission start time and an hour after mission start time with duration between 30 minutes and an hour. Additionally, the transmission area was set to vary in position only in the western half of the operational area. The no-go region was set to vary in position only in the southeastern quadrant of the operational area. The no-go barriers were constrained

Table 9.1 State space element details

Element	Description
Waypoint	A recommended target to pass through if not constrained by other factors like no-go areas or time. It is not required for mission success
Mission area	A 500 × 500 m region that the UUV must enter and remain inside for a predetermined amount of time. It must complete each mission area in the correct priority order. Completing all mission areas is required for mission success
Transmission area	A 700 × 750 m region that the vehicle must enter and surface for the open transmission time window. Completing this objective is required for mission success. Surfacing outside of this time window or transmission area is a safety failure
Large no-go area	A 400 × 500 m region that the vehicle cannot enter. If the vehicle enters this region, it will receive a safety failure, but the simulation continues
No-go barrier	If the vehicle collides with a barrier, it is considered a safety failure. The barrier areas are 700 x 40 m
Operational area	A 3 × 3 km region that the vehicle cannot leave. If the vehicle violates the boundary, it will receive a safety failure, and the simulation ends
Recovery point	A target circle with a radius of 15 m. The simulation ends when the vehicle reaches this point. If the vehicle does not reach this point, it receives a safety failure

to vary in position only within 400 m of the operational area vertical centerline. Finally, each scenario assigned a priority order for execution of the mission areas.

The mission and vehicle we utilized are similar to the one designed for initially evaluating our methods in simulation [13]. The major difference is that for the field tests we replaced physical obstacles with no-go regions. This was due to the practical limitations of introducing large-scale physical obstacles such as a shoal and fishing nets to the test range and manipulating them between trials. During earlier design iterations we considered leveraging the ability of the ATK software to insert them as live-virtual constructs using simulated sonar data. However, the ability to simulate this data relied upon the vehicle's odometry making the resulting behaviors no different than those for no-go regions.

Additional state space element information is listed in Table 9.1.

9.4.4 Scenario Generation

Using the techniques outlined in Sect. 9.2 and our previous work [12], we generated 150k scenarios for the November Keyport Demonstration, allocating 30k to global search and 120k to adaptive search methods. Due to scenarios pruned for being invalid (e.g., no-go area generated on top of a waypoint) or to runs failing on the cluster, only 123k completed successfully, and these constituted the final dataset.

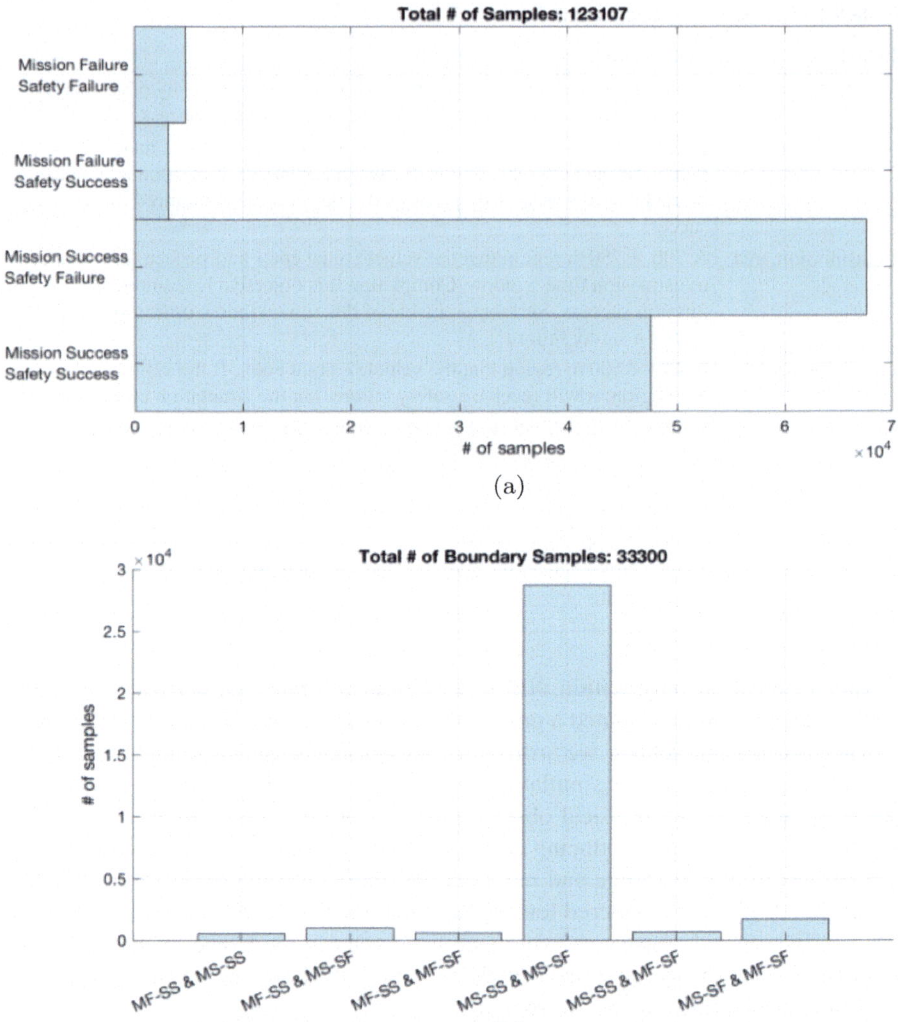

Fig. 9.5 Charts of number of completed runs in the pretest dataset for each performance mode (above) and boundary type (below). (**a**) Performance mode distributions. (**b**) Boundary type distributions

In this dataset, the Mission Success performance mode dominated the scenario outcomes, with 93% of all scenarios successfully completing the Mission Area and Transmission Area objectives. In contrast, 57% of all scenarios failed on the safety criterion. This effect can be seen in Fig. 9.5, where the Mission SuccessSafety Failure (MSSF) and Mission Success/Safety Success (MSSS) performance modes vastly outnumber those of all other modes and the boundary between them constitutes 85% of all boundary pairs.

The ease of the mission can be attributed to the size of the operational area. The physical size of the test range was smaller than some of those used in the original simulation studies of [13]; this reduced size also diminished the stress caused by the obstacle configuration and tidal current on the vehicle. To determine which mission parameters have the most impact, we extract the sensitivities computed by the S-NDV search algorithm [13]. As can be seen in Fig. 9.6, the primary input variables which affected the vehicle's ability to complete the mission successfully were the timing of the transmission area and the amount of starting battery capacity. Once the vehicle had more than 16% battery capacity, it nearly always completed the mission. The only exception was when the transmission window ended so early that it was impossible for the vehicle to arrive in time.

One of the objectives when selecting field tests was to find pairs of cases that represent different decisions made by the autonomy software. Therefore, it was necessary to analyze these performance boundaries to separate those where a change in performance is a result in a change in vehicle behavior from those where a change in performance is due to stochastic elements in the simulation. Changes in behavior are more easily replicated in the field and provide more salient information about the underlying decision-making of the autonomy software.

The primary reason for failing the safety criteria was colliding with a no-go area, which is illustrated quite clearly in Fig. 9.7. Upon further inspection we discovered that the primary factor for no-go area collision was the presence of a no-go area near one of the waypoints Fig. 9.8. This results in scenarios where the mission objectives and the safety criteria create conflicts between reaching a waypoint and safely avoiding an obstacle. It also implies that, any time an obstacle is placed directly in the desired path, the vehicle has a high probability of colliding.

This is an example of a "fuzzy" performance boundary where obstacles closer to the path result in higher collision probabilities due to deficiencies in the navigation software, errors in commanding the vehicle, deviations due to current, or inaccuracies in the estimated position. As scenarios approach the boundary, the uncertainty in the final score grows, with the scenario achieving either safety success or failure due to minor deviations in the boundary (Fig. 9.9). Thus, despite the prevalence of boundary pairs existing on a particular boundary, the majority of boundary pairs represent only minor variations in a trajectory rather than a change in the autonomy software's behavior.

These scenarios are useful for understanding a shortcoming of the autonomous vehicle software and that it does not leave enough room between itself and a no-go zone to account for state-estimation errors. However, since this is due to a stochastic effect, it would be difficult to replicate this boundary in the field without many repeated trials. Each of those repeated trials would result in the autonomy following the same trajectory and behaviors. As our goal is to capture as much variety in the vehicle's behaviors as possible using limited trials we chose to eliminate this type of safety failure from consideration for field tests.

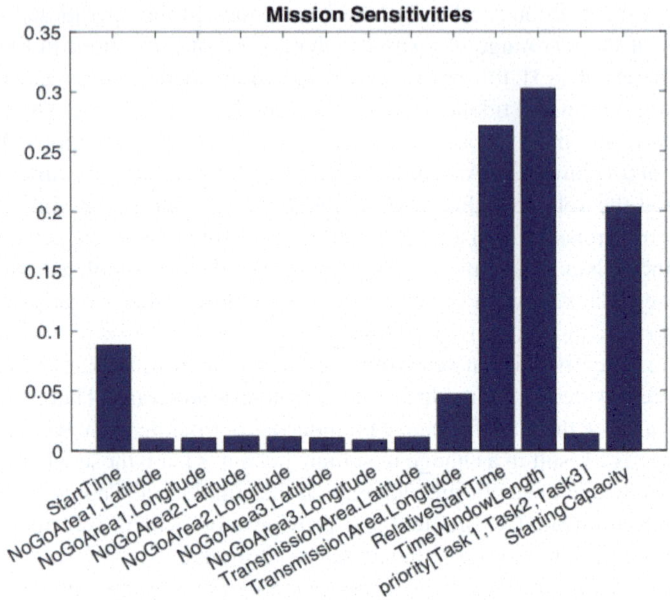

(a) Mission Parameter Response

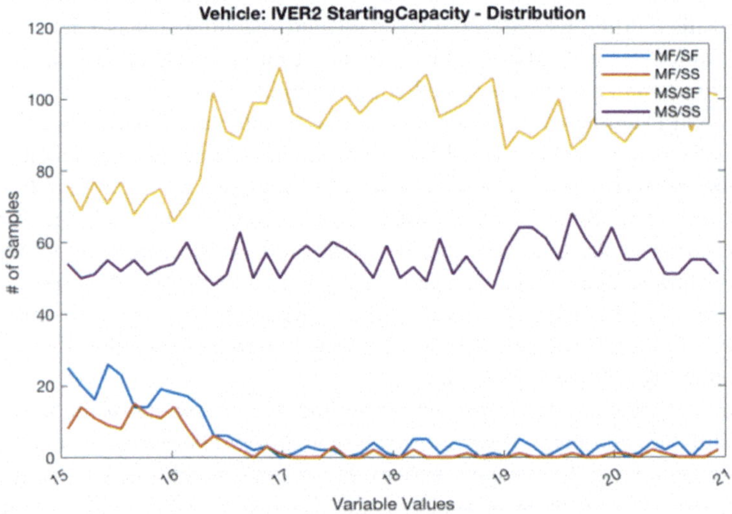

(b) Individual Parameter Distributions

Fig. 9.6 (Above) The input sensitives computed by the S-NDV search algorithm for the on mission success. (Below) Line histogram of the number of cases for each performance mode as the levels of starting battery capacity varies. (**a**) Mission parameter response. (**b**) Individual parameter distributions

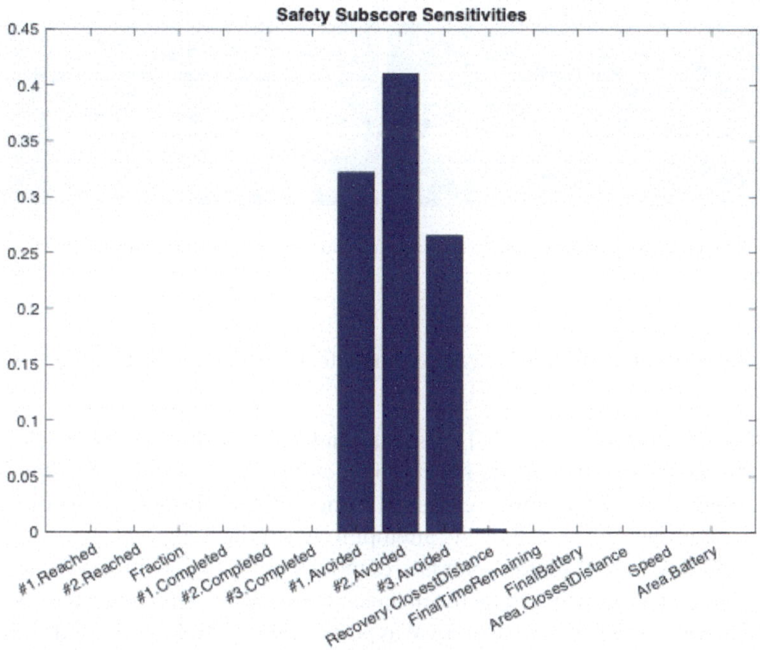

Fig. 9.7 The estimated sub-score response for the safety success criterion

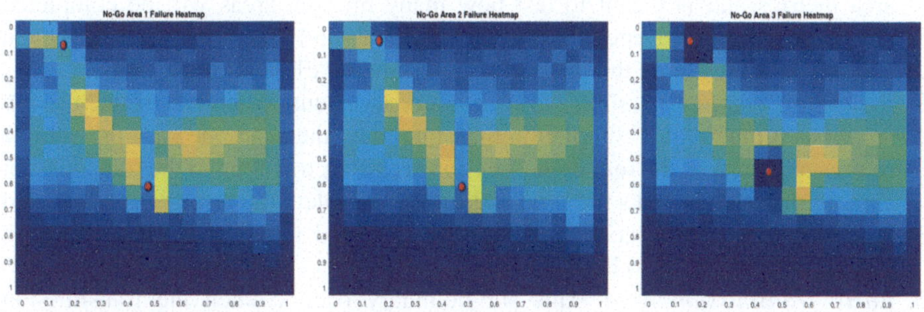

Fig. 9.8 Heatmap showing which no-go area positions were most likely to cause collisions. Lighter areas shown in yellow are more probable, while darker areas in blue are less probable. Red markers indicate the positions of the waypoints. Dark regions on top of the waypoints are result of these configurations being marked as "invalid" scenarios

9.5 Scenario Selection

Given the limited number of runs we could perform at the test range, we decided to create a testing suite of ten scenarios consisting of five boundary pairs. Each boundary was designed to test a different behavior of the system as identified using our sub-boundary algorithm. We utilized sub-score boundary generation to test the

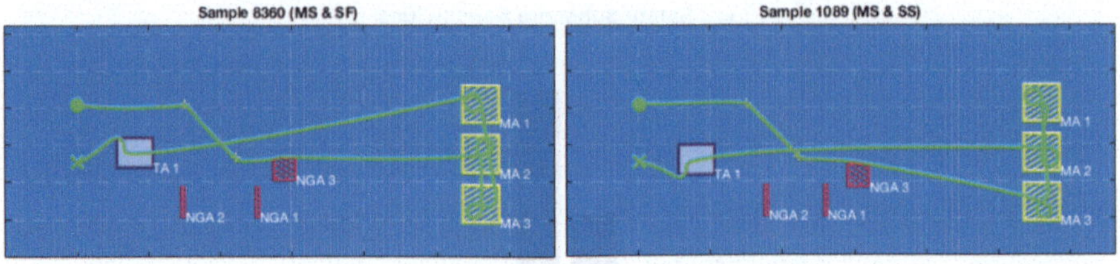

Fig. 9.9 Example of boundary pair between safety success and failure, illustrating the "fuzzy" boundary

sub-scores identified as the biggest contributors during the sub-score sensitivity analysis.

1. *Collision with No-Go Area 1.* This sub-boundary consists of the vehicle colliding with No-Go Area 1 when attempting to reach the first waypoint. In simulation if the no-go area was within a certain distance of the obstacle, it would neither avoid the no-go area nor abort its attempt to reach the waypoint.
2. *Waypoint 2 Abort.* This sub-boundary demonstrates a major shift in the trajectory the vehicle takes depending on the distance between No-Go Area 3 and Waypoint 2. If No-Go Area 3 is too close to Waypoint 2, the vehicle will abort; otherwise it will attempt to reach Waypoint 2 but will collide with the no-go area.
3. *Mission Area 1 Abort.* This sub-boundary occurs within the individual mission area metrics and is meant to test how many mission areas will be completed during a single trial. Depending on the fuel levels and time windows, the vehicle will either complete all the mission areas or abort early and return home.
4. *Return to Complete Mission.* This sub-boundary occurs when one of the mission areas has a much later completion time than the others. In one case the vehicle returns to complete the remaining mission area after traveling to the transmission area. In the other it lacks the time and fuel to complete the last mission area and instead returns to the recovery point. It also included a placement of No-Go Area 3 that was considered to be particularly difficult.
5. *Early Transmission Area.* This sub-boundary occurs when the transmission area is completed before any of the mission areas. This can occur when the transmission window ends very early in the mission. This also results in a more efficient path which can mean the difference between mission success and failure.

Of all the sub-boundary types these five were considered to be the most relevant to the performance of the system and the most reproducible in the field. The remaining sub-boundaries were primarily permutations of the same behaviors (e.g., mission areas in different orders and collisions occurring at the same time as a mission abort) or were results of software faults which were based on nondeterministic effects. While the latter of these are an important phenomenon to test more thoroughly testing, it in the water would require more time than we were allocated.

9.5.1 Dithering Study Results

Our search techniques achieve high-resolution sampling along the primary boundaries of our system (Mission Success, Safety Success), but as discussed in the previous section these did not always represent cases that would be interesting or easily reproducible on the water. We applied sub-boundary identification to identify pairs of scenarios which actually demonstrate changes in agent behaviors, but since these new criteria were only introduced during post-processing these regions were only sparsely sampled by the adaptive sampling algorithms. This means that we may have inaccurate estimates of how far our scenarios are from the desired sub-boundary or that there may even be sub-boundaries which we have failed to discover. Finally, uncertainty in the execution of the scenario may also cause the simulated result to fail to match the real-world test. Therefore we need to explore the regions of the testing space near our selected test scenarios to ensure that we understand the level of uncertainty associate with each, as well as determining whether any of these scenarios exist on previously unidentified sub-boundaries.

To address these concerns we performed a dithering study of the selected test scenarios. In this study each scenario in the testing suite was dithered with Gaussian noise to generate additional 2k scenarios in the neighborhood of test scenario. We then ran subclustering and sub-boundary identification scripts to identify all possible behaviors that could occur in the region of each boundary pair. The distances of the test scenarios to the boundaries discovered by the dithering study are listed in Table 9.2. For each boundary pair we provide the name of the sub-boundary type and the index for the score-tree element where the boundary occurs. For example, our first boundary pair is a collision sub-boundary which occurs on sub-score element 11, the metric that determines whether No-Go Area 1 was violated.

For a majority of the test scenarios we selected the dithering study revealed they were much closer to a sub-boundary than the original 120k dataset indicated. Boundary pairs 1 and 5 both resulted in the same sub-boundary type as indicated in the original set, simply with a closer distance. More concerning was the presence of previously undiscovered sub-boundaries in the vicinity of the other boundary pairs. We split these into two categories: orthogonal sub-boundaries and parallel sub-boundaries.

9.5.2 Orthogonal Sub-boundaries

Orthogonal sub-boundaries occur when multiple subclusters overlap and crossing one sub-boundary does not mean crossing the other. An example of this type of sub-boundary was found for boundary pair 2. The original sub-boundary we selected was the Waypoint 2 sub-score, which is Element 15 in our sub-score tree. When we ran the dithering study, we found that these test cases were also close to a Mission Abort sub-boundary, which is Element 21 in our sub-score tree. Both the original

Table 9.2 Dithering results

Boundary pair	Original boundary distance	Original boundary type	Scenario ID	Dithering boundary distance	Dithering boundary type
1	0.259	Collision (11)	65315	0.101	Collision (11)
			71349	0.256	Collision (11)
2	0.340	Waypoint (15)	55433	0.050	Mission Return (21)
			53451	0.042	Mission Return (21)
3	0.247	Mission Abort (26)	21063	0.050	Mission Abort (26)
			127386	0.042	Mission Abort (27)
4	0.2526	Mission Abort (35)	21063	0.052	Mission Abort (35)
			127386	0.073	Mission Abort (35)
5	0.2950	Transmission (8)	112122	0.054	Transmission (8)
			145263	0.073	Transmission (8)

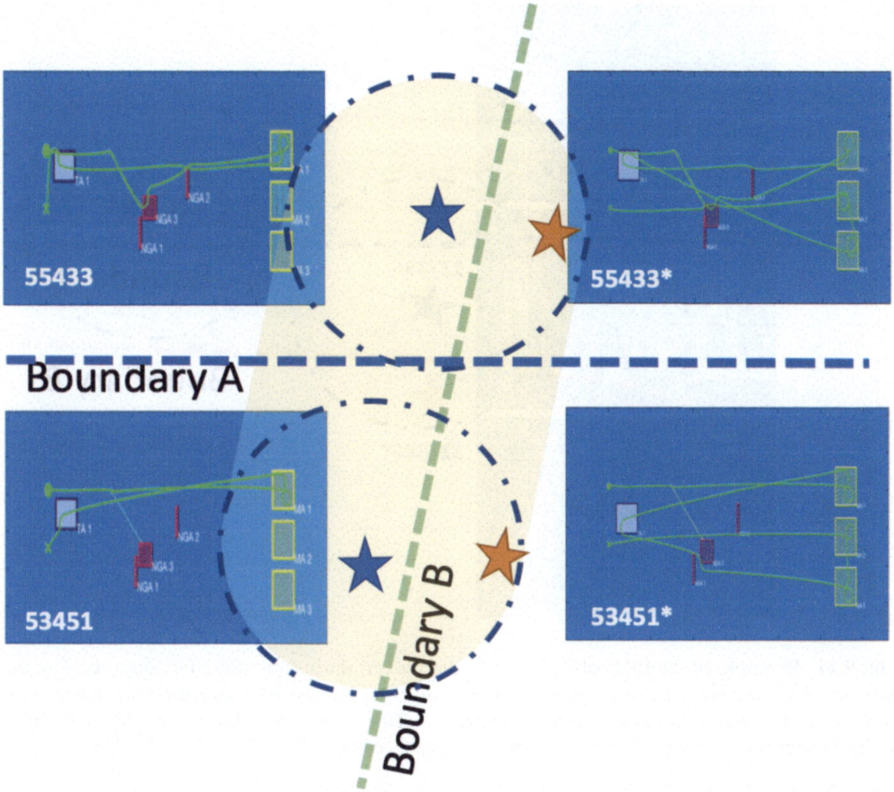

Fig. 9.10 Example of orthogonal sub-boundaries discovered during the dithering analysis. The top two scenarios are the original boundary pair, while the bottom two scenarios are the nearest neighbors discovered in the dithering analysis. While the right-left pairings provide an example of a Waypoint Abort sub-boundary, the top-bottom pairings provide an example of a Mission Return sub-boundary

boundary pair and the new sub-boundary pairs for these scenarios are shown in Fig. 9.10.

In this example the dithering results indicated that the scenarios were incredibly sensitive to the perceived fuel levels of the system. If the vehicle had sufficient fuel, it would return to complete the mission areas after surfacing in the transmission zone. Otherwise it would return to the recovery point immediately after transmitting the data. This decision to return to the mission is completely independent from the decision to abort the waypoint.

We selected this pair to test the sub-boundary caused by abandoning the waypoint, which still occurs even after the scenarios have been dithered. Therefore we were able to conclude that even if some bias occurred during test that we would still test the correct sub-boundary, albeit with the possibility that we would see

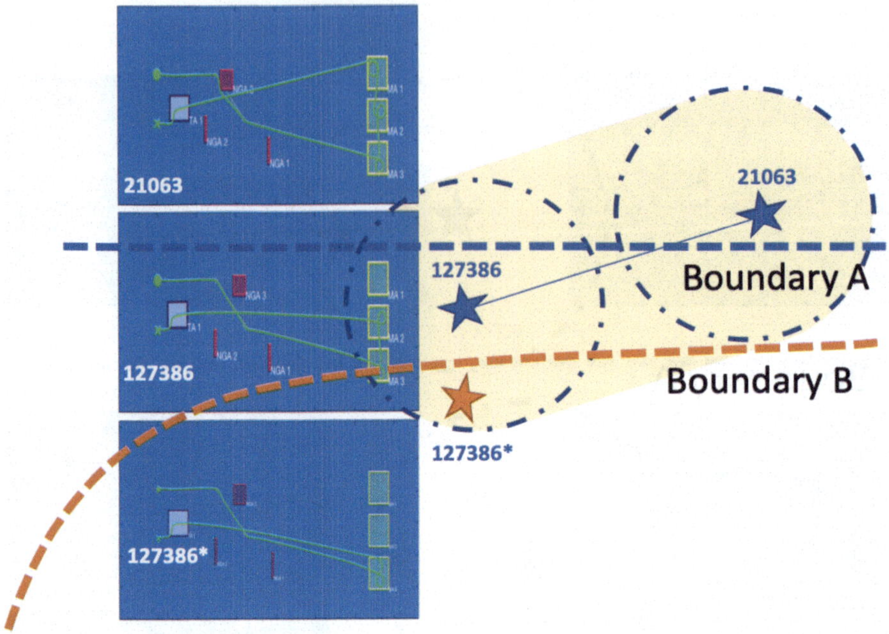

Fig. 9.11 Example of parallel sub-boundaries discovered during the dithering analysis. The top two scenarios are the original boundary pair, while the bottom two scenarios are the nearest neighbors discovered during the dithering analysis. These scenarios show examples from three different subclusters with varying levels of mission success

trajectories similar to Dithering Scenarios (5922) and (89933) rather than the ones predicted by Scenarios 53451 and 55433.

9.5.3 *Parallel Sub-boundaries*

Parallel sub-boundaries occur when multiple thin subclusters border one another, and their boundaries are aligned in the same direction in the state space. This is the case for boundary pairs 3 and 4 which were selected to test two highly sensitive regions of our state space. Our dithering study revealed that all of these boundary pairs were particularly close to the intended sub-boundary and nearby parallel sub-boundaries, meaning that any bias in the execution of the scenario could easily push one of these cases across the boundary. The original scenarios and the dithering generated scenarios for boundary pair 3 are shown in Fig. 9.11.

In this example a very small shift in the transmission window length causes the vehicle to complete anywhere from 1 to 3 of the mission areas. Our objective for this pair was to test the sub-boundary associated with the Mission Area 1 sub-score, which is Element 26 in our sub-score tree. The dithering study revealed not only that scenario 21063 is substantially closer to that sub-boundary than we originally estimated but also that scenario 127386 is adjacent to a parallel sub-boundary for the Mission Area 2 sub-score, which is element 28 in our sub-score tree.

These results indicate that our vehicle may exhibit behaviors from any of the subclusters represented in the dithering study for this pair. If the vehicle executes the mission more quickly or slowly than predicted in the simulation, we may miss the sub-boundary entirely. This results in either testing the wrong sub-boundary (e.g., the Mission Area 2 sub-boundary) or no boundary will be demonstrated at all due to both scenarios exhibiting the same behavior.

To ensure that the behaviors found in simulation can be replicated in the field the scenarios must be robust to variations. We achieved this by modifying our scenarios to move them further from one another in the state space. For boundary pair 3, this involved increasing and decreasing the transmission window length for scenarios 21063 and 127386, respectively. For boundary pair 4, this involved increasing and decreasing the starting battery for scenarios 61836 and 134243, respectively. Our test plan called for the unmodified scenarios to be executed first, and if they did not match their predicted behaviors, then the modified scenarios would be used in their place for any subsequent tests.

Running field tests to confirm the existence of performance boundaries is inherently difficult as the goal is to execute tests where small changes in the scenario cause large changes in the behavior. If successful, these tests provide substantial information about the decision-making process of the autonomy under test, but their sensitivity to changes in the vehicle or environment makes them more difficult to reproduce than standard field tests. Performing the dithering study therefore is critical to ensure the scenarios being selected will be robust to uncertainty while also predicting the full envelope of behaviors that can occur. It also provided us with valuable information about how to modify a scenario in the situation that it did not result in the predicted behavior.

9.6 Demonstration Results

The Range Adversarial Planning Tool (RAPT) final demonstration was held at NUWC Keyport November 14–17, 2017 using an IVER2 UUV running the JHU/APL Autonomy Toolkit for its surrogate autonomy.

The dataset utilized for the final test plan and analysis consisted of 120,000 RAPT-generated scenarios. We were constrained to only 24 in-water sorties for the field test period and so selected five boundary pairs from the 120k runs for execution during the in-water portion of the demonstration. The intention was to run each scenario twice with backup runs reserved for scenarios where unexpected behaviors occurred.

During the in-water trials at Keyport, it was quickly discovered that the tidal data files had almost no relationship to the actual currents experienced by the vehicle. This was due in part to our lack of control over when we could start each trial. In addition, the IVER2 executed the scenarios at a faster speed than was commanded within the simulation. These two factors taken together meant there was a strong execution bias in the results where the vehicle on the water behaved as if the time windows were later and the starting battery level was higher.

These results were fully encompassed and predicted by the dithering study and could be easily rectified by using the modified scenarios we had prepared in advance. In the cases where the field test did not result in the behavior predicted in the original simulated scenario, we performed the following steps. First, we searched the dithering set for the scenario that mostly closely matched the results we saw on the water. We then modified the original scenario by moving it directly away from that matching scenario in the state space, effectively doubling the distance between the pair. In every case after this modification we were able to reproduce the desired behavior. The rest of this section is devoted to a discussion of each of the boundary pairs.

The rest of this section is devoted to a discussion of each of the boundary pairs.

9.6.1 Boundary 1: No-Go Collision

This boundary pair tests the vehicle's navigation ability and how it handles waypoints near no-go areas and obstacles. In this boundary pair, the vehicle fails to abort a waypoint that is too close to a no-go area and ultimately violates the no-go region. This set of runs was an unmitigated success, both confirming the existence of the boundary and directly replicating the behavior of the simulation without any need to modify the scenarios. A side-by-side comparison of the predicted results versus the on-water results is shown in Fig. 9.12.

9.6.2 Boundary 2: Waypoint Abort

This boundary pair demonstrates the threshold at which the vehicle will abandon a waypoint due to the positioning of a no-go area. During the dithering study, it was discovered that these test scenarios were incredibly close to another sub-boundary where the vehicle continues the mission after completing the transmission area. Execution bias in the on-water tests caused the vehicle to complete the mission more quickly and with more fuel than originally predicted. However, it was still within the set of results captured by the dithering study. Since this change did not affect the behavior we were attempting to test, aborting the waypoint, we decided modification to match the original simulated result was not necessary. A side-by-side comparison of the scenarios is shown in Fig. 9.13.

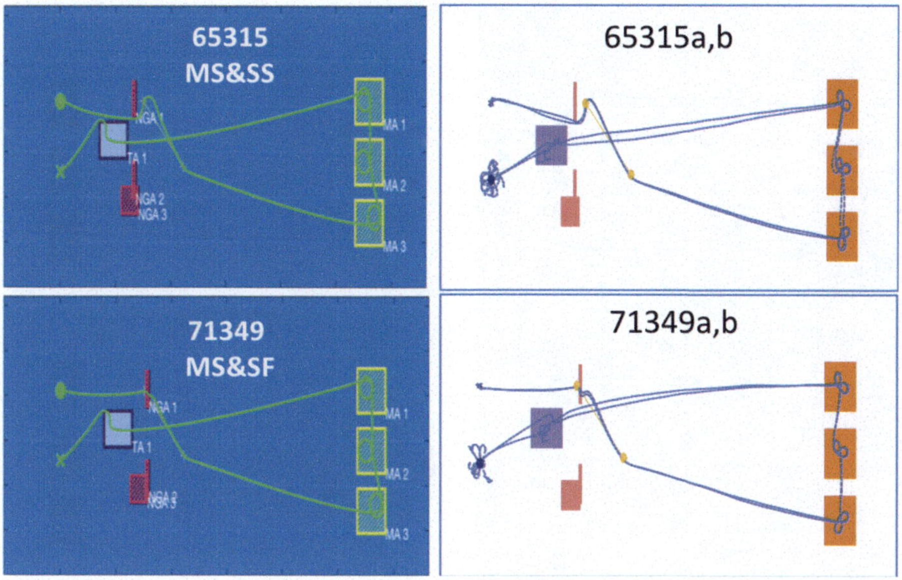

Fig. 9.12 Boundary pair 1 plots comparing the simulated result (left) to trajectory executed during in-water tests (right). Multiple trajectory lines are due to overlaying the results of all field tests on top of one another

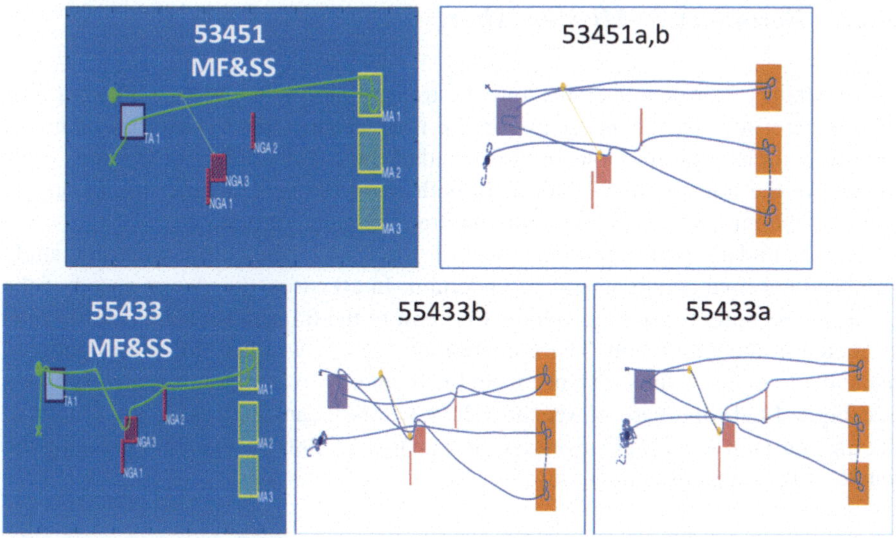

Fig. 9.13 Boundary pair 2 plots comparing the simulated result (left) to trajectory executed during in-water tests (right). Multiple trajectory lines are due to overlaying the results of all field tests on top of one another

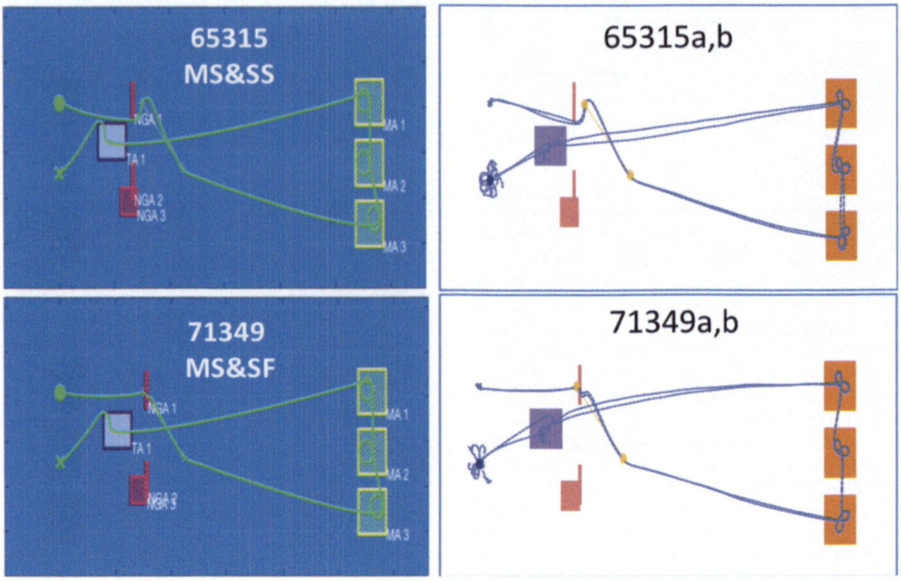

Fig. 9.14 Boundary pair 3 plots comparing the simulated result (left) to trajectory executed during in-water tests (right). Multiple trajectory lines are due to overlaying the results of all field tests on top of one another

9.6.3 Boundary 3: Mission Abort

This boundary pair demonstrates how the transmission window time affects transmission decisions by the autonomy. In the first scenario, the vehicle has plenty of time to complete the mission. In the boundary pair, the vehicle must return early to achieve the transmission window. In both cases the vehicle lacks the battery to continue the mission after completing the transmission window.

This boundary pair represents one of the cases where the dithering study indicated the final results would be uncertain. In all cases, the slower speed of the IVER in the water caused the vehicle to return to the transmission area earlier than predicted in the simulation. The scenarios designated with an "M" were modified manually to increase their distance from the boundary to achieve greater robustness. See Fig. 9.14. In the case of scenario 21063, this meant increasing the length of the transmission window. Conversely, in scenario 127386 this meant decreasing the length of the transmission window.

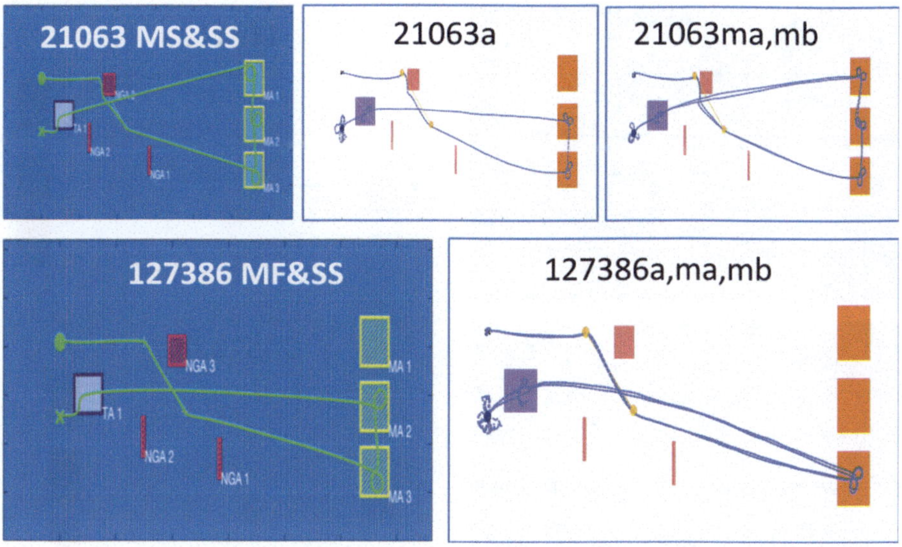

Fig. 9.15 Boundary pair 4 plots comparing the simulated result (left) to trajectory executed during in-water tests (right). Multiple trajectory lines are due to overlaying the results of all field tests on top of one another

9.6.4 Boundary 4: Mission Return

This boundary pair exercises the decision-making of the autonomy on whether to complete the mission after executing the transmission area. This is entirely determined by the starting battery level which is very close for both scenarios in the boundary pair. This boundary pair is incredibly close in all inputs, giving a good estimate of the threshold and also introducing risk due to noise and execution bias. A side-by-side comparison of the simulated result versus the on-water result is shown in Fig. 9.15.

In this case execution bias causes the results to shift away from what was previously predicted. Since this possibility was predicted during our dithering study of the selected runs, we created a set of modified scenarios that are designated with an "M" in Fig. 9.15 where the amount of fuel for scenario 13423 was reduced. However we underestimated the sensitivity of this boundary to both the transmission time and the starting fuel level, and while the modification did move scenario 13423 to the correct side of the sub-boundary which we were testing it returned earlier than was predicted in the original simulation.

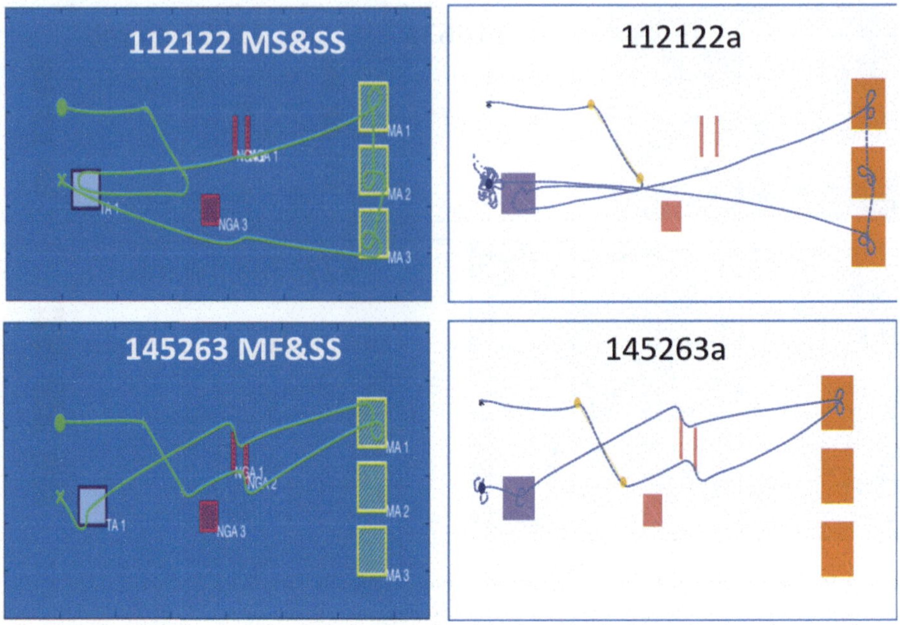

Fig. 9.16 Boundary pair 5 plots comparing the simulated result (left) to trajectory executed during in-water tests (right). Multiple trajectory lines are due to overlaying the results of all field tests on top of one another

9.6.5 Boundary 5: Transmission First

This boundary pair demonstrates vehicle behavior that will attempt the transmission area before completing all mission areas but always after the waypoints. In one scenario, this means that the vehicle achieves a total success since completing the transmission area first is the most fuel efficient. In the other, it fails the mission since there is not enough fuel to complete two full transits of the operational area. We were only able to perform a single run of this boundary due to unsafe weather conditions. This single pair managed to perfectly replicate the simulation results. A side-by-side comparison of the simulated result versus the on-water result is shown in Fig. 9.16.

9.6.6 Demonstration Results Discussion

For all the scenarios tested we were able to successfully predict the behavior either during the initial search or during the dithering study. The execution bias meant that the performance boundaries predicted by the dithering study were executed instead of those predicted by the 120k dataset. For boundary pair 2 this meant that the behavior being tested, aborting a waypoint, was still executed correctly, but the remainder of the trajectory executed on water differed from the simulation result. For boundary pairs 3 and 4 this execution bias meant that both scenarios would fall on the same side of the performance boundary. In order to account for this difference, we had to manually adjust the scenarios to start with less fuel in order. Once this adjustment was made, the original boundary was confirmed.

9.7 Conclusions

The objective for the RAPT program was to demonstrate that our simulation-based framework can be used to generate realistic and challenging test scenarios for in-water execution. Using hierarchical scoring and sub-boundary identification, we were able to identify five distinct types of performance boundaries for the system under test. During the November 14–17, 2017 demonstration at NUWC Keyport, these test scenarios were successfully executed in-water and validated the results of the simulation. This is both the first field demonstration of test scenarios generated using computational methods and the first demonstration of automatically generating a mission with more than ten input parameters.

Several challenges were encountered in the scenario generation and execution portions of the demonstration that are likely to occur in other domains. The first is a noisy performance surface which can throw off a search and boundary identification process if the noise is too high. Despite this, we were able not only to identify meaningful boundary cases but many of these cases were even closer to the boundary than originally predicted. The second was the transition of simulation results to in-water tests. Differences in both the dynamics and behavior of the hardware can skew performance of the vehicle, and this was seen with the slower speeds on the IVER. However, by using a combination of highly refined sampling in the region of the selected tests and using sub-boundary analysis to identify a greater variety of performance modes, we were able to predict the possible variations before execution in-water. As such, we were able to execute the tests with confidence and the vehicle's performance on water aligned with our predictions.

Simulation-based testing serves a vital role in the process of developmental and operational testing. It is not possible to run enough field tests to fully verify the system. Instead, field tests are best used to confirm and build confidence in the simulated results. Using adaptive sampling and performance boundary identification with software in the loop is the key to our success. It allowed us to identify the most prominent behaviors of the autonomy under test and the scenarios that would be capable of replicating those behaviors on the hardware platform. Every one of the field tests was designed to test a different behavior, maximizing the coverage of possible outcomes while minimizing the number of tests that needed to be run. It meant that we not only had a plan that optimized our time on the water but also had contingency plans prepared for a variety of outcomes.

Acknowledgments I would like to thank rest of the members of the JHU/APL RAPT team for supporting this study and the effort they put into designing the autonomous vehicle software, integrating it with the IVER UUV, and executing the field test: Paul Stankiewicz, Jordan Appler, Melissa Huntley, Doug Grantham, Johan Stewart, Michael Biggins, Chad R. Hawthorne, and James Horris.

This work was funded by the Test Resource Management Center (TRMC) and Test Evaluation/Science and Technology (T&E/S&T) Program and/or the U.S. Army Contracting Command Orlando (ACC-ORL-OPB) under contract W900KK-19-C-0004.

Any opinions, findings, and conclusions or recommendations expressed in this material are those of the author(s) and do not necessarily reflect the views of the Test Resource Management Center (TRMC) and Test Evaluation/Science and Technology (T&E/S&T) Program and/or the U.S. Army Contracting Command Orlando (ACC-ORL-OPB).

References

1. Bertaska, I. R., Shah, B., von Ellenrieder, K., Švec, P., Klinger, W., Sinisterra, A. J., Dhanak, M. & Gupta, S. K. Experimental evaluation of automatically-generated behaviors for USV operations. *Ocean Engineering* **106,** 496–514 (Sept. 2015).
2. Kuwata, Y., Wolf, M. T., Zarzhitsky, D. & Huntsberger, T. L. Safe Maritime Autonomous Navigation With COLREGS, Using Velocity Obstacles. *IEEE Journal of Oceanic Engineering* **39,** 110–119 (Jan. 2014).
3. Stern, R. E. *et al.* Dissipation of stop-and-go waves via control of autonomous vehicles: Field experiments. *Transportation Research Part C: Emerging Technologies* **89,** 205–221 (Apr. 2018).
4. Eriksen, C. C., Osse, T. J., Light, R. D., Wen, T., Lehman, T. W., Sabin, P. L., Ballard, J. W. & Chiodi, A. M. Seaglider: a long-range autonomous underwater vehicle for oceanographic research. *IEEE Journal of Oceanic Engineering* **26,** 424–436 (Oct. 2001).
5. Steinberg, M., Stack, J. & Paluszkiewicz, T. Long duration autonomy for maritime systems: challenges and opportunities. *Autonomous Robots* **40,** 1119–1122 (Oct. 2016).
6. Stankiewicz, P. G. & Mullins, G. E. *Improving evaluation methodology for autonomous surface vessel COLREGs compliance* in *OCEANS 2019-Marseille* (2019), 1–7.
7. Harris, A. & Conrad, J. M. *Survey of popular robotics simulators, frameworks, and toolkits* in *SoutheastCon, 2011 Proceedings of IEEE* (IEEE, 2011), 243–249.
8. Pepper, C., Balakirsky, S. & Scrapper, C. *Robot simulation physics validation* in *Proceedings of the 2007 Workshop on Performance Metrics for Intelligent Systems* (ACM, 2007), 97–104.
9. Torres-Torriti, M., Arredondo, T. & Castillo-Pizarro, P. Survey and comparative study of free simulation software for mobile robots. *Robotica* **34,** 791–822 (Apr. 2016).
10. Tuncali, C. E., Fainekos, G., Ito, H. & Kapinski, J. *Sim-ATAV: Simulation-Based Adversarial Testing Framework for Autonomous Vehicles* in *Proceedings of the 21st International Conference on Hybrid Systems: Computation and Control (Part of CPS Week)* (ACM, New York, NY, USA, 2018), 283–284.
11. Zou, X., Alexander, R. & McDermid, J. in *Computer Safety, Reliability, and Security* 33–48 (Springer, 2014).
12. Mullins, G. E., Stankiewicz, P. G. & Gupta, S. K. *Automated generation of diverse and challenging scenarios for test and evaluation of autonomous vehicles* in *Robotics and Automation (ICRA), 2017 IEEE International Conference on* (2017), 1443–1450.
13. Mullins, G. E., Stankiewicz, P. G., Hawthorne, R. C. & Gupta, S. K. Adaptive generation of challenging scenarios for testing and evaluation of autonomous vehicles. *Journal of Systems and Software* **137,** 197–215 (2018).
14. Corso, A., Moss, R. J., Koren, M., Lee, R. & Kochenderfer, M. J. A survey of algorithms for black-box safety validation. *arXiv preprint arXiv:2005.02979* (2020).

15. Underwood, S., Bartz, D., Kade, A. & Crawford, M. in *Road Vehicle Automation 3* (eds Meyer, G. & Beiker, S.) 91–109 (Springer International Publishing, Cham, 2016).
16. Alam, S., Abbass, H. A., Lokan, C. J., Ellejmi, M. & Kirby, S. *Computational Red Teaming to investigate Failure Patterns in Medium Term Conflict Detection* in *8th EUROCONTROL Innovation Research Workshop, EUROCONTROL Experimental Center, Bretigny-sur-Orge, France* (2009).
17. Alam, S., Lokan, C., Aldis, G., Barry, S., Butcher, R. & Abbass, H. Systemic identification of airspace collision risk tipping points using an evolutionary multi-objective scenario-based methodology. en. *Transportation Research Part C: Emerging Technologies* **35,** 57–84 (Oct. 2013).
18. Togelius, J., Yannakakis, G. N., Stanley, K. O. & Browne, C. Search-Based Procedural Content Generation: A Taxonomy and Survey. *IEEE Transactions on Computational Intelligence and AI in Games* **3,** 172–186 (Sept. 2011).
19. Althoff, M. & Lutz, S. *Automatic generation of safety-critical test scenarios for collision avoidance of road vehicles* in *2018 IEEE Intelligent Vehicles Symposium (IV)* (2018), 1326–1333.
20. Patrick, M., Alexander, R., Oriol, M. & Clark, J. A. Subdomain-based test data generation. en. *Journal of Systems and Software* **103,** 328–342 (May 2015).
21. Schwartz, A. & Do, H. Cost-effective regression testing through Adaptive Test Prioritization strategies. *Journal of Systems and Software* **115,** 61–81 (2016).
22. Hanh, L. T. M., Binh, N. T. & Tung, K. T. A Novel Fitness function of meta-heuristic algorithms for test data generation for Simulink models based on mutation analysis. *Journal of Systems and Software* **120,** 17–30 (Oct. 2016).
23. Arnold, J. & Alexander, R. in *Computer Safety, Reliability, and Security* 33–44 (Springer, 2013).
24. Dimjašević, M. & Giannakopoulou, D. *Test-case generation for runtime analysis and vice versa: verification of aircraft separation assurance* en. in (ACM Press, 2015), 282–292.
25. Sarkar, M., Homaifar, A., Erol, B. A., Behniapoor, M. & Tunstel, E. PIE: a Tool for Data-Driven Autonomous UAV Flight Testing. *Journal of Intelligent & Robotic Systems,* 1–18 (2019).
26. Koos, S., Mouret, J.-B. & Doncieux, S. The transferability approach: Crossing the reality gap in evolutionary robotics. *IEEE Transactions on Evolutionary Computation* **17,** 122–145 (2012).
27. Shorten, C. & Khoshgoftaar, T. M. A survey on image data augmentation for deep learning. *Journal of Big Data* **6,** 1–48 (2019).
28. Loquercio, A., Kaufmann, E., Ranftl, R., Dosovitskiy, A., Koltun, V. & Scaramuzza, D. Deep Drone Racing: From Simulation to Reality With Domain Randomization. *IEEE Transactions on Robotics* **36,** 1–14 (2020).
29. Chalmers, R., Scheidt, D., Neighoff, T., Witwicki, S & Bamberger, R. *Cooperating unmanned vehicles* in *AIAA 1st Intelligent Systems Technical Conference* (2004), 1–8.
30. Scheidt, D., Lutz, R., D'Amico, W., Kleissas, D., Chalmers, R. & Bamberger, R. Safe Testing of Autonomous Systems Performance.
31. Fossen, T. I. *Guidance and control of ocean vehicles* (John Wiley & Sons Inc, 1994).
32. Pals, C. R. & Newman, J. Thermal modeling of the lithium/polymer battery: I. Discharge behavior of a single cell. *Journal of the Electrochemical Society* **142,** 3274 (1995).
33. Tides, N. Currents. 2013. *NOAA tide predictions* http://tidesandcurrents.noaa.gov/tidepredictions.html. Accessed **21** (2014).

Chapter 10
Case Studies in Verifying Spacecraft Autonomy

Martin S. Feather, Lorraine E. Prokop, and Daniel J. Dorney

Chapter 10 Editorial Note

This chapter assesses the effectiveness of traditional verification methods against autonomous components through three case studies in spacecraft autonomy verification. It specifically addresses stakeholder questions related to certification issues, including security and effectiveness, and the definition of system limitations in the context of spacecraft autonomy verification. System usefulness and trust are key elements of this. Its description of how different techniques have been used to verify different systems supports improved understanding of how a system and its autonomous functions can be verified in practice. The OpNav system was verified based on artificial ground truth data generated by a simulation tool—the discussion of this approach addresses not only the elements of technical challenge 1 (creating an adequate system model) concerned with what level of fidelity can be used to determine an adequate model, but also the portion of technical challenge 6 (determining an adequate level of abstraction) concerned with environmental levels of abstraction and the portion of technical challenge 4 (identifying and modeling components) associated with defining the properties that are relevant to the task and behavior to be tested. This chapter, while not providing theoretical solutions to technical challenges 5 (appropriate levels of fidelity) or 10 (quantity of verification evidence), does empirically demonstrate that both the level of fidelity achieved and the amount of evidence gathered were sufficient to establish functionality of a

M. S. Feather (✉)
Jet Propulsion Laboratory, California Institute of Technology, Pasadena, CA, USA
e-mail: autonomyassurance@gmail.com

L. E. Prokop · D. J. Dorney
NASA, Langley Research Center, Hampton, VA, USA
e-mail: lorraine.e.prokop@nasa.gov; daniel.j.dorney@nasa.gov

S. Redfield et al. (eds.), *Verification of Autonomous Systems*,
https://doi.org/10.1007/978-3-031-88546-4_10

system in practice. Stakeholder questions are defined in the Chap. 1, Introduction, while technical challenges are presented and discussed in Chap. 2, Verification Challenges.

Abstract Autonomy has proven vital to many aspects of space missions. On crewed spacecraft, it is necessary when the response is needed faster than human reaction time can provide. On robotic (uncrewed) spacecraft, it is necessary when the delay caused by the round-trip light time precludes Earth-controlled operation. Autonomy goes beyond straightforward automation when the dynamics of a spacecraft's situation cannot be predicted in advance. Verification of autonomy, essential if we are to trust human lives and expensive missions to its control, is particularly challenging when there is a wide range of possible such situations. This chapter provides case studies of several successfully verified autonomous software systems acting in several phases of spacecraft flight (launch, transit, and entry and landing) and can serve as an example to future spacecraft in methods of successfully assuring spacecraft autonomy.

10.1 Introduction

This chapter summarizes several case studies of software successfully tested to perform spacecraft autonomy spanning comprehensive aspects of a space mission.

- During launch, the time-to-criticality (the length of time between first sign of a problem and that problem critically affecting operations) can be below human reaction time, necessitating the need for flight system autonomy. This is discussed in the context of the Atlas V Emergency Detection System (EDS), responsible for detecting when there is the need for a launch abort to save the human crew.
- During the transit phases of a crewed mission, loss of communication with ground control may threaten the successful return of the spacecraft and its crew to Earth. This is discussed in the context of the autonomous navigation software (OpNav) for the Orion Artemis-1 Mission, developed and verified to provide the ability of the spacecraft to self-navigate in the event of communication loss.
- During the entry and landing phase, as with launch, a short time-to-criticality can necessitate autonomy. This is discussed in the context of the Descent Image Motion Estimation System (DIMES) operating on two landings in 2004 of NASA's MER rovers on Mars, responsible for directing compensation for unpredictable cross-winds.

For the first two of these, human life is at stake. In such situations NASA requires the systems to be human-rated, defined in (NASA NPR 8705.2C n.d.) as follows: "A human-rated system accommodates human needs, effectively utilizes human capabilities, controls hazards with sufficient certainty to be considered safe for human operations, and provides, to the maximum extent practical, the capability to safely recover the crew from hazardous situations." For the third, they were purely robotic missions, once launched posing no hazards to humans—but their development and launch costs totaled $744 million at the time, so their successful landings on Mars were critical.

10.2 Launch Phase: ULA Atlas v Emergency Detection System (EDS)

10.2.1 Project Description

The launching of flight vehicles is a complex and energetic process necessitating many safety precautions and protocols. During launch, the time-to-criticality can be below human reaction time, necessitating the need for flight system autonomy. The Atlas V Emergency Detection System (EDS) is presented as an example of such automation.

An Atlas launch vehicle was used to propel John Glenn into space as he became the first American to reach orbit in February 1962. This early Atlas launch vehicle had an autonomous Emergency Detection System (EDS) known as the Abort Sensing and Implementation System (ASIS) (Holguin et al. 2010). This system monitored 13 basic measurements and could signal the Mercury capsule to abort if issues were detected.

In the nearly 60 years since Glenn's flight, many versions of the Atlas have been used to fly more than 300 commercial and government payloads to space. The most recent variant, the United Launch Alliance (ULA) Atlas V, had been successfully flown commercial and government payloads for nearly a decade when it was chosen to launch crew to the ISS as part of the Commercial Crew Program (CCP). As part of the efforts to human rate, the Atlas V, a modern EDS was designed and implemented on the launch vehicle. Human rating of the Atlas V was accomplished by verifying that the vehicle meets the requirements set forth in NASA design and construction standards (NASA CCP 2012). Information herein is drawn from multiple publications; in addition to those specifically referenced in this section, see also Chelkis et al. (2011), Kutter et al. (2010) and (FAA 2011).

10.2.2 Requirements and Autonomy Role

Designing and verifying launch automation are especially challenging given the short time-to-criticality, made even more so in the context of human space-flight's stringent needs for high reliability. To ensure a robust and reliable EDS,

"Groundrules" were established to guide selection of the parameters EDS would use for monitoring (Patton 2007). The following is a distillation of those rules:

- To reduce the potential for a false abort:
  - Keep the number of parameters to a minimum, thus lessening potential sources of anomalous indications of failures.
  - Require confirmation from a minimum of two redundant, independent sensors prior to initiating any abort/separation.
  - Use the highest system level detection possible (i.e., vehicle rates, tank pressures, etc.) for non-catastrophic faults, as these are the best indicators of overall vehicle health, and use component lower-level detection only when necessary.
- To detect conditions requiring an abort:
  - Include parameters necessary to detect catastrophic faults requiring immediate abort.
  - Leave in place the existing Atlas ground system that already monitors critical items and will safe the vehicle before engine start, i.e., retain use of a well-tested backup system.
  - Include parameters that, after engine start, support detection of early degradation in a critical Atlas booster or second stage (known as the Centaur, see Fig. 10.1) system.

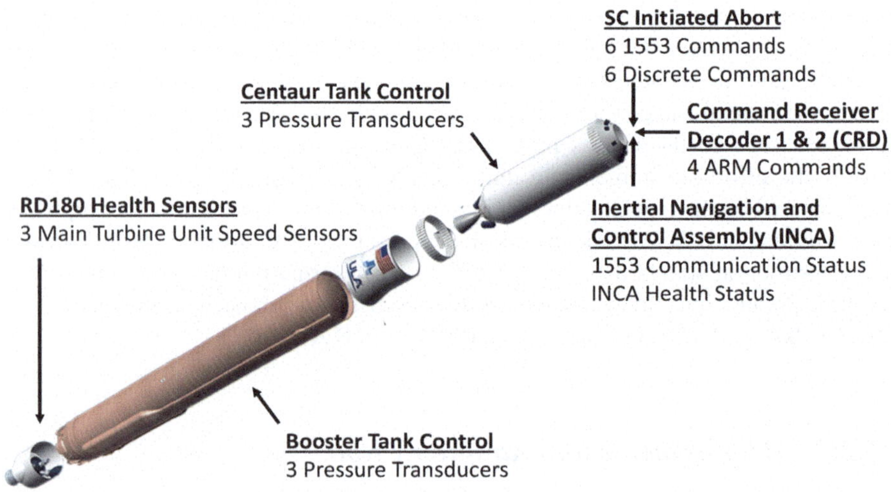

Fig. 10.1 Atlas V sensors utilized by the EDS. (Figure from https://mars.nasa.gov/resources/21443/atlas-v-401-rocket/?site=insight. Courtesy NASA/JPL-Caltech, with labeling information added from (Dawson and Mosley 2015))

Take into account prior knowledge and experience:
- - Manually compare the results of the parameter selection against previous crewed vehicle abort system experience (Mercury/Atlas, Gemini/Titan and Apollo).
  - Use the fault coverage assessment results from the Orbital Space Plane (Turner 2003; NASA OSP 2003), during which an extensive failure mode analysis was completed.
  - Passively monitor EDS parameters on as many uncrewed flights as possible prior to the first operational (crewed) mission, to refine vehicle characterization and determination of parameter limits, i.e., reduce risk by using uncrewed flights.

As stated above, the Atlas V EDS requires multiple measurements to initiate an abort. Figure 10.1 shows locations of the sensors used for the EDS (Dawson and Mosley 2015) to provide measurements. The sensors are part of the launch vehicle control system and are already redundant, reliable, and qualified to high standards. The Atlas V EDS health monitors include the following (Holguin 2011; Dawson and Mosley 2015):

- External monitors, providing additional situational awareness to the EDS:
  - Abort commands sent by the spacecraft
  - Directions from the flight termination system in the event that range safety detects an issue.
- Propulsion system monitors
  - Early detection of failure modes with the RD-180 liquid booster engine can be detected in the pump speed during ascent.
  - Loss of thrust from the engines will be detected by insufficient axial acceleration.
- Conversely, there are no monitors on:
  - The solid rocket motors (SRMs), since once ignited they cannot be shut down
  - The second-stage RL10 engine, because of its benign engine cycle and the ability to reach orbit even in the event of engine issues.
- Flight control monitors
  - These measure axial and lateral accelerations, body rates, and engine actuator commands.
- Structural integrity monitors
  - Ullage (pressurant gas filling the space above the liquid in the propellant tanks) pressure monitors the booster propellant (liquid oxygen and kerosene) tanks. The tanks are structurally stable and require pressurization to maintain structural integrity and to ensure the propellant reaching the RD-180 engine remains within the pressure/temperature operational run box.

  - Ullage pressure monitors the second-stage propellant (liquid oxygen and liquid hydrogen) tanks. Similar to the booster, the tanks require pressurization to remain structurally stable and to provide propellants within the operating box of the RL10 engines.

- Avionics monitors
  - The avionics systems involve bus communication between the flight control computer, control units, and engine controllers. The EDS monitors these systems for bus transmission errors and problems with communication between the units.
  - The flight computer itself also has health indicators of possible failed redundancy or the internal inertial measurement sensors. The EDS monitors the flight control computer health and inertial measurement health for loss of vehicle control due to a failed computer, and loss of accelerometer or gyroscope sensor accuracy.
- Staging monitors
  - The EDS monitors for an expected loss of communication to the booster avionics control unit, which occurs when the booster physically separates from the second stage and so doing severs the communication bus.

In total, the Atlas EDS baseline includes 76 parameters to be monitored, 37 on the booster, and 39 on the second stage (Patton 2007).

10.2.3 Architecture and Design

To meet the human safety requirement for single (at a minimum) fault tolerance, the EDS system is architected as follows (details distilled from Holguin et al. (2010)):

EDS consists of two independently powered identical units, physically separated, each of which receives the same set of commands, inhibits, and sensor inputs. Either unit can independently detect a failure and generate an abort signal. Note that the failure of one of these units does not necessitate an abort, since the other unit continues to provide the EDS function.

To reduce the potential for a false abort, each unit has a fail-safe design such that an internal failure of the unit will *not* generate an abort signal, since an internal failure would not be an indication of off-nominal vehicle performance.

The units also convey real-time telemetry of their health status to the spacecraft and to ground control.

10.2.4 Verification and Validation

In 2010, the ULA Atlas team demonstrated a prototype of the EDS avionics box operating within the high fidelity Systems Integration Laboratory (SIL). As described in (ULA 2011) (Holguin 2011), demonstration involved abort simulation runs covering a number of representative failure scenarios in which the EDS was fed inputs to see that the EDS detected anomalous conditions and issued the command to separate and initiate the abort sequence. The failure scenarios included loss of thrust, excessive rates or attitudes, off-nominal trajectory performance, and propellant tank leaks. The outcomes were positive in that abort conditions were correctly recognized and there were no false triggers, as described in (Herbella et al. 2011): "The software output the abort signal when the correct quantity of sensors, with persistency, was out of pre-defined allowable limits. The overall demonstration performance was excellent and with architectural and hardware/software designs highly relevant to the planned configuration." ULA's (formerly Lockheed Martin's) SIL is a high fidelity "test like you fly" testbed providing hardware-in-the-loop simulation of launch and flight operations (Winn 2003). More recent testing included an integrated Boeing/ULA integrated abort test of the CST-100 Starliner software and the EDS at Johnson Space Center (JSC) (Gerbhardt 2017). The simulations at JSC utilized the actual flight software.

10.2.5 Hazards and Uncertainty with Mitigations and Resiliency

The function of the EDS is to monitor precursors of catastrophic launch vehicle failure in order to direct an abort if need be. Thus there are two classes of hazards—those that could cause launch vehicle failure, and those that could cause the EDS itself to fail and thus be unable to perform its function. Around 50 years of launch vehicle experience has contributed to the understanding of the former class and the evolution of how to monitor for them (Herbella et al. 2011).

The possible failure causes and controls for the EDS avionics and systems are contained within NASA Commercial Crew Program (CCP) launch vehicle hazard reports (Canfield 2016). "Failure causes" are conditions that can result in catastrophic consequences; "controls" are design features, processes, inspections, etc., that can mitigate the failure causes. The hazard reports cover power failures, software failures, avionics failures, and battery failures. NASA surveillance of the controls also includes process assurance actions (PAAs) such as record reviews, process witnessing, product examinations, and process audits.

10.2.6 Application and Summary

The EDS flew in shadow mode during the uncrewed Orbital Flight Test (OFT) of the Boeing CST-100 Starliner spacecraft on December 19, 2019. In shadow mode, the EDS produces flight data but the data is not used to make operational flight decisions. Although the CST-100 spacecraft encountered issues, the launch vehicle and the EDS met all requirements (the data matched that from other launch vehicle systems, no system values exceeded abort thresholds and there were no false triggers) (Baylor 2019; Gebhardt 2019). The EDS will be operated on all subsequent Boeing CCP missions.

10.3 Transit Phase: Orion Optical Navigation (OpNav)

10.3.1 Project Description

During the Orion spacecraft's Artemis I mission, a future uncrewed 3½ week mission to, around, and safely back from the Moon, there are to be approximately 15 scheduled trajectory corrections. These trajectory corrections are to be accomplished via navigation state update from the ground Mission Control Center. The Orion Optical Navigation (OpNav) software is to serve as backup in the event of loss of communication with ground. In such circumstances, OpNav is to derive navigation information onboard the spacecraft accurate enough to complete the mission around the Moon and land safely on Earth. Similar to the manual space sextant (a device to measure the angular distance between two objects) used on the Apollo missions (Holley 1976) as a navigation backup, the OpNav system on Orion will determine corresponding information, but do so autonomously.

The autonomous navigation concept has a history of use on robotic missions. It was demonstrated in the Deep Space 1 mission (Bhaskaran et al. 2000), computing its own navigation solutions during a portion of its deep space cruise. Another example is that of the Deep Impact mission, where it was used for terminal guidance to impact a comet nucleus (Deep Impact (Frauenholz et al. 2008)) and later to track the nucleus during a comet flyby (Stardust (Bhaskaran 2012)). With the Orion Artemis-1 Mission, this will be the first application of autonomous navigation as Class A Safety-Critical software. This is NASA's most stringent classification, the software for which is subject to the process governed by the NASA Software Engineering Requirements document (NASA NPR 7150.2B n.d.).

The OpNav system on Orion will use still images of the Earth or Moon to analyze the radius of those bodies and their obliqueness to determine range and bearing (Christian 2015). It will use the stars to determine and compensate for lens distortion and to adjust for attitude fluctuations between itself and the Orion star trackers (Christian et al. 2016). For Artemis II, the OpNav software is to provide self-attitude determination through optical images of a starfield even partially obscured by Earth/Moon using an onboard star catalog (Samaan 2003). The algorithms for determining this information have been developed and refined over years of research, but were finalized in a NASA government effort to build the Orion OpNav software and deliver to the Orion program in 2018.

Verification of the OpNav system faced a challenge common to many forms of spacecraft autonomy, namely creating or accessing a test environment that correctly represents expected operating environment. In the optical navigation system case, there were no available test data sets of Earth/Moon optical imagery, timestamped to the accuracy required, to serve as truth. So the question became, how could we test with synthetic images with confidence such that we would be assured the answers are correct? In light of this question, the test, verification, and validation campaign for the optical navigation software is explored as a case study.

10.3.2 Requirements and Autonomy Role

The OpNav system in Artemis I is to autonomously compute range/bearing/attitude instrument information onboard the spacecraft in the event of communication loss. It must be accurate enough to allow the primary flight computer, based on that data, to successfully complete the mission, including performing a successful entry and landing.

A subset of key performance requirements that formed the basis of the OpNav verification tests is shown in Table 10.1. Testing OpNav to verify that all output measurements were within those tolerances posed a challenge; also challenging was to establish truth data for comparison. Note: below, *Attitude* refers to three-dimensional orientation in space, *boresight* is a frame of reference centered with the optical camera view, *about* being about the yaw (focal) axis, and *cross* being in the focal plane.

Table 10.1 Selected OpNav performance requirements

Title	Shall statement
Calibration accuracy	The optical navigation software shall correct for misalignments between the optical navigation sensor boresight and star tracker boresights to within 5 arc seconds
Undistortion accuracy	The optical navigation software shall undistort images to within 0.2 pixels of truth
Range and bearing accuracy	The optical navigation software shall resolve range and bearing to within an accuracy of truth as defined in NASA EG Technical Brief FlyDyn-CEV-17-35, May 26, 2017
Attitude accuracy—cross boresight	The optical navigation software shall resolve attitude to within an accuracy of 18 arc seconds cross boresight (three sigma) of truth assuming a maximum 0.2°/s body rate per axis
Attitude accuracy—around boresight	The optical navigation software shall resolve attitude to within an accuracy of 100 arc seconds about boresight (3 sigma) of truth assuming a maximum 0.2°/s body rate per axis

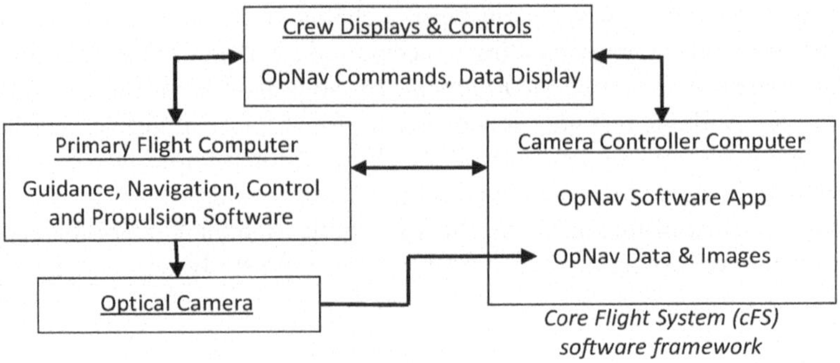

Fig. 10.2 Orion optical navigation (OpNav) system architecture

10.3.3 Architecture and Design

A diagram of the top-level Orion OpNav architecture is shown in Fig. 10.2. A commercial optical camera is used to take greyscale still images. Images are received and processed by the OpNav application, with results provided to the onboard navigation computer. For the Artemis II crewed mission, the crew may inspect and select the OpNav data sensor output as a navigation instrument. The OpNav application itself is approximately 8800 C-language Source Lines of Code (SLOC) that operates as a single application within NASA's Core Flight System (cFS) software framework (cFS n.d.) on a Linux-based operating system.

10.3.4 Verification and Validation: The OpNav Test Campaign

A primary challenge in verifying the optical navigation autonomous software was that synthetic imagery representing truth was used to test the software. To gain confidence in this method, not only were the image generation tools and star catalog accredited, but the images themselves were validated against truth. Additionally, the software underwent an extensive test campaign consisting of nominal and off-nominal images.

To gain confidence in the image generation tool and synthetic images, the first step was to understand the accuracy of the star catalog. Fortunately, accreditation of the Hipparcos star catalog (Hipparcos Star Catalog n.d.) was already performed by the European Space Agency (ESA), so a peer review of that process was performed and that catalog selected. Synthetic images were validated against the star catalog as described under "model and tool accreditation."

To gain confidence in the OpNav software itself, a test campaign was undertaken, summarized below, and quantified in Tables 10.2 and 10.3. Of note is that the test

Table 10.2 OpNav test summary statistics

Item	Quantification
Flight Software Size (SLOC)	8800
Test Software Size (SLOC)	35,400
Individual Pass/Fail Test Cases (Total)	21,322
Unit Test Cases	2769
Verification Test Cases (Nominal)	16,618
Verification Test Cases (Off-Nominal)	1935
Nominal Synthetic Test Images generated	3382
Off-Nominal Test Images	2299

Table 10.3 Peer review statistics

Review type/product	Major issues	Minor issues
Software Development Plan (SDP)	1	6
Software Requirements Specification (SRS)	4	21
Critical Design Review (CDR)	7	11
Verification Test Review (VTR)	0	7
Test Readiness Review (TRR)	0	0
Code inspections	49	305

software was approximately four times the size of the flight software in Source Lines of Code (SLOC). The development of the OpNav software followed a waterfall software development lifecycle against an aggressive schedule. In addition to major reviews, the project employed incremental peer reviews throughout the process to review the code, refine and approve individual verification tests, and to qualify the nominal and off-nominal image test set. Unit, verification, and off-nominal testing were run as nightly regressions on a flight representative processor.

Unit Testing—Unit testing provided 100% line, branch, and condition permutation coverage (unless specifically waivered), parameter boundary, and nominal path tests. Nine individual unit tests comprising 2769 individual pass-fail cases were run, written with a test framework called UTAssert (n.d.). Deviations from 100% coverage were peer reviewed and documented.

Static Analysis and Code Inspections—This analysis consisted of both manual team inspections and static analysis tool use, with manual inspections primarily checking for logic/algorithmic errors, unhandled off-nominal conditions, or causes for processor exceptions. While there were only nine static analysis tool findings, there were 517 manual inspection findings, of which 49 of those were considered major and the rest either minor or considered "nits." With good participation and expertise, the manual inspections were regarded as one of the most highly value-added processes in increasing confidence in this complex software.

Peer Reviews—In addition to using peers for code inspections, peer reviews were an integral aspect of routine project cadence. Formal reviews were held with stakeholders at major milestones (Table 10.3), and informal reviews were held weekly to monitor the progress of development, to identify off-nominal conditions, and approve associated test data. Incremental test peer reviews were held upon the completion of each verification test. The formal peer review process followed a NASA work instruction that identifies roles such as leader, moderator, reader, and participants and requires tracking documentation for such things as time spent and severity of findings.

Verification Testing—Verification tests were written to verify against the OpNav Software Requirements Specification (SRS) document, specifically to ensure all results returned were within tolerances. Each verification test was individually peer reviewed to ensure proper test cases, coverage, supporting data, platform, completeness, fidelity, and correctness of code against performance tolerances. Because of the lack of available truth data precisely correlating time and optical imagery, verification images were primarily synthetically generated using a NASA-developed tool called Engineering DOUG (Dynamic Onboard Ubiquitous Graphics) Graphics for Exploration (EDGE) (EDGE n.d.). Forty-three separate verification tests designed to test 51 requirements were written comprising 16,618 individual pass/fail test cases using 21,500 Python SLOC.

To determine what images to generate, the entire planned Orion vehicle mission trajectory information (i.e. position, velocity, and orientation) at one-second intervals was received from the Orion Program. Fifteen segments of time called "OpNav Passes," each usually 2 h in duration, are planned for the mission. Given this, approximately 3400 images were generated of the Earth, Moon, and Stars, during the expected OpNav passes across the nominal Artemis I trajectory.

Off-Nominal and Robustness Testing—A series of off-nominal tests were designed against unexpected behavior such as errors in inputs and poor imagery or established performance limits in "robustness." Tests with off-nominal data were required to either pass within tolerance or to fail in expected ways returning an error code such that there were no "false positives," i.e., the instrument should never return data out of established performance tolerances. The team routinely brainstormed and augmented the list of off-nominal conditions throughout the project. As a result, approximately 2300 off-nominal images were tested within 30 separate off-nominal verification tests and 1935 individual pass-fail test cases. "Robustness tests" were conceived and performed to characterize the limits of the sensor outside of defined requirements. For example, tests were run to characterize the greatest distance away from the Earth or Moon that it would still provide an accurate result. A list of the off-nominal and "robustness" categories of conditions tested included the following, with several example images shown in Fig. 10.3:

- Errors in inputs: Wrong heliocentric velocity, wrong target, wrong state data, late launch, no time from Orion, no star tracker from Orion, time/attitude mismatch, bad attitude from star trackers, wrong size image, images without accurate timestamp, Lunar Reconnaissance Orbiter (LRO), and images from Earth-based observation
- Poor imagery due to stray light, earthshine, specular limb, over- or under-exposure, sun in image, sun burn, partial limb out of field of view, black image, white image. Earth and Moon both in image, no limb with lit planet, specular limb, oversaturated limb, and planetary body phase variation

Fig. 10.3 Sample off-nominal images (Earth-and-Moon, varied exposures, lens scratch). (Images courtesy of NASA)

- Imagery confounders: Hot pixels, lens scratch, large streaks, dust on lens, ice crystals, floating debris, radiation damage, and solar panel in image
- Equipment problems: Camera settings reset, radiation damage, misalignment, and cold camera
- Conditions outside of required design: Large vehicle angular velocity, extreme Earth and Moon distances outside of Orion mission trajectory

Integration Testing—Once the application software was verified, it proceeded to integration testing in progressively higher fidelity Orion program integration test facilities. The performance parameters established during verification testing were incorporated into procedures and tolerances for OpNav input into Orion navigation system. At least six integrated runs and two half-mission runs were conducted (outbound mission and inbound mission) as joint test runs with the Orion Integrated Test Lab (ITL) and the Houston Mission Control Center.

Validation/Certification Testing—Although verified prior to delivery to the program, the formal validation and certification of OpNav are planned during the outbound leg of the Artemis I mission. This is because the laboratory validation of OpNav, even though conducted on flight hardware, used synthetic imagery. The system will be inspected for accurate results and certified during two dedicated OpNav Certification passes on Artemis I—one imaging the Earth and the other the Moon. If successful, it will be activated for use during loss-of-comm, and certified for use on the next crewed mission.

Model and Tool Accreditation—Accreditation of tools used in the generation of OpNav truth data was carefully considered. These tools included the Anteres trajectory generation tool, FreeFlyer (FreeFlyer n.d.), EDGE for synthetic image generation, the Hipparcos Star Catalog (n.d.), and Star Catalog generation software. For all items other than EDGE and the Star Catalog Generation Code, previous accreditation results were reviewed and accepted by the team. For EDGE-generated star images, validation was done by comparing centroid residuals between undistorted images and the Hipparcos star catalog. For EDGE-generated synthetic Earth and Moon images, EDGE was validated against the FreeFlyer tool and peer reviewed with the team.

10.3.5 Hazards and Uncertainty with Mitigations and Resiliency

Most of the challenges for the OpNav application lay in generating all the necessary test cases and imagery needed to gain confidence in the operation of the software. In addition to considering the entire nominal mission, a concerted off-nominal test campaign effort was taken to identify the limiting operating parameters or "robustness" of the system. Examples included the determination of:

- How far away from the moon, or earth, or earth-moon system, the sensor can still operate and return a value within tolerance.
- How bad the star tracker data can be and still have the sensor return a correct value.
- How many pixels can be corrupted and still get a good measurement.
- How much of the image can be obscured and still get a good measurement.

Resilience to these off-nominal situations was considered part of normal software regression testing. An overall software requirement was for any measurement returned to be within specified tolerances, and this was verified to be the case for all tested situations. Expected behavior and OpNav error tolerances were incorporated into the overall Orion navigation system error tolerance budget.

However, in addition to returning sensor values within tolerance, additional "safety-gate" mitigations were designed into the system. First, the onboard navigation filter rejects values that are too far out of family, so if the OpNav state information is too far from the known state, its data is rejected by the primary GNC flight software. Additionally, there are human (crew and ground) configurable "gates" in the onboard software to enable which, if any, of three onboard navigation filters would accept OpNav data as input. In a crewed mission, the crew has control of these gates and can enable/disable the OpNav data after inspection. In the loss-of-comm case in an uncrewed mission, these gates will be set after successful system certification and confidence established.

10.3.6 Application and Summary

The OpNav system for Orion is an application of autonomous onboard navigation in the transit of space. It was verified for the Artemis I mission and will be validated in flight. The verification effort for this software demonstrates a comprehensive test campaign ranging from coverage testing at the lowest level through nominal and off-nominal verification, assured with model accreditation and end-to-end mission integration tests in high fidelity facilities. The ratio of the amount of test code to flight code was approximately 4:1 and should be a consideration in the planning effort for this type of software. The team employed active peer reviews of the algorithms, analysis, code, test strategy, and off-nominal conditions throughout the development process. Through this concerted effort, a case for confidence was built to achieve flight certification.

10.4 Entry Descent and Landing Phase: Descent Image Motion Estimation System (DIMES)

10.4.1 Project Description

Entry, descent and landing (EDL) are irreversible activities—entry especially, once started must be carried through to the end. Critical decisions must be made under time pressure, often faster than a human crew (if present) can manage. Landing robotic spacecraft on Mars exemplifies these challenges—the light-time delay in communicating with them from Earth eliminates anything other than a completely automatic landing performed by the spacecraft themselves. The Descent Image Motion Estimation System (DIMES), an autonomous machine vision system used during the two NASA Mars Exploration Rover (MER) landings in 2004, is presented as an example of such automation.

Figure 10.4 illustrates the landings. A supersonic parachute slowed the descent, and once the heat shield was jettisoned, the lander separated from the back shell and descended to the end of a tether (left image). Close to the surface rockets fired to bring the lander to zero vertical velocity nearly 40 feet off the ground (center image). The tether was then cut, and the airbag-cocooned lander dropped and bounced several times on the surface (right image) to come to a complete stop.

DIMES' role was to estimate the horizontal velocity imparted on the lander by steady-state horizontal winds during descent through the Martian atmosphere. This information was then used by the landing system to counter that velocity by,

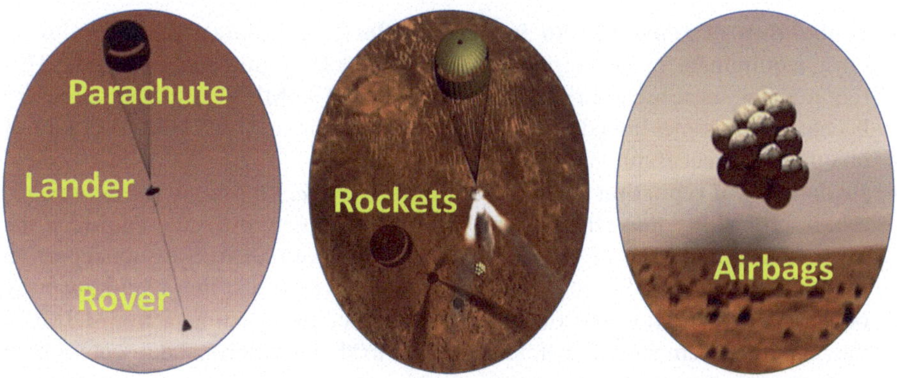

Fig. 10.4 How the MERs were landed. (Artist impressions from https://mars.nasa.gov/mer/mission/timeline/edl/steps/ Courtesy NASA)

if necessary, firing the transverse rocket system in the appropriate direction and magnitude. This was crucial to a successful landing, since the addition of a wind-induced horizontal velocity to the vertical drop could potentially exceed the airbags capacity to survive and cushion the lander. The core of DIMES was a procedure for computing horizontal velocity from the same ground feature as tracked between successive images. To make this robust, DIMES combined radar, image, and inertial data available from the existing sensors to yield its velocity estimate.

DIMES had to be autonomous—the magnitude and direction of Martian winds were not predictable in advance of the descent itself; light-time delays precluded any kind of real-time involvement from Earth, since at the time of the landings, the one-way speed-of-light time was over 9 min, while the entire Entry, Descent and Landing, from hitting the top of the Martian atmosphere to the airbag-cushioned bounces on the ground would take lass then 7 min (NASA 2004).

Because a failed landing would have been mission-ending, DIMES served to reduce a critical mission risk. However, its incorrect operation could not only fail to reduce that risk, but potentially introduce risk by directing an erroneous rocket firing, leading to a failed landing that otherwise would have been successful.

Figure 10.5 illustrates DIMES' operation. It had time to take three images of the ground, at approximately the altitudes shown. It picked two ground features in the first image to then identify in the second image, and two (not necessarily the same) features in the second image to then identify in the third image. By comparing the change of position of those features from one image to the next, it estimated the horizontal velocity of the lander.

10.4.2 Requirements and Autonomy Role

Two probabilistic requirements were levied on DIMES' velocity estimation (Cheng et al. 2005):

- **Probability of reporting an incorrect velocity estimate < 0.1%** (with an error requirement of no more than 5 m/s at 1600 m altitude (3 sigma)). This requirement set a high bar for the correctness of DIMES' estimate, necessary because as was mentioned above, an incorrect estimate could lead to failure of an otherwise successful landing.
- **Probability of reporting a velocity estimate ≥ 90%.** This requirement imposed a high probability that DIMES would report a velocity estimate, but left it the low-probability option of not doing so. This allowed DIMES to not report an estimate unless it could produce one that would satisfy the first requirement by being almost certain to be correct. In the absence of an estimate from DIMES, the landing system would operate as normal to slow the vertical descent without performing any compensation for wind-induced horizontal velocity, and thus still have a chance of success depending on the magnitude of such winds.

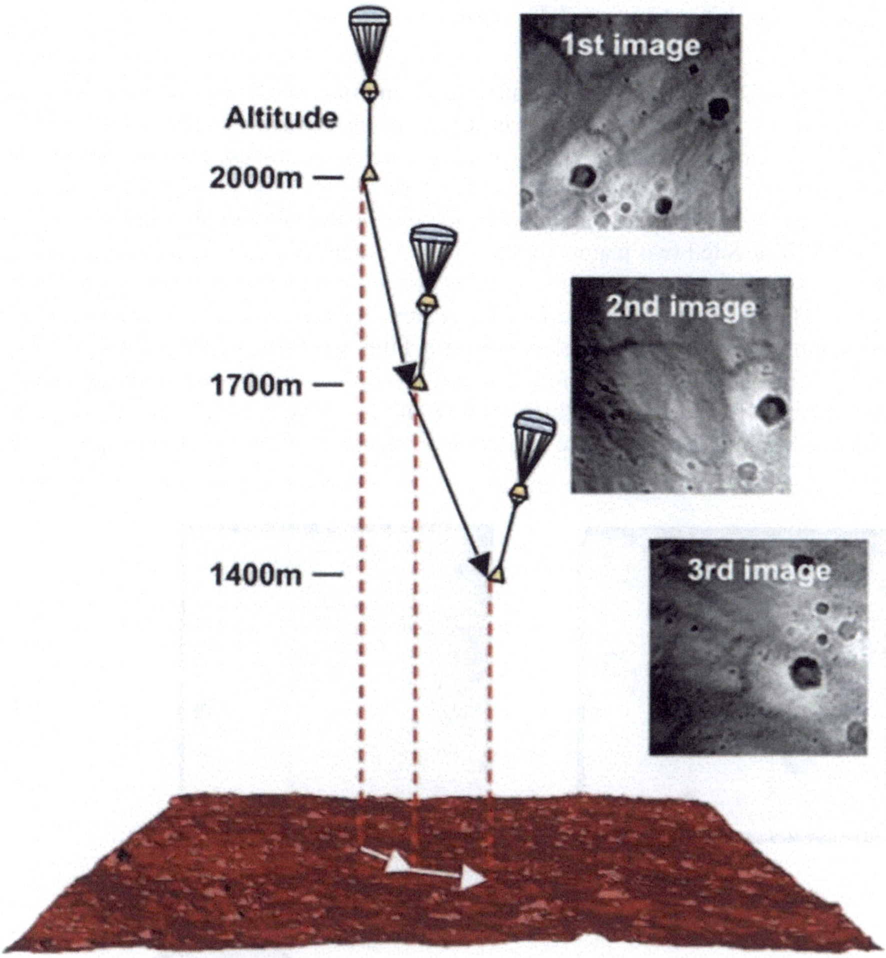

Fig. 10.5 Descent images taken by DIMES on MER-A. (Image from https://www-robotics.jpl.nasa.gov/images/mer1-540.jpg Courtesy NASA/JPL-Caltech)

This pair of requirements stemmed from concerns given the novelty of the approach and the short time (approximately 2 years) available for its development and testing. As stated in (Cheng et al. 2004), "we designed DIMES to 'do no harm'; the DIMES algorithm will produce no answer rather than one that cannot be trusted." The referenced papers do not discuss how the determination was made of balance between the two probability thresholds. What they do describe is the way their design achieved them, and how that was validated.

In addition, DIMES' computations be accomplished within an allocation of 40% of the not-very-powerful flight processor in approximately 20 s, in time to inform the direction of compensatory thruster firings as necessary. This requirement had a major effect on the subsequent development, driving many of the optimizations in the algorithm.

10.4.3 Architecture and Design

The core of DIMES was a procedure for computing horizontal velocity from the same ground feature as identified in a pair of successive images. Johnson et al. (2007)]'s related work section lists prior research tasks focused on various vision-based approaches and states "During development we took concepts and in some cases code from all of these approaches to develop the DIMES algorithm."

DIMES selected two features in an image to identify in the next image. The top of Fig. 10.6 shows the first pair of images taken during MER-A's landing at Gusev Crater. The left image was the first one taken. Superimposed on it are two small red squares, the two features selected for tracking, while the small green squares on the right image show where those features were found in the second image taken. The lower half of the figure illustrates how the two images as a whole correspond, with the left image shown here shifted and rotated to fit on top of the right image

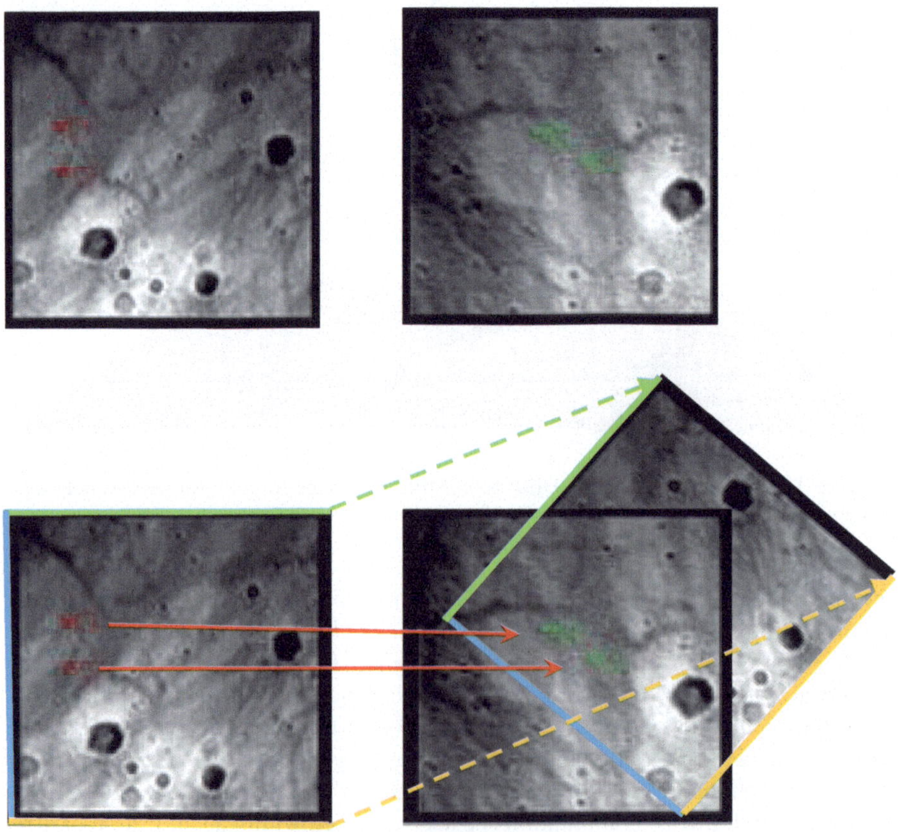

Fig. 10.6 The first pair of images and feature matches by DIMES on MER-A. (Formed from https://www-robotics.jpl.nasa.gov/images/mer1-540.jpg images. Courtesy NASA/JPL-Caltech)

and match the tracked features. DIMES had rotation information from the lander's inertial management unit (IMU).

To meet the real-time performance requirement, DIMES designers incorporated optimizations and approximations that speed up the vision processing steps but did so without compromising the validity of the resulting velocity estimate. For example, selecting a feature within an image to then track was done first on a coarse grid (64×64 pixels instead of 256×256), and then a 9×9 region of the fine grid, centered on the coarsely identified location, was explored to pick the exact location. The net result of these speedups was sufficient to allow DIMES to select and track two features between the first and second images, and two more between the second and third images, and still be in time to produce its estimate. This proved important to allowing DIMES multiple opportunities to yield a trusted estimate, as explained in the Hazards and Uncertainty with Mitigations and Resiliency subsection to follow.

10.4.4 Verification & Validation (V&V): The DIMES Test Campaign

Given the mission-critical use of DIMES' velocity estimate, extensive V&V of its performance was necessary to convince the mission to allow its in-flight use. DIMES illustrates two key challenges to V&V of spacecraft autonomy: its terrestrial testing cannot completely reproduce the conditions the spacecraft will experience, and its V&V must show that the autonomy will react appropriately to a range of conditions, since the exact conditions it will experience are not known in advance. To address these challenges, the following were used in combination to assure DIMES:

- Flight system testing to show the flight software would work as expected on the flight system, including showing that it would satisfy its timing requirements. The latter was done by running DIMES on the flight processor in the absence of other significant processing to determine the longest processing time DIMES would take. Then, the idle time of the EDL software was measured without DIMES running. This showed there to be more than enough idle time for DIMES to run (Cheng et al. 2005).
- Field testing to show the camera and algorithm would produce reasonable velocity estimates when imaging Mars-like terrain at representative altitudes.
- Monte Carlo simulation, to subject DIMES to a sufficiently large number of test cases to show satisfaction of its probabilistic requirements.

A primary challenge in DIMES' field testing and Monte Carlo simulation testing was to provide the software with conditions and imagery representative of that it would experience on Mars.

Field testing was made as flight-like as possible, using the same model of IMU as would be used in flight, and an engineering model of the flight camera fitted with a filter to replicate the dimmer lighting on Mars (further from the Sun than is Earth) and cooled to Mars temperatures. The tests were done from a helicopter, covering the range of flight dynamics expected during landing, with the exception of vertical velocity. That exception was justified through analysis, showing the vertical velocity of the lander had much less effect on DIMES performance than did attitude rates and altitudes.

Since field testing could only be done on Earth, the approach taken was to locate terrestrial terrain representative of that known to be present in the landing sites picked for the two rovers. Three test sites close to Southern California were selected, based on inputs from the Mars science community: a lava field, representative of the darkest and least-contrast MER landing sites, a dune field, representative of greater terrain relief, and a dry lake bed representative of the smoothest and smallest vertical relief landing terrains. Together these encompassed the range of conditions (albedo, contrast, and slope) crucial to DIMES (Johnson et al. 2005).

For **Monte Carlo testing**, a camera simulator specific to DIMES was developed and used to generate examples of the imagery that DIMES would experience during descent (Johnson et al. 2007). Its simulated images served three purposes: to tune DIMES from early proof-of-conception, to show that DIMES would meet its probabilistic requirements in a large number of simulations, and to test DIMES' sensitivity to off-nominal conditions, for example, dust on the camera lens.

For all these purposes, it was important that the simulator faithfully generates the imagery effects that DIMES' camera would see. To this end, the simulator used images previously taken from orbit of candidate landing sites, predicted lander descent profiles and operating conditions, and camera parameters and calibration data, to produce geometrically and radiometrically accurate descent camera images. Furthermore, the simulator incorporated critical relevant effects from the following categories:

- Scene appearance. For example, albedo and shading.
- Descent dynamics, themselves derived from landing simulations. For example, image scene scale and orientation depending on lander altitude and attitude respectively.
- Camera optical effects, determined from calibration data. For example, optical intensity fall-off (vignetting) and dust on the lens.
- Camera electronic effects derived from analysis and calibration data. For example, dark current and dead pixels in the CCD.

Monte Carlo testing was used to quantify DIMES' performance for each of the two MER landing sites. However, within the areas of both landing sites were wide variations in appearance, with correspondingly differing effects on DIMES performance—in particular, bland terrain impeded DIMES ability to confidently generate valid velocity estimates. To emerge with accurate statistics on performance, representative images of each appearance class were tested separately, and the results combined in proportion to their fraction of the predicted landing ellipses.

DIMES development also underwent reviews, a hint of the intensity of which is seen in the acknowledgment statements of several of the DIMES papers, where they say "We also thank the DIMES review board for the withering but useful reviews during DIMES development."

10.4.5 Hazards and Uncertainty with Mitigations and Resiliency

DIMES' design and V&V hinged on thorough identification and treatment of the hazards that could potentially confound its velocity estimation. This identification drew from expertise in the areas of computation (vision processing algorithms), hardware (optics and electronics), Mars descent (the dynamics of the vehicle and expected Martian conditions), and scene (features and characteristics of the landing sites). Potential hazards were identified in the following categories:

- Computation hazards: Executing too slowly to return a result in time—a concern given the short timeframe in which DIMES had to operate, and the limited computational capability available.
- Environmental challenges: Inability to find a suitably distinctive feature in one image to unambiguously locate in the next, with the danger of matching to an incorrect one, could arise from conditions of:
  - Reduced image contrast due to dust in the atmosphere.
  - Smooth, relatively featureless landing sites (which had already been picked for landing safety and science return, not for the convenience of DIMES).
  - Repetitive terrain (e.g., dunes) or multiple features of similar appearance (e.g., craters).
- Relatively standard impediments and confounders from the vision system hardware—the camera's optics and electronics, for example, bad pixels in the camera's CCD. These could have interfered with locating a feature, or themselves be mistaken as a ground feature.
- Impediments and confounders stemming from descent:

  The shadow cast by the lander's large parachute, and the heat shield (which, released shortly before image acquisition, would fall past the lander still slowed by its parachute) could enter the camera's field of view and be mistaken for static ground features.

  Because of rotation and/or horizontal shift, it was possible that a feature selected for tracking from the first image would not be in the second image.

Each of these hazards was either eliminated by design or addressed by having DIMES assess the trustworthiness of its velocity estimates, and reporting only one with sufficient trustworthiness. Since DIMES' design allowed for four opportunities to calculate a velocity estimate; provided at least one of these was trustworthy, it had a suitable one to report. The two ways DIMES established trustworthiness were through internal credibility criteria, and through an independent plausibility check against an external computation. These are summarized below; for details see (Johnson et al. 2007).

- Internal credibility criteria, based on characteristics of the feature selected in the first of a pair of images and its match in the second image. To be trustworthy, all the following conditions had to hold:
  - The feature had to have sufficiently high contrast and have a sufficiently close match—helps avoid image noise causing a spurious match.
  - The feature had to be a localized feature—helps avoid using an edge as a feature, which might match anywhere along its length.
  - The best match had to be significantly better than the second-best match—this helps ambiguous matches.
- External plausibility check: There was enough time for DIMES to take and process three successive images, yielding two velocity estimates from two features selected and tracked between the first pair of images, and later, two more velocity estimates from two features selected and tracked between the second pair. The acceleration computed from the difference between the first and second image pairs' velocity estimates was compared to the acceleration as accurately measured by the onboard IMU. A mismatch above a preset threshold was taken as an indication of the untrustworthiness of one or both of the velocity estimates.

10.4.6 Application and Summary

DIMES operated on the successful landings of both the MER rovers.

For the first landing, that of MER-A (the rover named Spirit), DIMES produced a velocity estimate sufficient to trigger compensatory action to reduce the horizontal velocity. Had it not done so, "the [landing] velocity would have been just on the threshold of airbag performance" (Cheng et al. 2005). The data also showed that DIMES discarded use of the second feature of the second image pair because it failed one of the internal credibility criteria, and correctly tracked the other features.

For the second landing, that of MER-B (Opportunity), DIMES also produced a velocity estimate, but in this case, it was benign enough to not trigger any compensatory action. In this landing too, the data showed that DIMES discarded one feature for failing an internal credibility criteria, and correctly tracked the other features.

Additionally, images collected from the landing were subsequently analyzed, confirming that the camera simulator generated imagery comparable to the real images "indicating that realistic descent images can be simulated for algorithm development and testing in future flight programs" (Johnson et al. 2007).

Overall, DIMES exemplified the challenges faced by designers of many kinds of autonomous space systems intended to operate correctly the first time, in conditions not predictable nor perfectly replicable on Earth. Its V&V of DIMES entailed identifying the hazards that could confound the velocity determination, designing where possible to address them, and a regime of testing hinging on simulation-based verification where the simulation was validated with Mars-analogue-like tests. Since DIMES, similar capabilities have been developed for what is termed "Terrain Relative Navigation" as a means to achieve pin-point landing through precise position estimation gauged from imagery taken during descent (Johnson and Montgomery 2008).

10.5 Conclusions

These three case studies illustrate typical challenges to the assurance of autonomy on space missions, and approaches taken to address those challenges. Each case study is of an application that augments normal operations to handle unusual, unpredictable conditions. The Atlas V's EDS system is there to monitor the launch and ascent for problems, and to take action if need be to abort the mission to save the crew. Orion's OpNav system is to be a backup in the event of loss of communication with the ground. The MERs' DIMES system was there to detect and direct compensation for potential cross-winds in the Martian atmosphere. Complexity is present in all three cases. Both OpNav and DIMES derive their critical information from imagery, which is more complicated than deriving threshold information from a pressure sensor. Conversely, EDS has multiple sensors' parameters (76) to monitor, the need to "require confirmation from a minimum of two redundant, independent sensors," and the need to also worry about the computer itself.

Autonomy is needed in these situations because a straightforward pre-scripted response is insufficient for the range of conditions that may be experienced. However, the root of the challenge is to assure that the autonomy will perform its safety- or mission-critical actions correctly across this range of conditions.

Furthermore, because these are space mission applications, it is infeasible to exactly recreate on Earth the conditions that will be experienced, and opportunities to test in space those applications prior to use are rare (EDS was tested in "shadow mode" on one CCP flight; OpNav will be validated on the outbound leg of the Artemis I mission) or completely absent (DIMES). In all three applications, the response has been to rely extensively on modeling and simulation, against which the autonomy is tested. This posed two fundamental questions: how to be confident those models and simulations would be sufficiently accurate, and how to be confident that the testing would be sufficiently comprehensive of off-nominal conditions. EDS testing used the well-established high fidelity systems integration laboratory, while the test efforts of OpNav and DIMES devoted efforts to establish the confidence in their respective simulations. Each application drew upon past experience in its domain—decades of past experience with launch vehicle abort control systems informed the development of EDS, while past experience with the hazards that optical systems used in space informed the development and testing of both OpNav and DIMES.

In other respects, these systems were architected and tested using techniques long honed in NASA's projects. As is typical for critical systems, not just autonomous ones, their redundant architectures protect against internal and external anomalies. NASA's stringent procedures and policies requiring reviews, audits, etc., applied to these systems' developments, and standard software development practices (code inspections, unit testing, regression testing, hardware-in-the-loop testing, and static analysis) were diligently applied.

Finally, note that these three cases do not encompass all forms of autonomy. None of them are based on run-time reasoning over models, as was exemplified by autonomous planning and scheduling seen on the successful Earth Observing One (EO-1) spacecraft, which was able to autonomously respond to dynamic science events, including volcanic eruptions, flooding, ice breakup, and cloud cover change (Chien et al. 2005). A precursor to this was the Remote Agent Experiment (Rajan et al. 2000), which flew on NASA Deep Space 1 (DS1) spacecraft as a demonstration of autonomy, combining onboard planning and scheduling together with a model-based diagnosis and recovery system. EO-1's autonomy also used onboard machine learning techniques to identify valuable context within the voluminous data gathered by its instruments, and so know which data to prioritize for download (Wagstaff et al. 2017). As yet these kinds of autonomous capabilities are not widely seen on space missions, although they feature in numerous operational concepts, for example, see the survey (Tipaldi and Glielmo 2017) and the discussion in (Kothari et al. 2020). For their use to become widespread, means to verify them will be needed.

Acknowledgments This research was carried out at NASA—Langley Research Center, and the Jet Propulsion Laboratory, California Institute of Technology under a contract with the National Aeronautics and Space Administration (80NM0018D0004).

References

Baylor (2019) https://www.nasaspaceflight.com/2019/10/boeings-starliner-two-pivotal-flights-years-end, *Boeing's Starliner Set for Two Pivotal Flight Tests Before the Year's End*, Baylor, M., October 17, 2019.

Bhaskaran, S., Riedel, J., Synnott, S. and Wang, "The Deep Space 1 autonomous navigation system-A post-flight analysis," *Astrodynamics Specialist Conference* (p. 3935), August 2000.

Bhaskaran, S., "Autonomous navigation for deep space missions," *SpaceOps 2012* (p. 1267135), 2012.

Canfield, A., "*The Evolution of the NASA Commercial Crew Program Mission Assurance Process.*" KSC-E-DAA-TN31363, May 18, 2016.

core Flight System (cFS) (n.d.). Goddard Space Flight Center, https://cfs.gsfc.nasa.gov/

Chelkis, F. Y., Chwiedor, T. D., Connolly, P.T., and Hanson, J. P., "Incorporation of RD-180 Failure Response Features in the Atlas V Booster Emergency Detection System," *AIAA 2011-7120, AIAA SPACE 2011 Conference and Exposition*, Long Beach, CA, September 27–29, 2011.

Cheng, Y., Goguen, J., Johnson, A., Leger, C., Matthies, L., San Martin M., and Willson, R., "The Mars Exploration Rovers Descent Image Motion Estimation System," *IEEE Intelligent Systems*, 19(3), (May) 2004, pp. 13–21.

Cheng, Y., Johnson, A., and Matthies, L., "MER-DIMES: a planetary landing application of computer vision," *2005 IEEE Computer Society Conference on Computer Vision and Pattern Recognition* (CVPR'05), vol. 1, (Jun) 2005, pp 806–813.

Chien, S., Sherwood, R., Tran, D., Cichy, B., Rabideau, G., Castano, R., Davis, A., Mandl, D., Frye, S., Trout, B. and Shulman, S., "Using autonomy flight software to improve science return on Earth Observing One." *Journal of Aerospace Computing, Information, and Communication*, 2(4), pp.196–216, 2005.

Christian, John A., "Optical Navigation Using Planet's Centroid and Apparent Diameter in Image", *AIAA Journal of Guidance, Control, and Dynamics*, Vol. 38, No. 2, Feb 2015.

Christian, John A., Benhacine, L., and Hikes, J. et al., "Geometric Calibration of the Orion Optical Navigation Camera Using Star Field Images," *The Journal of the Astronautical Sciences*, Vol. 63, No. 4, Dec. 2016.

Dawson, I., and Mosley, B., "Launch Vehicle Abort System Design Evolution and Validation," *AIAA 2015-4680, AIAA SPACE 2015 Conference and Exposition*, Pasadena, CA, August 31–September 2, 2015.

EDGE (n.d.), NASA Technology Transfer Program, https://software.nasa.gov/featuredsoftware/edge

FAA (2011) *U.S. Commercial Space Transportation Developments and Concepts: Vehicles, Technologies, and Spaceports*, Federal Aviation Administration, January 2011.

Frauenholz, R.B., Bhat, R.S., Chesley, S.R., Mastrodemos, N., Owen Jr, W.M. and Ryne, M.S., "Deep impact navigation system performance," *Journal of Spacecraft and Rockets*, 45(1), pp.39–56, 2008.

FreeFlyer (n.d.), https://ai-solutions.com/freeflyer/

Gerbhardt, C. (2017) https://www.nasaspaceflight.com/2017/08/spacex-boeing-home-stretch-commercial-crew-readiness/, *SpaceX and Boeing in Home Stretch for Commercial Crew Readiness*, August 11, 2017.

Gebhardt, C. (2019) https://www.nasaspaceflight.com/2019/12/boeing-ula-momentous-starliner-uncrewed-test-flight, *Boeing, ULA Launches Starliner, Suffers Orbital Insertion Issue—Will Return Home Sunday*, December 19, 2019.

Herbella, C., Mingee, R. and Heinsheimer, T., "Evolution of Abort Management of Crewed Launch Vehicles from Mercury ASIS to Commercial Crew EDS," *AIAA SPACE 2011 Conference & Exposition* (p. 7128) (2011).

Hipparcos Star Catalog. (n.d.). http://vizier.u-strasbg.fr/viz-bin/VizieR-3?-source=I/311/hip2

Holguin, M., Herbella, G. and Mingee, R., "Commercial Crew Launch Emergency Detection System—The Key Technology for Human Rating EELV," *AIAA 2010-8670, AIAA SPACE Conference and Exposition*, Anaheim, CA, August 30–September 2, 2010.

Holley, M. D. "Apollo Experience Report—Guidance And Control Systems: Primary Guidance Navigation And Control System Development, NASA TN D-8287" Lyndon B. Johnson Space Center, United States, National Aeronautics and Space Administration, May 1976.

Holguin, M., "Commercial Crew Development: Enabling Launch on EELV," *AIAA 2011-7309, AIAA SPACE 2011 Conference and Exposition*, Long Beach, CA, September 27–29, 2011.

Johnson, A., Willson, R., Goguen, J., Alexander, J., and Meller, D., "Field Testing of the Mars Exploration Rovers Descent Image Motion Estimation System," *Proceedings of the 2005 IEEE International Conference on Robotics and Automation*, (Apr) 2005, pp 4463-4469.

Johnson, A., Willson, R., Cheng, Y., Goguen, J., Leger, C., San Martin M., and Matthies, L., "Design Through Operation of an Image-Based Velocity Estimation System for Mars Landing," *International Journal of Computer Vision* 74(3), 2007, pp 319–341.

Johnson, A.E. and Montgomery, J.F., 2008, March. "Overview of terrain relative navigation approaches for precise lunar landing," *2008 IEEE Aerospace Conference*, March 2008, pp. 1–10.

Kothari, V., Liberis, E. and Lane, "The final frontier: Deep learning in space," *Proceedings of the 21st International Workshop on Mobile Computing Systems and Applications* (pp. 45–49), March 2020.

Kutter, B. F., Zegler, F., Barr, J., Gravlee, M., Szatkowski, J., Patton, J., Ward, S., "Ongoing Launch Vehicle Innovation at United Launch Alliance, IEEE 2010-1020, *2010 IEEE Aerospace Conference*, Big Sky Montana, March 6-13, 2010.

NASA "Mars Exploration Rover Landings—Press Kit" January 2004, available from https://www.jpl.nasa.gov/news/press_kits/merlandings.pdf

NASA CCP (2012). https://www.nasa.gov/exploration/commercial/crew/crewsafety.html, *Commercial Crew Transportation—Safety Requirements Shape Commercial Crew Designs.*

NASA NPR 7150.2B. (n.d.) NASA, "*NASA Software Engineering Requirements*" NASA procedural Requirements NPR 7150.2B

NASA NPR 8705.2C. (n.d.). NASA, "*Human-Rating Requirements for Space Systems*" NASA Procedural Requirements NPR 8705.2C

NASA OSP (2003) *Orbital Space Plane (OSP) Level I Requirements Program Interpretation Document*, NASA, February 18, 2003.

Patton, J. A., "Atlas Emergency Detection System (EDS)," *AIAA 2007-6127, AIAA SPACE 2007 Conference and Exposition*, Long Beach, CA, September 18–20, 2007.

Rajan, K., Bernard, D., Dorais, G., Gamble, E., Kanefsky, B., Kurien, J., Millar, W., Muscettola, N., Nayak, P., Rouquette, N. and Smith, B., "Remote Agent: An autonomous control system for the new millennium," *Proceedings of the 14th European Conference on Artificial Intelligence* (pp. 726–730). IOS Press, August 2000.

Samaan, M.A., "*Toward Faster and More Accurate Star Sensors Using Recursive Centroiding and Star Identification*," Texas A&M University, 2003.

Tipaldi, M. and Glielmo, L., "A survey on model-based mission planning and execution for autonomous spacecraft," *IEEE Systems Journal*, 12(4), pp.3893–3905, 2017.

Turner, S., "Flight Demonstrations of Orbital Space Plane (OSP) Technologies," *AIAA 2003-2710*, January 2003.

ULA (2011) https://www.ulalaunch.com/about/news-detail/2011/02/08/ula-completes-ccdev-emergency-detection-system-abort-demonstration

UTAssert (n.d.) *UTAssert-Based Unit Test Tools,* Goddard Space Flight Center https://opensource.gsfc.nasa.gov/projects/ut-assert/index.php

Wagstaff, K.L., Altinok, A., Chien, S.A., Rebbapragada, U., Schaffer, S.R., Thompson, D.R. and Tran, D.Q., " Cloud filtering and novelty detection using onboard machine learning for the EO-1 spacecraft," *Proc. IJCAI Workshop AI in the Oceans and Space*. 2017.

Winn, R., "Atlas V-First to Flight," *21st International Communications Satellite Systems Conference and Exhibit* (p. 2257), April 2003.

Chapter 11
Evaluating Autonomous Systems

David Scheidt, Craig Lennon, and Donald Sofge

Chapter 11 Editorial Note

This chapter focuses on tools, techniques, and processes for evaluating autonomous systems on the basis of usefulness as well as safety. It addresses stakeholder questions regarding best practices for how to verify a system and its functionality. Its comprehensive overview of autonomous system components and design highlights the different ways in which autonomous systems can violate requirements and need additional verification. In addition to explicitly calling out the three main roles of verification and identifying how the different tools and techniques support them, this overview effectively addresses technical challenge 21 (gaps in tool and technique development). Learning components and their failure modes are specifically called out enabling the chapter to highlight the different tools and techniques available to address them (technical challenge 27—verifying learning systems). Although metric development and evaluation of autonomous system behavior are generally poorly understood (technical challenge 16—describing the capability embodied by a robot), this chapter includes a detailed breakdown of how different aspects of an autonomous behavior can be easily verified through testing and which elements pose particular challenges to verification. This also plays into technical challenge 25 (disambiguating between different assumption failures) by connecting these aspects to the different methods used to assess system performance under different circumstances. The integration of performance, safety,

D. Scheidt (✉)
Weather Gage Technologies, Annapolis, MD, USA
e-mail: dscheidt@weathergagetech.com

C. Lennon
Army Research Laboratory, United States Army, Bel Air, MD, USA

D. Sofge
Navy Center for Applied Research in Artificial Intelligence, U.S. Naval Research Laboratory, Washington, DC, USA
e-mail: donald.sofge@nrl.navy.mil

S. Redfield et al. (eds.), *Verification of Autonomous Systems*,
https://doi.org/10.1007/978-3-031-88546-4_11

and security considerations (technical challenge 9) is addressed via a thorough guide to testing and evaluation methods in the context of verification. Examples enable the reader to identify cases where simulations are inadequate to test specific aspects of hardware and where physical testing is inadequate to address certain verification performance assessment needs (technical challenge 11). This chapter does not address technical challenge 22 (verifying the fitness of a given robot for a given purpose), instead assuming that the system has been developed for the task its performance is being assessed against. Stakeholder questions are discussed in more depth in the Chap. 1, Introduction. Technical challenges are explored in full in Chap. 2, Verification Challenges.

Abstract Evaluating autonomous systems requires consideration of their intended use, architecture, components, and operational environment. This chapter presents examples of methods and tools for verification and evaluation of autonomous systems, designing systems to support verification, and providing assurance of system properties such as safety.

11.1 Introduction

This chapter provides an overview of methods and tools for evaluating autonomous systems. It provides examples that will inform the reader as to the challenges that exist in evaluating such systems, but also provides methods and tools that might help in overcoming these challenges. The term autonomous system brings different things to mind for different audiences. Accepted definitions of autonomous system include:

- "A network that can own or announce network addresses to the Internet" [1]
- "A robot performing a given task in which the robot solves the task without human intervention, while adapting to operational and environmental conditions" [2]
- "An autonomous system is capable of understanding higher-level intent and direction. From this understanding and its perception of its environment, such a system is able to take appropriate action to bring about a desired state. It is capable of deciding a course of action, from a number of alternatives, without depending on human oversight and control, although these may still be present" [3].

In the next section, we present examples of different types of autonomous systems before narrowing the focus to a particular class of such systems. Then in subsequent sections we consider the relevance of autonomous system architecture to evaluation of the system, varieties of artificial intelligence used in autonomous systems, selected methods of test and evaluation, methods of design that facilitate evaluation, and methods of assurance for an autonomous system.

11.2 Autonomy Use Cases

11.2.1 Structures

Immobots are immobile structures that incorporate autonomous and robotic elements. These include smart homes, agricultural systems like the voluntary milking system, the autonomy and automation used on board ships and submarines to control the internal state of the host system (as opposed to autonomy and operation used to control the navigation or mobility of the vessel), and large interconnected cyber-physical systems. A particular challenge in evaluating such systems comes from the fact that you cannot evaluate the whole system separately from its intended use environment. Autonomous vehicles or robotic arms are designed to operate within an environment and should be tested in a designated test facility that is representative of the target environment. Autonomous factories represent both the system under test and the operating environment and may be only testable as a factory, although its components may be testable in isolation. The structure has the advantage of being fixed, but testing is costly, since time spent on testing is time the facility cannot be used.

11.2.2 Vehicles

An autonomous vehicle has the advantage of being testable at a designated test site, but the challenge is that the vehicle may be used in a variety of operational environments different from the test environment. This creates uncertainty in how the results of tests generalize to operational environments. A robotic factory or farm will deal with unexpected environmental interactions (e.g., from workers or cows), but the variety of such interactions is far greater with a vehicle when it can be used in many different environments. With the factory, testing is costly. With the vehicle, testing is less costly, but it may be hard to know how much and what type of testing you have to do.

11.2.3 Component Level

Many systems consist of autonomous components mounted on or within an automated or teleoperated framework. For example, a system that is primarily remotely operated—a spacecraft or underwater vehicle—may have specific autonomous operations or functions that are called only under circumstances in which operator intervention is not possible. If communications latency is high, an operator may not have the ability to intervene before the window for intervention closes. If the same operation must be done many times to the same degree of precision, an autonomous behavior may be more suited to the task than a human operator who can become tired or inattentive. In this use case, the operational environment may be hard to get to, or poorly understood, if the robot is being sent on an exploration mission. By contrast with the vehicle case, you may be confident that you know what the operational environment will be but may not be able to adequately represent that environment for tests.

For the rest of this chapter, we consider autonomous systems that (1) sense their environment, (2) plan based on an internal state that includes a representation of their environment, and (3) physically move. Within this class of autonomous systems, we focus on systems developed to perform tasks that are traditionally performed by humans. By focusing on previously human-controlled systems, we can use prior human performance as a baseline which enables us to identify the boundary between human operator performance and autonomous performance.

11.3 Autonomy Architecture

There are many approaches to the design of an autonomous system. Functional approaches focus on design components—architecting around the core elements needed to support autonomous operation. Scope approaches focus on environmental and mission constraints to impose structure on the design of specific behaviors. Temporal approaches look at the different interactions between the environment, the platform, and the mission during different design and operational stages. Each approach will lead to an architecture by which subsystems (e.g., sensing, planning, and moving) interact to produce system behaviors. Whichever architecture is chosen presents challenges and opportunities for evaluating the system. As an example, the architecture in Fig. 11.1 is for an autonomous ground robot [4].

The results of perception and localization pass through a higher-level planning layer (symbolic) and then a lower-level planning layer (metric). This separation allows the planners to be tested in simulation. The lower-level planner receives goal points from the high-level planner and could be tested without the high-level planner by sending it waypoints chosen by the tester, but it could not be tested without the simultaneous localization and mapping (SLAM) module, which provides a map of terrain over which to plan. By contrast, the high-level planner only gets object locations from SLAM and from the perception block, and therefore might be testable with only a stream of object types and locations provided to it. The architecture of an autonomous system determines how its subsystems interact, which determines how and what tests can be conducted on the system.

11.4 AI and Autonomy

A key property of AI algorithms is their ability to map patterns of input to patterns of output when the input–output correspondence is too complicated to be programmed as a set of simple rules. To understand the role of AI in autonomous systems, we

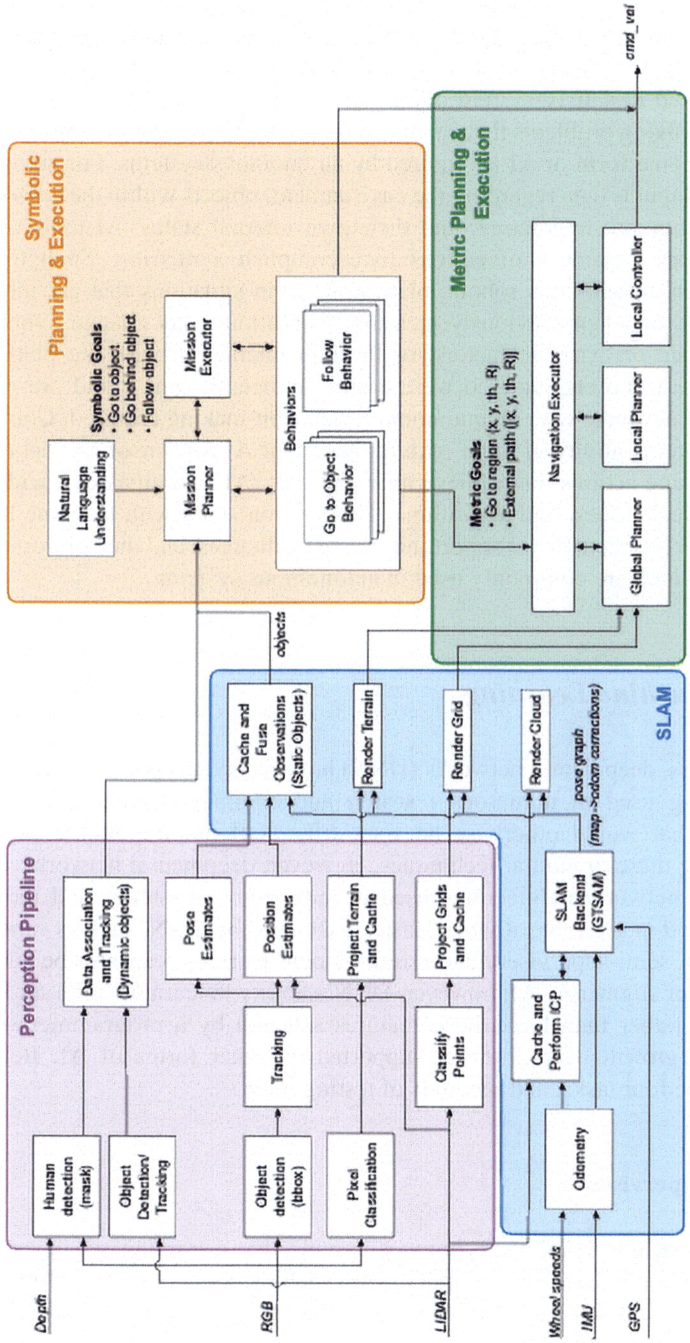

Fig. 11.1 Example architecture for an autonomous robot

distinguish automatic systems from autonomous systems. Both autonomous and automatic systems are capable of operating without human intervention. *Automatic* systems respond to external stimuli using a predefined stimulus response matrix. *Autonomous* systems respond to stimuli by determining an appropriate response that is expected to satisfy system objectives. Because, by definition, autonomous systems are solving problems that cannot be solved by closed form, control theoretic approaches, some form of AI is required by autonomous systems. For autonomous systems, the input is data regarding the environment, objects within the environment with which they are interacting, and their own internal states. AI tools take this input and plan sequences of actions to accomplish a mission. Such tools are required when autonomous robotic platforms are in situations that are inherently unpredictable, or when previously measured conditions may change dynamically due to intrinsic or extrinsic factors, or through interaction with the platform. If the conditions to be encountered were easily predictable, one could pre-program everything in advance, with no autonomous decision-making required. Complexity and lack of predictability of input require the use of AI for sensor and data fusion, and for choosing actions that achieve mission goals. AI encompasses a wide range of methods, techniques, and algorithms. This section starts with machine learning methods, which present novel opportunities and challenges, and then discusses other types of AI which are commonly used in autonomous systems.

11.4.1 Machine Learning

In recent years, deep neural networks (DNN) have received a great deal of attention and are being used in numerous research and commercial settings to provide capabilities that would otherwise be very difficult if not impossible to achieve without using these or similar techniques. However, deep neural networks are only one class of network architectures used in autonomous systems, and these may even be trained in a variety of ways. Tasks performed by DNN, such as supervised, unsupervised, semi-supervised, and reinforcement learning, can also be addressed by symbolic or stigmergic AI; however, DNN's ability to learn the relevant features of the input rather than have those features selected by a programmer supports performance growth over time not supported by other forms of AI. Below we describe these four tasks, and methods of testing them.

11.4.1.1 Supervised

Supervised machine learning techniques assume that a training set of exemplars (patterns or records) has been compiled, with each record containing an *input* value or vector, and a desired *output* value or target (or target vector). This set is referred to as the *training set*. In addition, a second separate set of exemplars (designated the *test set*) is set aside and not used for training. The algorithm is then trained to match input to output by using the examples in the training set, typically over thousands of iterations, with the network weights modified in each iteration, until the error of the network across the entire training set falls below some previously chosen threshold. The network is then tested using the inputs from the test set, and the network outputs are compared with the targets from the test set. The cumulative error across the entire test set is measured to evaluate the quality of the network mapping. Prior to deployment, the network should be checked against a set of data representative of the input it will receive when deployed, for the purpose of verifying that the network will perform as expected in its intended environment. This verification data should be separated from the training and test data used during the development process.

There are numerous techniques for adjusting the weights of the network based upon error feedback (e.g., backpropagation of error, gradient descent, and evolutionary optimization). The weight update or training method is generally chosen to fit the network representation model and the problem space to which it is being applied. The number of elements in the input and output vectors, as well as the complexity of the mapping of the problem space, affect the data requirements for training. Very large networks with thousands or millions of weights may require extremely large amounts of data for training and testing, while smaller networks require less but may not be capable of capturing the intricacies required for adequate modeling of the process.

Since supervised learning maps certain types of inputs to certain types of outputs, evaluation of such learning requires a comparison of whether representative inputs are matched to their appropriate type of output. This consists of (1) selecting which inputs are representative of the input the system will encounter in operation and (2) assessing the mapping of input to output given representative inputs. An example of assessing the mapping of input to output is given in the Experimental Methods section, in which an algorithm that assigns labels to pixels of an image is evaluated by comparing whether the assigned label is correct or incorrect. Deciding which inputs are representative, however, is a challenge. To see why, consider Fig. 11.2 taken from [5]. The pair of images comes from a driving simulator, in which an autonomous vehicle was steering based on the input shown in the images. The right image was modified in a realistic but adversarial way as an attack on the vehicle, and results in the vehicle steering left when it should steer right.

Most humans examining the right image of Fig. 11.2 would not expect it to lead to behavior different from the image on the left, but it does. Our selection of training, test, and verification data is influenced by our beliefs about what variations in the environment need to be included in the data, but this adversarial attack shows that our human intuition as to what constitutes representative data may lead us astray.

11.4.1.2 Unsupervised

Unsupervised learning methods have been used in autonomous systems for a variety of purposes, typically for matching observed patterns of data (e.g., based upon the current state of the platform and environment) with previously encountered patterns.

Fig. 11.2 Adversarial attack where, from left to right, the simulator provides: (**a**) input which should and does result in a right turn, (**b**) input with an adversarial attack that should, but does not, lead to a right turn

Clustering algorithms, K-means or Expectation Maximization (EM) for example, may map an observed state to a neighborhood of previously encountered similar states, and the actions taken in those previous states may be used to determine an appropriate action for the autonomous system in its current state. The advantage of unsupervised learning methods is that it isn't necessary to provide labeled data (that is, exemplars' output or target values) for training. Generation of labels (or targets) can be quite costly and time-consuming and may not be possible for very large training sets. The disadvantage of unsupervised techniques is that how the data are organized is harder to control, and this may affect the quality of the learned mapping.

As an example of unsupervised learning, consider masked language models, which are unsupervised learning algorithms trained to predict a masked (hidden) word in a string of text. The unsupervised training of such models provides a pre-trained algorithm that can then be trained more quickly on subsequent tasks, for example, translation between languages, as in [6]. In evaluating unsupervised learning done to support other tasks, e.g., translation, one can evaluate it based on the improvement in performance it enables on the translation as compared to a baseline of translation without the unsupervised learning.

11.4.1.3 Semi-supervised

Semi-supervised machine learning techniques combine both labeled and unlabeled data. These techniques may be useful when some labeled data is available, but not all, and it is possible to create classes or categories based upon the labeled exemplars. A supervised learning process may be used to train a model with a small amount of data, and then this model is used to bootstrap another model trained with a far larger number of exemplars of unlabeled data. Concept classes may be applied to the unlabeled exemplars based upon similarity or proximity to the concept classes of the labeled exemplars.

11.4.1.4 Reinforcement Learning

In reinforcement learning (RL), it is assumed that the autonomous decision-making process receives some form of feedback directly from the environment as a result of its actions rather than mediated through selected exemplars. The goal of the RL system is to consider its previous actions based upon that reinforcement signal, and to update (or optimize) its plan for future states. For example, a game-playing robot may make a series of moves resulting in a win (or loss) of a point, multiple points, or the entire game. The challenge of mapping the outcome back to prior state-action pairs is referred to as the temporal credit assignment problem. It is somewhat similar to structural credit assignment as found in supervised learning, mapping errors back to weights and inputs, except that the autonomous system (and the environment in which it operates) may have experienced multiple changes over the course of the decision-making sequence leading up to the current outcome. While RL methods have existed for decades in the ML community, and arguably longer in the control systems community, recent progress has allowed use of powerful RL methods with advanced deep network architectures to provide highly capable control systems that can learn through experience. RL training, much like training deep networks, is fraught with many pitfalls, and therefore considerable experience with use of RL to train systems is required to overcome effects such as premature convergence, overfitting, or failure to progress. As with creating effective deep network architectures, the current state of practice is as much art as science.

To see the challenges of evaluating RL, consider the areas in which RL has made the most recent dramatic progress—playing games. The RL algorithm from Silver [7] learned via self-play to achieve high-level performance at chess, shogi, and Go.

Other RL algorithms have been demonstrated to be effective at playing video games [8]. In both cases, the evaluation was by comparison with human performance. This suggests that when a task has a fixed and specified set of rules and a clear reward criterion (e.g., winning and losing), the performance of the RL agent can be evaluated with respect to its ability in achieving rewards while maintaining the specified rules. This would imply that one aspect of the evaluation process must be to determine whether the reward criteria are indeed the only relevant property to be evaluated, and whether the specified rules include all rules the system needs to follow. If criteria other than reward or specified rules are relevant, the challenge will be to understand how to evaluate the RL agent with respect to the other criteria.

11.4.2 Symbolic AI

Symbolic AI methods use high-level symbolic representations, generally human-readable text names, symbols, or variables, to represent decision-making processes. Examples of these are production rules (as used in expert systems), logical operations on Boolean arguments, semantic nets, frames, and blackboards. An advantage of symbolic AI over neural network representations is that the decision-making process may be easily understood by a human AI expert by simply examining the model's representation (e.g., list of rules), whereas it is extremely difficult for a human to casually examine a connected network of weights and activation functions and determine what it does and doesn't do.

Recent efforts in artificial intelligence and machine learning have attempted to create explainable AI such that a human may easily audit the decision-making processes that the AI will make in the future or has made in the past. The ability to audit the decision-making process can be an important step, if not a requirement, in achieving verification of autonomous systems using artificial intelligence.

11.4.2.1 Search-Based Techniques

Search-based techniques rely upon rapid but often computationally expensive searches of a state or action space to explore many (if not all) possible actions that may be taken for the best action to be selected. An early example of search-based AI was chess playing computers that would explore possible moves 1, 2, 3, . . . , up to n moves ahead, and then choose the move that would produce the best result according to some measure of quality of board position. Lacking a more nuanced ability to evaluate board positions, the chess computers that could look farthest ahead would have an advantage. But since the number of possible moves increases exponentially with the number of look-ahead steps, this method is fundamentally limited by the raw processing power of the processors available.

Search-based methods may be combined with other more sophisticated AI and machine learning techniques, such as reinforcement learning, tree-based search, and optimization, etc. In such cases, the system may explore some number of steps ahead in order to optimize its next move, but it makes no attempt to exhaustively explore a state space in order to map every state and possible step in order to find the optimal solution.

Search algorithms are often used in path planning, especially for autonomous systems. An example of an evaluation of vehicle path planning is given in the Autonomy Requirements and Metrics section. Here we will point out that search-based algorithms may be selected as a component for a variety of reasons, two of which include: a) the algorithm is provably optimal under some set of assumptions, and b) the algorithm has performed well on evaluations of similar tasks. In each case, the applicability of assumptions and the similarity of the tasks would need to be evaluated, but so would constraints on the algorithm such as how long it needs to find a route. For some tasks, such as path planning or sensing, the amount of time needed to perform the task is on equal footing with other characteristics of the algorithm.

11.4.2.2 Probabilistic Methods

Input to an algorithm, for example, from a sensor, can have errors. One approach to dealing with such input is to assume a certain amount of randomness to it. Commonly used research techniques, such as regression, take this approach. In autonomous systems, simultaneous localization and mapping often involve a probabilistic method which places the system at a certain point, with an error distribution, and uses noisy sensor estimates to update the position. A key assumption is that the noise, being random, will cancel out on average, allowing repeated measurements to increase the certainty with which the system's position is known.

A Kalman filter is a probabilistic method sometimes used in robotic control, and it models the robot's trajectory as a deterministic dynamical system with normally distributed noise added to it. In Agarwal [9] such filtering was used in a test of localization of a small drone. The drone was flown around, and estimated its own position, which was also recorded from external cameras, and the two compared to evaluate the drone's localization. Evaluating probabilistic methods requires consideration of how well the probabilistic model represents the system, including how well errors fit the distribution assumed by the model.

11.4.2.3 Hybrid Systems

Recall that AI techniques generally use symbolic representations or sub-symbolic representations as found with trained neural networks. However, there are advantages to combining both symbolic and sub-symbolic representations in the same system. Traditionally, AI has been approached as a top-down process, trying to reason from high-level concepts and plans, and then decomposing those into lower-level representations and actions. The connectionist school which produced many of the neural network-based solutions found today, including deep networks, assumed a more bottom-up approach to AI, and used sub-symbolic representations in the form of network weights and connections between them. While sub-symbolic ML techniques often rely on inductive reasoning (reasoning by pattern matching or similarity), top-down methods may utilize deductive reasoning (logic chains) to make decisions about the results of possible future states or actions. Humans and animals clearly use sub-symbolic representations for learning pattern association (e.g., recognizing voices, audible calls, and faces through visual stimulation), but humans may also use deductive means, facilitated by symbolic representation, when there is not enough experiential data available for inductive inference. Many groundbreaking AI systems in recent history including IBM's Watson (Jeopardy-playing AI) and Google's AlphaGo (world-class computer Go player) utilized hybrid AI systems.

As an example of a hybrid system, we will consider Handelman's robotic system that combined symbolic planning to outline tasks [10], and an RL algorithm to execute specific subtasks. The example task in [10] was for a human and robot to cross a hallway together, with the robot checking the hallway first to make sure it is clear of threats before its human teammate enters the hall. In this system, a high-level symbolic plan is produced, either by a human or by a robot, for example:

1. Robot makes sure hallway is clear.
2. Human crosses hallway.
3. Robot follows human.

Subtasks consist of steps in the plan, e.g., robot makes sure hallway is clear. Making sure the hallway was clear, in this case, meant moving down the length of the hallway and detecting threats. The subtask of searching the hallway was executed by robot employing a RL trained algorithm, which achieved positive reward when it oriented the robot toward a threat, and negative reward when a line of sight existed between the threat and the RL agent. Key points of evaluation in a symbolic method that plans like this one include: a) establishing that the planner will produce a plan (a list of subtasks that complete the task) in an acceptable amount of time, b) establishing that a plan which is produced must complete the task successfully, c) establishing that any plan produced will be possible to execute in reality. Challenges with evaluating RL algorithms are discussed in the RL section, but in this hybrid system we add the challenge of ensuring that the trained RL algorithm meets the constraints assumed by the planner. This means identifying conditions the planner expects to be met at the beginning (precondition) and end of a task (postcondition), and verifying that the RL algorithm does meet each precondition and postcondition.

11.5 Testing and Evaluating Autonomy

Test and evaluation are activities designed to provide information about the system under test. During developmental testing, this information can be used to inform requirements for the final system. Once the requirements are established, testing is used to determine if the system meets those requirements, and potentially to identify system limitations not considered in the requirements. The process requires:

- *System under test*: Identifying the system to be tested. This may be more than a single platform, as the system should include whatever external support is needed to make the object work as intended in its operational environment.
- *Requirement*: Identifying what property of that system is to be tested, and clearly specifying that property and what it is expected to be. The property is often documented in greater detail in established standards. In the example given in Table 11.1, the property is the ability of the vehicle to traverse a road that has holes in it, and the characteristics of the road and holes are described in a standard test procedure.
- *Verification method*: Identify a method by which one can determine whether the requirement is met. Some requirements, like a delivery date, or using a certain quality control process, can be checked by inspection. Others involve a function of the system during operation, and tests of the system in use are needed.

Table 11.1 Joint light tactical vehicle [12] requirement example

Requirement	Verification method
The JLTV ... shall traverse staggered 24 in (61 cm) diameter potholes with depths alternating between 6 in (15 cm) and 12 in (30 cm) and 45° edge at 10 mph (16 kph)	Testing shall be conducted over the obstacle described ... IAW TOP 2-2-611 to verify compliance

- *Test*: A test is something done with the system to elicit system behavior verifying that the system has the property required of it. Tests generally have *metrics* (standards of measurement) associated with the property being tested to allow the property to be quantified.
- *Evaluation*: The results of tests need to be evaluated, but so do other system properties. For example, the system may use the required amount of fuel, but when in use with other existing systems, this collective fuel demand may be too great. Or the system may have safety or security implications when used with other systems.

The test and evaluation of an autonomous system imply evaluating not only the hardware and software of the system, but also the behavior of the system is expected to exhibit when in autonomous operation. A challenge with autonomous systems has been to specify requirements related to desired system behavior in a way that allows for metrics that are both clear and testable. For example, less than one collision per million miles driven over interstate would be a clear metric for an autonomous vehicle, but it is not easy to test. By contrast, it is easy to test the behavior of the same vehicle on a test track, but unclear that this implies safe autonomous operation on a highway. The next section examines challenges related to autonomy requirements.

11.5.1 Autonomy Requirements and Metrics

System requirements are an agreement between the entity acquiring a system and the entity supplying or developing the system as to what the system should be or do. To understand the challenge of autonomous system requirements, we will consider requirements as being in one of two interdependent categories:

- Requirements related to the physical system and software.
- Requirements related to autonomous behaviors—the actions the system is expected to take without human intervention.

Requirements for complex physical systems and software have been challenging for decades, so standards and best practices have evolved for dealing with the challenges [11]. Some autonomous behaviors, such as self-directed navigation toward a heat source or sonar signature, have been around for decades. Such behaviors, which are guided by a deterministic response to a single and identifiable physical property of the environment, have been successfully implemented on missiles and torpedoes. However, recent AI research has suggested the possibility of replacing traditionally human abilities (e.g., driving). Using autonomous behaviors to replace traditionally human behaviors poses a challenge, because we cannot define a requirement for an autonomous system as success on the criteria by which we currently judge human performance at the behavior.

To illustrate the challenge, consider the level of detail that goes into requirements for an Army ground vehicle, the Joint Light Tactical Vehicle (JLTV), as compared with the test which evaluates human driving performance. There are 143 pages of requirements in the purchase description for the joint light tactical vehicle [12], one of which is shown in Table 11.1.

The phrase "IAW TOP ..." means in accordance with (IAW) test operation procedure (TOP), referring to a specific test procedure established for obstacle-related requirements.

Compare that requirement and verification method with the test that would be used to evaluate the ability of a human to drive the system. Army drivers are graded by the road test scoring sheet [13], a single page sheet, with criteria to check during a road test. Example criteria are shown in Fig. 11.3.

The requirements for a human driver to be acceptable at using brakes are less well defined than what constitutes a reasonable array of potholes for testing the JTLV. Vague criteria for testing humans are widely accepted because a human tester judging a human candidate has a conceptual model of that human, and thus can extrapolate from a few observations and ask questions to clarify. Furthermore, the implicit requirement that the driving candidate be a human of certain maturity and explicit requirements on the human having a certain level of eyesight imply the presence of other abilities, which then do not need to be explicitly outlined. The

GENERAL DRIVING BEHAVIOR

- Use clutch properly *(to shift, double clutched, didn't ride)* .... ☐
- Used gears properly *(not over-rev/lug engine, clash gears, coast)* .. ☐
- Used brakes properly *(no hard braking, no riding or pumping brake)* .... ☐
- Proper steering *(both hands on wheel, not over/under steer)* .... ☐
- Obeyed all traffic signs and signals .... ☐
- Drove without an accident .... ☐
- Never put vehicle over sidewalks, lane markings, stop lines, etc. ... ☐
- Examiner was never thrown to left, right, or forward .... ☐
- Driver was never forced to take evasive action .... ☐
- Wore seat belt .... ☐
- Yielded right of way to pedestrians .... ☐
- Yielded right of way to other vehicles, as appropriate .... ☐
- No errors .... ☐

Fig. 11.3 Road test scoring sheet [13] example criteria

same small amount of data is unlikely to generate the same level of certainty for autonomous system behavior. Autonomous driving cannot be tested by giving it a driver's test intended for humans but requires metrics for driving related behaviors.

In the early stages of autonomous driving research, attempts were made to assess the performance of an "Experimental Unmanned Vehicle" (XUV) navigating off road. One evaluation assessed the technology in cross country driving across arid, vegetated, and urban terrain, driving over 1100 km to conduct 588 test runs [14]. In the test runs, the vehicle would navigate autonomously when possible, but be teleoperated by a human when the autonomy failed. The assessment included metrics such as the speed of the vehicle in autonomous operation and the percentage of the runs conducted in autonomous operation (versus teleoperation) by distance and time. The metrics for percent time in autonomous operation were an assessment of how successful the autonomy was in terms of how much human attention it would demand. The metric of speed during autonomous operation provided a means of analyzing the performance at the task the behavior would be expected to accomplish. Both speed and percentage of time in autonomous operation were metrics attempting to measure autonomous behavior against a human performance baseline to determine if the technology was worth developing further.

As autonomous driving has matured to the point of testing on public roadways, a variety of metrics have been developed for components of the system. This includes metrics for (1) evaluating the detection of objects and their position relative to the vehicle, (2) the tracking of these objects, and the risk of collision with them, and (3) similarity of vehicle behavior to human driving behavior [15]. Using metrics such as these, one can evaluate an autonomous vehicle over a period of time and under a variety of test conditions. Such a test, however, will not provide confidence that the same vehicle will have the same test results at a different time and place. As we saw in the supervised learning section, two scenarios that look similar by a human may result in different decisions by the autonomous system. If we do not understand components (e.g., object detection) well enough to predict their performance in new environments, then we cannot predict the performance of a system using these components. Therefore, autonomous system testing generally follows testing of components. In the next section, we see examples of methods used for such testing.

As discussed earlier, autonomous machines are capable of devising and executing a course of action that achieves predefined objectives in response to unexpected circumstances. Autonomy requirements involve formally defining:

- The goals and objectives that the autonomous system could be asked to perform
- The scope of unexpected circumstances that the autonomous system could encounter
- The operating conditions the autonomy will be required to operate within

Autonomy goals and objectives consist of two primary structures: (1) tasks the system could be asked to perform and (2) constraints that place limitations on its decisions. Each task definition should include the parameters that define the scope of the task and a fitness criterion that can be used to evaluate task performance.

Task fitness criteria may be satisfiable or open ended. Satisfiable criteria produce clear cut Boolean evaluations that the task was or was not accomplished, such as "robot, mow the lawn by noon today." Open-ended criteria produce evaluations without a defined success criterion, such as "robot, mow the lawn as soon as possible." Constraints define "laws" that shall not be violated in performing the task. These are encoded versions of the traffic regulations for autonomous automobiles, or rules of engagement for military robots. Constraints may be task specific and defined by an operator at run time. When specifying requirements, the autonomy engineer should specify the language, scope, and parameters that are used to define constraints, formally expressing the possible constraints that could be issued without enumerating individual constraint examples.

To define the uncertainty space that an autonomous system could be required to address, the requirements engineer must define the classes of unexpected occurrences that the autonomy could encounter. It may be useful to decompose uncertainties by their temporal permanence. Three categories of uncertainties are as follows:

- Static uncertainty—These are objects or conditions that are unknown at task inception but, once discovered, do not change. For example, an autonomous apple picking robot will not know of the location of each apple at the start of a harvesting mission but, once an apple is detected, it is likely to remain until harvested.
- Dynamic uncertainty—These are objects or conditions that are unknown at task inception and, once discovered, are subject to change. For example, an autonomous car will continually identify and track other cars whose locations are constantly changing.
- Terminological uncertainty—Autonomous systems will use their understanding of the expected behaviors and capabilities of objects they interact within the performing of a task. For example, an autonomous car will utilize predictive motion models of other cars to generate a safe route through traffic. Usually the internal model of "how things work," called terminological knowledge, is static during task performance; however, robots that incorporate in situ learning have the capability to learn about the behavior or capabilities of novel objects or characterize changes in behaviors over time. When these robots are deployed, the requirements must express terminological uncertainty expected to be addressed by the autonomy.

Having defined the goals/objectives and uncertainty requirements for a robot's autonomy, it is useful to decompose the autonomy requirements with respect to the autonomy architecture. The requirements engineer should define a cognitive model that defines component-level specifications that include the role each component has in producing an autonomous decision, the information required, information and/or decisions produced, and uncertainty space over which the component functions. Effective cognitive models provide a principled basis for component-level testing and analysis.

11.5.2 Assurance, Test, Evaluation, Verification, and Validation (ATEVV) Methods

Evaluations of an autonomous system in development are an attempt to understand how a system is likely to perform in an operational scenario before the system is placed in service, and sometimes before the system exists. This challenge is exacerbated by constraints of time and budget, which limit testing. The challenge is met in two ways: (1) by modeling the system to understand functionality which has been outlined in requirements but does not yet exist and (2) by considering the system as an integrated set of components which can be tested early and at lower expense than the complete system. The modeling approach (1) includes processes of system design such as formal methods and analysis of failure modes. Testing (2) involves tradeoffs between the cost of the testing and the insight gained into system performance. In this section, we highlight some methods of design, test, and simulation that have been applied to autonomous systems. These methods include:

- Formal methods used during the design phase to avoid defects.
- Testing component algorithms for defects, which can be done early in development and at low cost.
- Testing a simulated version of the system, which can allow for early system software testing and for user tests before the physical system is built.
- Testing on a prototype system with an operator, which allows for greater understanding of the human–system interaction, but at a point when the system is costlier to change.
- Testing the prototype in live-virtual-constructive tests, which combine a physical system with simulated environmental conditions, to assess how the system performs in an operational context before the system is complete.
- Testing of the completed system within an operational environment provides the greatest insight into its capabilities, but at a point in time at which deficiencies are difficult and expensive to fix. This gold standard level of system test is too late to fundamentally change system design because it can only occur after the system is built.

In the next section, we examine methods by which defects can be avoided or discovered. The following section provides illustrative examples of applications of these methods.

11.5.2.1 Formal Analysis

The process of translating the desired properties of a system into software or hardware involves moving in stages from a vague but understandable system description to a precise description, to eventual implementation, in a way that is difficult for humans to understand. Each phase of this process has methods of analysis to support its correct execution. In this section, we provide examples of such methods, focusing on those based on automated analysis of formal system models and code.

We begin with a method of analysis for identifying how a system can fail and the consequences of such a failure. This method is sensible to a human and can be conducted without automation, but is most applicable to systems where failure can be traced to a single specific cause, for example, a circuit shorting out or a joint breaking. Autonomous systems may have more complicated modes of failure and generally require more complicated methods of assurance, such as those presented in the System Assurance section. Those methods of assurance then require evidence that certain system functions will be executed properly, and the formal methods of analysis presented in the Formal Methods sections would provide pieces of evidence to support such an argument. In these sections, we consider how a formal specification of system requirements or properties can allow one to check for the consistency of high-level requirements, desirable system properties of an abstract system model, correctness of algorithm design, and securely implemented source code.

Failure Modes, Effects, and Criticality Analysis

Failure Modes, Effects, and Criticality Analysis proceeds by examining each component (or subsystem) of a system to determine: (1) in what ways (modes) the component could fail, (2) what effect each failure mode would have on system operation, and (3) what the severity of the effect would be. In Fig. 11.4 are two examples from a technical manual [16] showing how the analysis traces from system to component level. This analysis may be done in early design stages by analyzing an abstract model of the system based on the required functionality of the components, as in the left side of Fig. 11.4. It can also be done once software and hardware have been selected to understand overall system risk resulting from the selection, as in the right side of Fig. 11.4. Such a method can guide testing by identifying what assumptions regarding component performance are most critical to overall system operation.

This method requires failure to be traceable to a part of the system in way a human can understand. In Fig. 11.4, the condenser has a clearly defined role as part of the chiller, making its relation to the cooling process easy to trace. Compare this with Fig. 11.1, in which objects can be placed into the robot's map of the environment by several interrelated methods. The perception pipeline combines camera (RGB) with LIDAR and other depth estimates and fuses this with objects from the simultaneous localization and mapping (SLAM) module. In addition, objects can be placed into the robot's map by human command (not shown in the diagram). Finally, these processes are operating in parallel, and the timing of when the map is updated by the different components could potentially alter the resulting map. When the robot's map fails to align with ground truth, it may be difficult to trace that failure back to a specific component or algorithm.

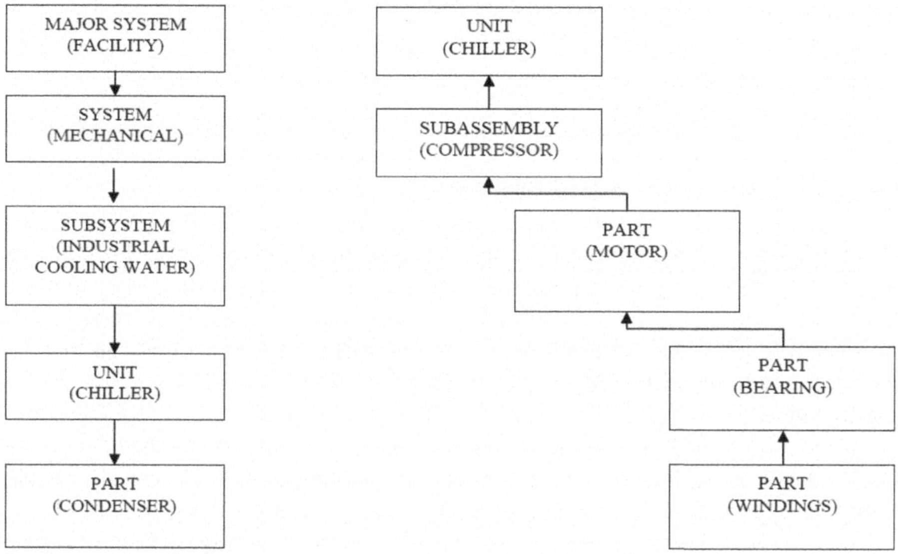

Fig. 11.4 Examples of functional (left) and hardware (right) decompositions

The methods of analysis we turn to below do not rely on a human to do the checking, but they do rely on humans to specify what properties should be checked, and to represent the system in a way that is checkable. In the next section, we see an example of formal analysis used to check requirements for consistency, and in the section on Formal Specifications, we consider formal means of modeling a system and its desirable properties, and how the presence or absence of such properties can be checked.

Checking Requirements by Theorem Proving

As an example of formal analysis applied to high-level requirements, consider the case study in Cofer [17] in which theorem proving is used to support the safety certification of an aircraft flight guidance system. In this system, there are two parallel channels (sides) that can control the aircraft, a primary, and an alternate, allowing control to be transferred between pilot and co-pilot.

We rephrase Cofer's high-level requirements here as:

- The system shall start with primary side in control.
- At least one side will have control at any given time.
- At most one side will have control at any given time.
- Control shall be transferred from one side to another if and only if the transfer switch is pressed.

```
R1. At least one side shall be the pilot flying side.
```

```
R1: THEOREM
    Reachable_State(s) =>
      Left_Pilot_Flying_Side(s) or Right_Pilot_Flying_Side(s)
```

Fig. 11.5 Example theorem for a high-level requirement [17]

Theorem proving, for example with the Prototype Verification System [18], allows for the definition of system operations (e.g., transfer), along with axioms that are given to be true (e.g., the primary side starts in control). Desirable properties of the system (e.g., at least one side will have control) are theorems that can then be proven via automated theorem proving, along with the consistency of the set of requirements.

In Fig. 11.5, also from Cofer, the side in control is referred to as the pilot flying side, and below the requirement is a provable theorem that every state reachable within the system shall have the property that one of the two sides is in control.

Thus, the high-level requirement that exactly one side is always in control can be recast as a provable theorem. After requirements are analyzed, a system model can be developed. In the next section, we see an example of how to formally specify the system model and its desired properties.

Formal Specifications

We will define formal specification as "the expression in some formal language and at some level of abstraction, of a collection of properties some system should satisfy" [19]. To be useful, this formal language must have rules for expressing and interpreting the properties, and for using the properties to infer useful information. A survey of methods of formal specification and verification for autonomous systems can be found in Luckuck [20], which presents a variety of methods of describing systems and their properties. As a simple example from Mohammed [21], we consider a state-transition model describing a water tank having a water level between 0 and 99, augmented by an automated controller which can open or close to let in more water. The controller has a 3 s delay, meaning that if the valve is set to open, it must remain in that state for at least 3 s before a close command can be issued, and the same holds true for the open to closed transition. The state of the system consists of the state of the water level ($0 \leq$ level ≤ 99) combined with the state of the controller, which is opened or closed (open, close). In Fig. 11.6, the system specification is shown, with nodes q_a and q_d indicating states at which the valve can change between open and closed, and the other nodes indicating transition states the system must go through due to the 3 s delay.

The desired properties of the system are that it never be empty and not overflow, which can be expressed as never enter the state (level < 1) and never enter the state ($99 <$ level). There are many possible systems which could, based on a measurement of the water level and control over the input valve, keep a given level of water in

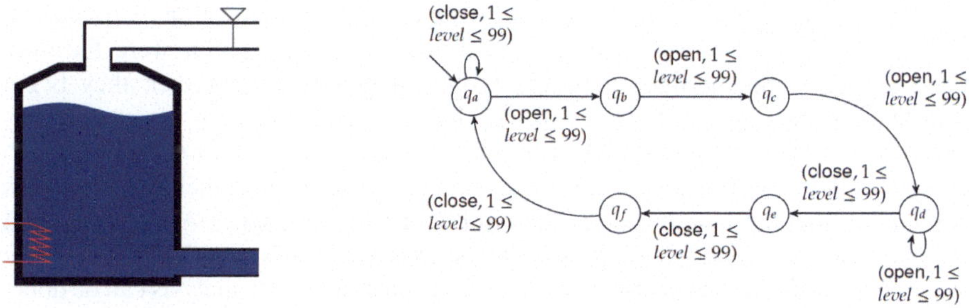

Fig. 11.6 An automated water tank (left) and its state-transition specification (right)

the tank. With these formal specifications of the system and desired properties, it is possible to check to see if a particular control system will conform to the specifications.

Model Checking and Synthesis

Given a formal specification, as in Fig. 11.6, the next step is to design an abstract controller, which can be verified as conforming to the specification, and can itself be implemented in code. The process of checking an abstract controller against this type of specification is called model checking. An overview of model checking methods is available in [22]. Alternatively, one could use the specification to synthesize a controller (model) from the formal specification itself. Such a correct-by-construction controller can be guaranteed to conform to the specification but has no guarantees with respect to desirable properties that are not expressed in the specification. A technique like this was used in Mohammed [21] to synthesize a model of the system showing all states from which one could conform to the water tank formal specification. This model was then used as a shield for a reinforcement learning algorithm. Whenever the algorithm, in the course of exploration, chooses a behavior that violates desired system properties, the shield would substitute an action which would not violate these properties. This allowed the reinforcement learning algorithm to learn safely, where safety is defined as conformance to formally specified system properties.

In the water tank example, model checking provided for formal analysis of an abstract system. Next, we examine how it can be used on algorithms which determine system behavior.

Proving Algorithm Properties

As an example of formal analysis applied to an algorithm, we consider the decentralized perimeter surveillance system, a protocol that allows multiple unmanned air vehicles (UAV) to coordinate patrol along a linear boundary under conditions of uncertainty with regard to the patrol area and communication between the UAVs [22]. In the algorithm, each UAV starts patrolling as if it were the only UAV patrolling the perimeter. As the UAVs encounter one another, they begin coordinating their patrol efforts until they have divided up patrol responsibilities. Davis et al. [23] subjected the algorithm to formal analysis via model checking to prove that the length of time it takes for all UAVs to have coordinated their patrols is no more than the length of time it would have taken a single UAV to fly around the perimeter five times. Such proofs can be expensive, however, in terms of time required to formulate the problem in a manner such that formal analysis can be done, and in terms of computation. Davis's analysis of the UAV coordination protocol took 18 days to complete using a computer with 256 GB RAM and 80 cores. Since the cost of the method is high in terms of time and computation, the benefit would need to be substantial, either in terms of avoiding a large risk at one point in time, or small risks over a long period of time.

Checking Source Code

Algorithms must eventually be implemented in source code, which provides another way for errors to be introduced into the system. Some common errors can be detected via automated analysis of the source code. For example, consider buffer overflow errors, in which input creates data beyond the memory area allocated to it, potentially altering data that should be off limits to the program [24]. A similar error was behind the Heartbleed bug, which allowed remote attackers to obtain off limits information from computer memory by reading information beyond the limits the area allocated to the buffer [25]. Some source code errors such as these can be detected by automated source code security analyzers [26]. A limited set of tools that check for errors in compiled code are also available [27].

11.5.2.2 Oracles

Oracles, a standard tool in software testing, are procedures to distinguish correct from incorrect system response. A variety of methods of handling test oracles are surveyed by Barr [28]. Oracles can provide feedback during development and testing but can also be used during system operation to determine if a system is behaving within acceptable limits. As a simple example, consider an aircraft which has a lower bound to its altitude during operation. During development, an autonomous control system could be tested in simulation to see if it goes lower than the allowed altitude. But this threshold can also be used in runtime to provide assurance that the aircraft will not violate it during flight.

Test Oracles

A software test consists of (1) a system under test, (2) input to the system, (3) a response from the system, and (4) a map from the response to acceptance criteria. The acceptance criteria could be pass/fail or could be a real number between 0 and 1, but in either case, the map described in (4) is the test oracle [28]. As an example of a test time oracle for AI, consider an algorithm designed to detect the presence of an object in an image. For the purposes of training and testing the algorithm, a set of images is labeled by a human. These images, without labels, constitute the test input, and are fed to the classifier which produced labels for each object detected in the image. The set of labels output by the classifier is the test response. The object class labels are compared to the label assigned by the human labeler to produce performance metrics (e.g., precision), making the human labels and metric calculation the test oracle.

Runtime Assurance

Runtime assurance is the ability to ensure the safe operation of a system containing components which have not themselves been sufficiently verified [29]. It may be part of system assurance, which will be described further in the assurance section. As an example, the Automatic Ground Collision Avoidance System (GCAS) is a runtime assurance mechanism designed to prevent humans from crashing a plane into the ground [30]. Although a pilot is not a machine learning system, the same methods used by GCAS to guard against pilot error could be applied to a machine learning system. The GCAS system detects potential collisions of fighter aircraft with the ground, and if criteria related to trajectory and altitude are met, the GCAS system alters the aircraft trajectory to avoid collision. The GCAS system has been installed on operational aircraft and has saved the life of pilots who lost consciousness during flight [31].

In this section, we have described tools and methods that can reduce the number of defects of a system. In the next section, we examine experimental methods, which can both identify defects in the system, and provide better understanding of system performance.

11.5.2.3 From Algorithm to System

The previous sections introduced formal methods for design and analysis of a system, and introduced oracles, which are used to evaluate the behaviors of algorithms used by the system. In this section, we use four example systems to illustrate stages of testing. We start with the testing of an individual algorithm, then examine simulation-based testing done without a physical system. We then look at examples of operator in the loop testing, live-virtual-constructive testing, and operational testing.

The first example system is a research platform from the Robotics Collaborative Technology Alliance (RCTA) which has substantial, but immature, autonomous capabilities. During experiments in 2016, the system demonstrated its ability to map its environment in semantic terms, receive verbal instructions in structured language, ground these instructions to elements of the environment, and execute simple missions, like "Stay to the right of the car; screen the back of the building that is behind the car" [32]. The system was designed for the demonstration of cutting-edge research, and never intended for production, but provides an example of advanced artificial intelligence for scene segmentation on a physical platform and illustrates *software component* and *simulation-based testing*. The architecture of this system was provided as an example in Fig. 11.1. The second system is the Robotic Wingman [33], which has less autonomous capability than the RCTA system, but is substantially more mature. It can autonomously navigate cross country while a human operator remotely operates its weapons system to engage targets. It serves as an example of *simulation-based testing*, and of *operator in the loop testing*. The third system, of equal maturity, is a drone under test in the DARPA Collaborative Operations in Denied Environments (CODE) program. The drone has the capability to navigate autonomously with a group of similar drones, sharing information to locate its position in a GPS denied environment and illustrates the use of *live-virtual-constructive testing*. Finally, for our example of *operational testing*, we use the MK 54 homing torpedo. This weapon is currently on US Navy ships and is capable of autonomously classifying, tracking, and attacking targets.

Software Component Testing

The lowest cost method of testing the software running on an autonomous system is to test the software for functionality, reliability, or other desirable properties before it is deployed to the physical system. Testing the software independent of the operation of the physical system will not find all bugs, but those that are found this way will be easier to identify and less expensive to fix. There are many books outlining methods of traditional software testing, e.g., [34, 35]. Here, we will consider testing of the scene segmentation algorithm used on the RCTA system [36]. This algorithm was critical to the navigation behavior of the RCTA robot and was tested as a component by labeling individual images. Each pixel in an image was labeled as being in one of eight classes, for example, tree, building, or grass. An example of a labeled image, with a color key indicating the label, is shown in Fig. 11.7.

Correct behavior for the algorithm was determined to be correct pixel level labeling of images provided to the algorithm, with the test oracle being a set of images labeled by a human. Overall performance of the algorithm was assessed by examining confusion between pixel classes, which is when a pixel from one class is labeled as being in another class. Pixel confusion is represented by a confusion matrix, as in the one shown in Table 11.2. Pixels on the diagonal are correctly classified.

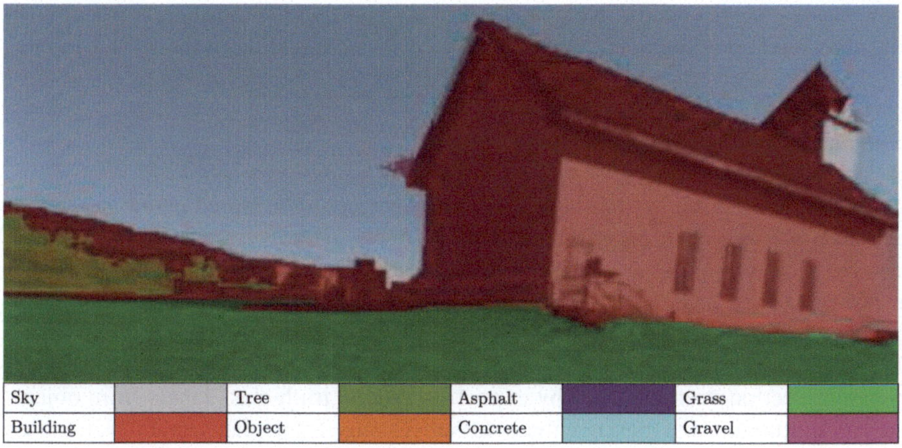

Fig. 11.7 Scene segmentation example [36]

Table 11.2 Confusion matrix—entries represent counts of image pixels jointly classified through ground truth (rows) and semantic labeling (columns)

	Sky	Tree	Asphalt	Grass	Building	Object	Concrete	Gravel
Sky	17,881	0	0	0	341	0	0	0
Tree	1	1096	0	17	787	0	0	0
Asphalt	0	0	0	0	0	0	0	0
Grass	0	38	0	40,163	331	0	0	0
Building	565	51	0	114	14,357	0	0	0
Object	0	14	0	12	368	0	0	0
Concrete	0	0	0	0	0	0	0	0
Gravel	0	0	0	0	0	0	0	0

Aggregate measures of performance for a given class on an image were established by class recall and precision:

$$\text{Recall}_{\text{class}} = \frac{\text{Number of class pixels correctly identified}}{\text{Total number of class pixels in the image}}$$

and

$$\text{Precision}_{\text{class}} = \frac{\text{Number of class pixels correctly identified}}{\text{Total number of pixels labeled as class}}$$

and an overall assessment of classifier performance was made by averaging of precision and recall across classes. The separate testing of the semantic labeling algorithm allowed for better understanding of what was being integrated into the system, and for quick retraining and retesting as the testing season progressed and new classes, such as snow, were needed for navigation [37]. Further, corrections to the algorithm could be made without the need for field testing of the RCTA robot.

Simulation-Based Testing

While the perception system for the RCTA robot could be tested using a selection of saved sensor input, e.g., images, some aspects of the system, such as control, need motion for test. However, the motion need not be on a live system. The RCTA robot was tested in a physics-based simulation (RIVET) in order to assess control-based behaviors subject to assumptions about system perception [38]. As shown in Fig. 11.8, the operation of the semantic perception algorithm mentioned in the previous section was emulated by extracting ground truth class labels from objects in the simulation and passing them to subsystems using the semantic classifier. The result was the ability to check navigation related behaviors for bugs under the assumption of perfect perception in order to disentangle perception challenges from bugs in the navigation algorithms under test.

Another use of simulation is to allow early and inexpensive user interaction with a system. In 2017, the Robotic Wingman program [39] used a simulation which employed actual system software without hardware (software in the loop) to test the user interface for the Wingman system as shown in Fig. 11.9. Such testing allows for early and quick feedback on system usability, potentially before the system exists.

Fig. 11.8 Simulated scene from RIVET with and without semantic labeling [38]

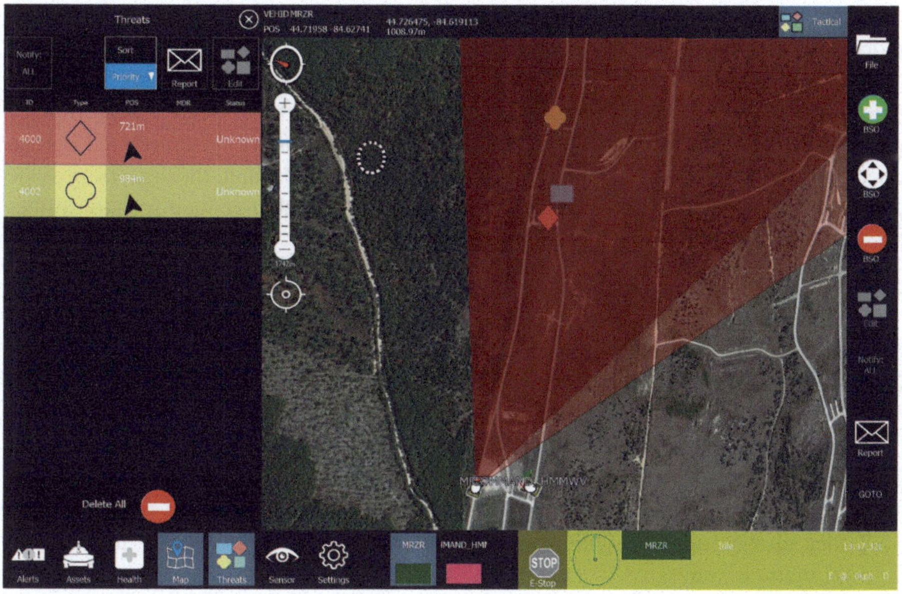

Fig. 11.9 Early Robotic Wingman user interface under test

The simulation-based testing uncovered issues related to the size of icons and graphics, and the amount of information being conveyed to users. The report detailing these issues [39] also shows the challenges of integration of such a system, and the need for integration testing to determine if system components are communicating properly.

Operator-in-the-Loop Testing

Effective interaction between a human operator and a robotic system can increase situational awareness, reduce communication load, and maintain operator situational awareness [40]. Operator in the loop testing, whether in simulation during early-stage development or on a prototype system, provides insight as to whether the system can and will be used as expected during design. However, some features of the expected operational environment cannot be tested with a single prototype system. Behaviors related to teams of autonomous systems engaging hostile systems, for example, cannot be tested by a single system on a test range unless the system's teammates and enemy are simulated.

Traditional human-factors evaluations are designed to optimize operator performance, focusing on operator workflow evaluation. Autonomy-operator evaluation requires a departure from human-centric evaluation as there will exist circumstances in which the delays associated with keeping the operator "in-the-loop" will negatively impact mission effectiveness.

When an unmanned vehicle is autonomous, and capable of making decisions independently, the relationship between the operator and the machine fundamentally changes. The operator becomes more of a supervisor than a pilot. Effective autonomy-operator interactions should seek to optimize the collective performance of the decision-making team, which includes the autonomous system(s) and human operators. Evaluating operator–robot interaction requires that the testers provide evidence that system is avoiding sub-optimal practices that reduce mission effectiveness including:

1. Under trust—In which mission effectiveness is diminished because the pilot failed to authorize an autonomous robot to perform a task that the robot was capable of performing.
2. Over trust—In which mission effectiveness is diminished because the operator tasked the robot to perform a task the robot's autonomy was not capable of performing.
3. Over monitoring—In which an operator fails to perform more important tasks because he/she was unnecessarily monitoring a robot.
4. Under monitoring—In which a robot fails to perform a necessary task due to pilot inattentiveness.

In designing operator-autonomy experiments, it is useful to view the system under test as a Command and Control (C2) system. Command and Control research shows us that, for any engagement, there is an optimal amount of attention and authority that a commander should provide to a subordinate. Sharing knowledge and deliberating on decisions have a cost-benefit trade-off. Commanders that spend time acquiring better situational awareness and making more detailed plans make decisions that appear to be smarter but at the cost of slower reaction time. We say "appear to be smarter" because decisions made based on old situational awareness may, in fact, be poor decisions if the facts used to decide are no longer true.

In [41] it was shown that the optimal C2 strategy is dependent upon three operational characteristics:

1. The rate at which team members are capable of sharing knowledge/decisions, which in some cases is derived, in part, by operator workload
2. The complexity of the problem being solved
3. The rate at which situational awareness is changing due to unpredictable combatant actions.

These operational attributes are subject to change over the course of an engagement, and when they change, the operator-robot team must be trained/designed to adapt their C2 strategy to match the new operating conditions. Tests of this adaptation should include tactical complexity, situational dynamics, and communications availability as factors. The test engineer can then examine how well the operator-robot team was able to adapt the level of human supervision in order to maintain performance as the factors change.

Returning to the Robotic Wingman example, experimentation with military operators was conducted in 2018 using the robotic vehicle in Fig. 11.10. The

Fig. 11.10 Robotic Wingman (right) with the human operated equivalent vehicle on the left

methods used by Schaeffer [39] to support operator in the loop experimentation centered on the human teams operating the vehicle and included questionnaires given to the operators, physiological measurement of the operators during operation, and analysis of the flow of crew communication.

The physiological measurement, conducted via wearable devices on the operators' wrists and chests, provided real-time data which could be monitored for indications of operator stress during the gunnery exercise. Such indicators can then inform subsequent analysis to understand if the stress is the normal result of high pressure activity, or if it results from flaws in system design. The analysis of communication flow during operation provides insight into whether the crew is interacting in a way anticipated during system design, and operator questionnaires provide a means of soliciting direct input from the operators.

Live-Virtual-Constructive Testing

Combining physical systems operating in the physical world with simulated systems allows for testing situations which would otherwise be prohibitively risky or expensive. Live testing involves testing of real systems, virtual testing creates an operational environment not physically present, and constructive testing adds computer controlled agents. As an example of all three, consider an actual drone (live) leading virtual drones that follow in formation (virtual) against a simulated hostile air defense system with AI-based decision-making (constructive and virtual). The White Force Network is such a live-virtual-constructive (LVC) test system, created as part of the DARPA CODE program to allow for testing of collaborative autonomous behaviors in a communications denied environment. The architectural diagram in Fig. 11.11 [42] outlines how information flows between the system under

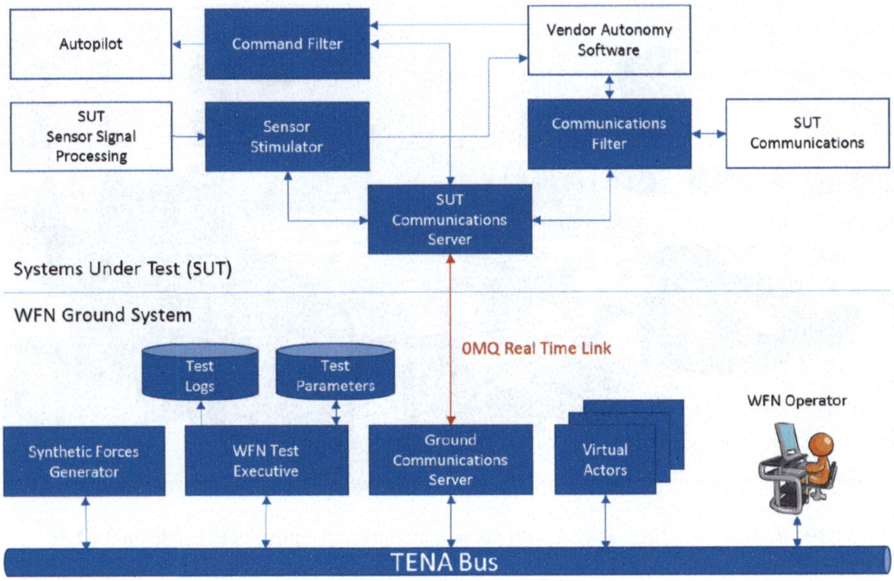

Fig. 11.11 White Force Network Architecture [42]

test and the ground system, which can manage virtual actors and inject them into the perception pipeline of the autonomous system to evaluate how the autonomy software makes decisions.

The system can, for example, set a virtual aircraft on a collision course with a drone to test whether the drone will avoid it, or inject the effects of electronic warfare jamming into the systems. Such tests may be too risky or expensive to carry out in live operation but are reasonable in an LVC test environment.

Operational Testing

A test of an autonomous system under operational conditions allows for the best understanding of the operational potential of the system, although usually at a point in the development cycle at which fixing system deficiencies is costly. The MK 54 torpedo is an example of a mature autonomous weapon, capable of classifying, tracking, and attacking targets, under operational test in 2014 [43]. Figure 11.12 shows an MK 54 being launched during operational test. The MK 54 testing resulted in recommendations for improvement of the system's capabilities, methods of testing which might have both resolved problems sooner and improved training and documentation for operators.

The testing also pointed out missing test capabilities, such as the need for tests of how the weapon would respond to the type of measures an adversary would take in combat.

Fig. 11.12 MK 54 Torpedo, an autonomous weapon under operational test in 2014

The operational test lies on the end of the fulcrum opposite the software coverage testing with which this section began. Testing of algorithms provides weak insight into the operational effectiveness of the system on which they will eventually be used, but deficiencies are easy and inexpensive to fix. Operational testing provides great insight into operational effectiveness, but at a point where deficiencies are difficult and expensive to fix. Thus to the extent that deficiencies can be found or avoided by software testing, or by formal methods of design mentioned in the Formal Analysis section, these testing methods pay for their own implementation.

11.5.3 System Assurance

The design and test methods of the previous section provide evidence that a system will have certain properties, for example, meeting the specifications in the system requirements. Further assurance is needed for the people purchasing or using the system to have confidence that the system is safe, or will do what they expect of it. Such confidence is often provided by certification that a system meets certain standards. But the development of standards generally lags technological development, and such standards will not exist for systems which have never been developed before. In this section, we introduce assurance arguments, which are one means of developing confidence in systems for which no certification standards exist.

Assurance arguments are structured arguments that explain why a system has a certain property. A method of building such an argument based on claims, arguments, and evidence (CAE) is shown in Fig. 11.13 [44].

In the CAE format, the arguments consist of:

- *Claims* asserting that a system or subsystem has some property

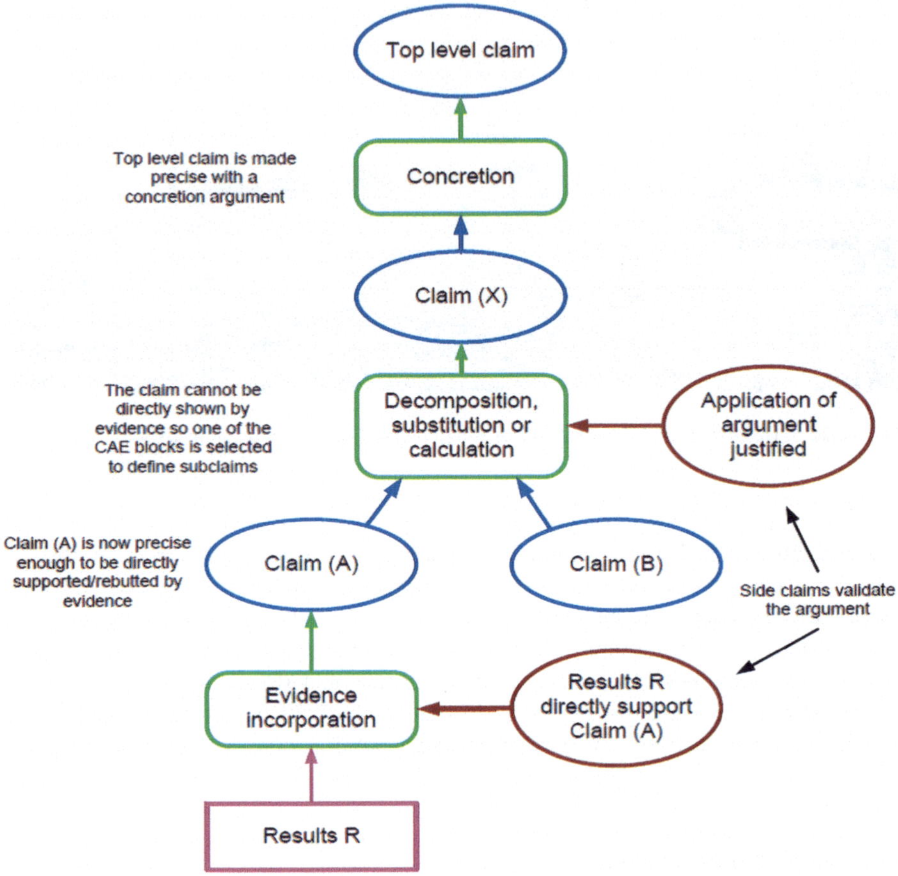

Fig. 11.13 An example structure for claims, arguments, and evidence (CAE) based reasoning

- *Arguments* that link evidence to the claim
- *Evidence* which is factual and nature and provides the basis of the claim

A top-level claim that a system is safe might be argued to require subclaims of safety in snow, rain, and no precipitation. Each of these subclaims might then be either further subdivided or supported by evidence. Tools such as the Assurance and Safety Case Environment [45] support the development of assurance arguments in claims, argument, evidence format.

An alternative structure of reasoning is Goal Structured Notation (GSN), as shown in Fig. 11.14 [46]. Here a top-level goal (G0) of safety has additional context describing the system (Ca) and defining safety (Cb) and a strategy for achieving the goal (S1). This strategy requires further subgoals, which refine the larger goal to identify what needs to be done to satisfy the top-level goal.

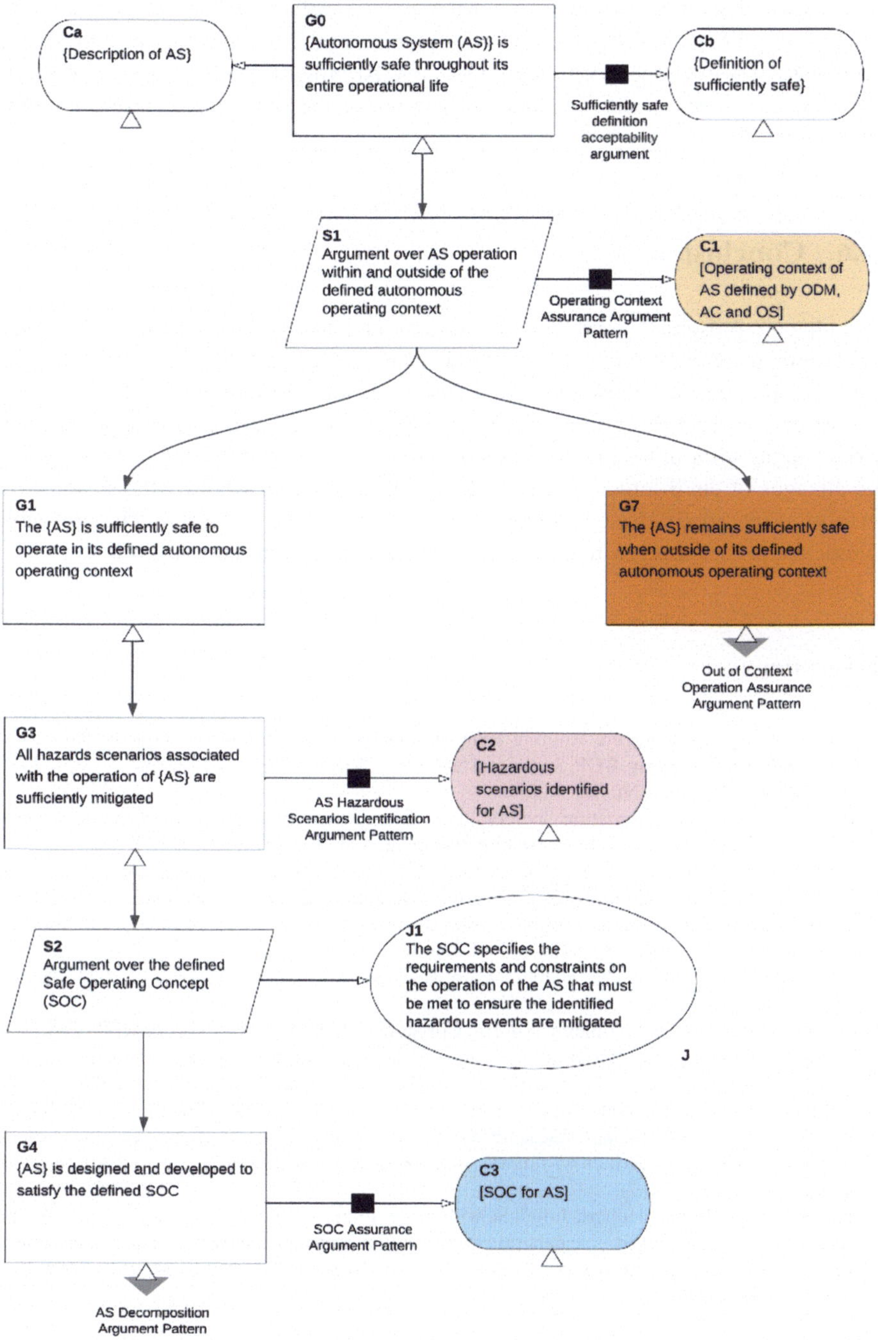

Fig. 11.14 Example argument with goals, context, and strategy [46]

Tools such as the Assurance of Machine Learning for use in Autonomous Systems Tool [47] and the AdvoCATE [48] facilitate the construction of GSN-based safety cases for machine learning components. Arguments such as these can be laid out in general form as the development process starts and refined as development continues.

11.6 Conclusion

This chapter presented a variety of methods and tools for verification, test, and evaluation of autonomous systems. It also presented challenges, for which tools and methods need development. The promise of autonomous systems is that they may be able to perform complicated behaviors while adapting to changes in their environment, without being controlled by a human. But the complexity of behaviors and changes of environment pose challenges in evaluating whether the system will perform as intended. These challenges need to be overcome to establish reasonable confidence in the safety, reliability, and performance of autonomous systems.

References

1. Borchet et al. "BGP Secure Routing Extension (BGP-SRx): Reference Implementation and Test Tools for Emerging BGP Security Standards," NIST Technical Note 2060, Sept. 2021. https://doi.org/10.6028/NIST.TN.2060
2. IEEE Robotics and Automation Society. "IEEE standard ontologies for robotics and automation." *IEEE Stan.* 1872 (2015): 1-60. https://standards.ieee.org/ieee/1872/5354/
3. United Nations Institute for Disarmament Research. "The Weaponization of Increasingly Autonomous Technologies: Concerns, Characteristics and Definitional Approaches." UNIDIR, 2017. https://unidir.org/publication/the-weaponization-of-increasingly-autonomous-technologies-concerns-characteristics-and-definitional-approaches/
4. ARL Autonomy Stack, "Scalable, Adaptive, and Resilient Autonomy", 2023. https://arl.devcom.army.mil/cras/sara-cra/sara-overview/
5. Boloor, Adith, Karthik Garimella, Xin He, Christopher Gill, Yevgeniy Vorobeychik, and Xuan Zhang. "Attacking vision-based perception in end-to-end autonomous driving models." *Journal of Systems Architecture* 110 (2020): 101766. https://doi.org/10.1016/j.sysarc.2020.101766.
6. Conneau, Alexis, Kartikay Khandelwal, Naman Goyal, Vishrav Chaudhary, Guillaume Wenzek, Francisco Guzmán, Edouard Grave, Myle Ott, Luke Zettlemoyer, and Veselin Stoyanov. "Unsupervised cross-lingual representation learning at scale." *arXiv preprint arXiv:1911.02116* (2019).
7. Silver, David, Thomas Hubert, Julian Schrittwieser, Ioannis Antonoglou, Matthew Lai, Arthur Guez, Marc Lanctot et al. "A general reinforcement learning algorithm that masters chess, shogi, and Go through self-play." *Science* 362, no. 6419 (2018): 1140-1144. https://doi.org/10.1126/science.aar6404.

8. Kaiser, Lukasz, Mohammad Babaeizadeh, Piotr Milos, Blazej Osinski, Roy H. Campbell, Konrad Czechowski, Dumitru Erhan et al. "Model-based reinforcement learning for Atari." *arXiv preprint arXiv:1903.00374* (2019).
9. Agarwal, Ankit, Jacob R. Crouse, and Eric N. Johnson. "Evaluation of a commercially available autonomous visual inertial odometry solution for indoor navigation." In *2020 International Conference on Unmanned Aircraft Systems (ICUAS)*, pp. 372–381. IEEE, 2020.
10. Handelman, David A., Corban G. Rivera, Robert St Amant, Emma A. Holmes, Andrew R. Badger, and Bryanna Y. Yeh. "Adaptive human-robot teaming through integrated symbolic and subsymbolic artificial intelligence: preliminary results." In *Artificial Intelligence and Machine Learning for Multi-Domain Operations Applications IV*, vol. 12113, pp. 145–157. SPIE, 2022.
11. ISO, IEC. "Ieee: Iso/iec/ieee 29148, systems and software engineering, life cycle processes." *Requirements engineering* 600 (2011).
12. Purchase Description for Joint Light Tactical Vehicle Version 3.0, January 20, 2012.
13. DA form 6125, Army Publishing Directorate, September 2019. https://armypubs.army.mil/ProductMaps/PubForm/Details.aspx?PUB_ID=1007869
14. Bodt, Barry A., and Rick S. Camden. "Technology readiness level six and autonomous mobility." In *Unmanned Ground Vehicle Technology VI*, vol. 5422, pp. 302-313. SPIE, 2004. https://doi.org/10.1117/12.547450
15. Sharath, Mysore Narasimhamurthy, and Babak Mehran. "A literature review of performance metrics of automated driving systems for on-road vehicles." *Frontiers in Future Transportation* 2 (2021): 759125. doi: https://doi.org/10.3389/ffutr.2021.759125
16. Army, U. S. "Failure modes, effects and criticality analysis (FMECA) for command, control, communications, computer, intelligence, surveillance, and reconnaissance (c4isr) facilities." *Department of the Army, Technical Manual No. TM* (2006): 5–698.
17. Cofer, Darren, and Steven Miller. "DO-333 certification case studies." In *NASA Formal Methods: 6th International Symposium, NFM 2014, Houston, TX, USA, April 29–May 1, 2014. Proceedings 6*, pp. 1–15. Springer International Publishing, 2014. https://doi.org/10.1007/978-3-319-06200-6_1
18. Crow, Judy, Sam Owre, John Rushby, Natarajan Shankar, and Mandayam Srivas. "A tutorial introduction to PVS." Wift, 1995. http://pvs.csl.sri.com/doc/wift-tutorial.pdf
19. Lamsweerde, Axel van. "Formal specification: a roadmap." In *Proceedings of the Conference on the Future of Software Engineering*, pp. 147-159. 2000.
20. Luckcuck, Matt, Marie Farrell, Louise A. Dennis, Clare Dixon, and Michael Fisher. "Formal specification and verification of autonomous robotic systems: A survey." *ACM Computing Surveys (CSUR)* 52, no. 5 (2019): 1–41.
21. Alshiekh, Mohammed, Roderick Bloem, Rüdiger Ehlers, Bettina Könighofer, Scott Niekum, and Ufuk Topcu. "Safe reinforcement learning via shielding." In *Proceedings of the AAAI conference on artificial intelligence*, vol. 32, no. 1. 2018. https://arxiv.org/abs/1708.08611
22. Baier, Christel, and Joost-Pieter Katoen. *Principles of model checking*. MIT press, 2008.
23. Davis, Jennifer A., Laura R. Humphrey, and Derek B. Kingston. "When human intuition fails: Using formal methods to find an error in the "proof" of a multi-agent protocol." In *Computer Aided Verification: 31st International Conference, CAV 2019, New York City, NY, USA, July 15–18, 2019, Proceedings, Part I 31*, pp. 366–375. Springer International Publishing, 2019.
24. Black, Paul, Kass Michael, Michael Koo, and Elizabeth Fong. "Source Code Security Analysis Tool Functional Specification Version 1.1" NIST Special Publication 500-268. NIST, 2011. https://www.nist.gov/system/files/documents/2021/03/23/source_code_security_analysis_spec_SP500-268_v1.1.pdf
25. Mitre, C. V. E. "Heartbleed OpenSSL Bug [CVE-2014-0160]." (2014).
26. NIST, Software Quality Group. "Source Code Security Analyzers". https://www.nist.gov/itl/ssd/software-quality-group/source-code-security-analyzers
27. NIST, Software Quality Group. "Binary Code Scanners". https://www.nist.gov/itl/ssd/software-quality-group/binary-code-scanners

28. Barr, Earl T., Mark Harman, Phil McMinn, Muzammil Shahbaz, and Shin Yoo. "The oracle problem in software testing: A survey." *IEEE transactions on software engineering* 41, no. 5 (2014): 507–525.
29. Clark, Matthew, Xenofon Koutsoukos, Ratnesh Kumar, Insup Lee, George Pappas, Lee Pike, Joseph Porter, and Oleg Sokolsky. "A study on run time assurance for complex cyber physical systems." *Air Force Research Lab, Tech. Rep. ADA585474* (2013). https://apps.dtic.mil/sti/pdfs/ADA585474.pdf
30. Swihart, Donald E., Arthur F. Barfield, Edward M. Griffin, Richard C. Lehmann, Shawn C. Whitcomb, Billie Flynn, Mark A. Skoog, and Kevin E. Processor. "Automatic ground collision avoidance system design, integration, & flight test." *IEEE Aerospace and Electronic Systems Magazine* 26, no. 5 (2011): 4–11.
31. Thuloweit, Kenji. "416th FLTS testers meet with Auto GCAS survivor." 412th Test Wing Public Affairs Sept. 1, 2016. https://www.edwards.af.mil/News/Article/932081/416th-flts-testers-meet-with-auto-gcas-survivor/
32. Oh, Jean, Thomas M. Howard, Matthew R. Walter, Daniel Barber, Menglong Zhu, Sangdon Park, Arne Suppe et al. "Integrated intelligence for human-robot teams." In *2016 International Symposium on Experimental Robotics*, pp. 309-322. Springer International Publishing, 2017.
33. Schaefer, Kristin E., Anthony L. Baker, Ralph W. Brewer, Debra Patton, Jonroy Canady, and Jason S. Metcalfe. "Assessing multi-agent human-autonomy teams: US Army Robotic Wingman gunnery operations." In *Micro-and Nanotechnology Sensors, Systems, and Applications XI*, vol. 10982, pp. 353–372. SPIE, 2019.
34. Spillner, Andreas, Tilo Linz, Schaefer, and Hans Schaefer. "Software Testing Foundations: A Study Guide for the Certified Tester Exam." Fourth Edition, Rocky Nook, Inc. 2014.
35. Patton, Ron, "Software Testing." Second Edition, Sams Publishing, 2005.
36. Lennon, Craig, Barry Bodt, Marshal Childers, Rick Camden, Arne Suppe, Luis Navarro-Serment, and Nicoleta Florea. "Performance evaluation of a semantic perception classifier." *Army Research Labs, Tech. Rep. ARL-TR-6653* (2013).
37. Lennon, Craig, Barry Bodt, Marshal Childers, Jean Oh, Arne Suppe, Luis Navarro-Serment, Robert Dean et al. "An integrated assessment of progress in robotic perception and semantic navigation." *Aberdeen Proving Ground (MD): Army Research Laboratory (US)* (2015).
38. Dean, Robert Michael S., and Charles A. DiBerardino. "Robotic collaborative technology alliance: an open architecture approach to integrated research." In *Open Architecture/Open Business Model Net-Centric Systems and Defense Transformation 2014*, vol. 9096, pp. 96–102. SPIE, 2014.
39. Schaefer, Kristin E., Ralph W. Brewer, E. Ray Pursel, Michelle Desormeaux, Anthony Zimmermann, and Eduardo Cerame. "US Army Robotic Wingman Simulation: June 2018 Integration Workshop." (2018)
40. Frey, Matthias A., and Axel Schulte. "Towards a tactical UAV assessment system to support transport helicopter crews in MUM-T scenarios." In *2020 IEEE International Conference on Systems, Man, and Cybernetics (SMC)*, pp. 3206–3213. IEEE, 2020.
41. Scheidt, David, and Kevin Schultz. "On optimizing command and control structures." In *Proceedings of the 16th international command and control research and technology symposium*. Quebec City, CA, June 21–23, 2011.
42. Scheidt, David, Robert Lutz, and James Sturim. "The DARPA CODE White Force Network." *Interservice/Industry Training, Simulation, and Education Conference*, 2017.
43. Gilmore, Michael, "Director, Operational Test and Evaluation FY 2014 Annual Report." OFFICE OF THE SECRETARY OF DEFENSE WASHINGTON DC OPERATIONAL TEST AND EVALUATION. (2015).
44. Bloomfield, Robin, Gareth Fletcher, Heidy Khlaaf, Luke Hinde, and Philippa Ryan. "Safety case templates for autonomous systems." *arXiv preprint arXiv:2102.02625* (2021). https://arxiv.org/ftp/arxiv/papers/2102/2102.02625.pdf
45. ASCE Software Overview, https://www.adelard.com/asce/

46. Hawkins, Richard, Matt Osborne, Mike Parsons, Mark Nicholson, John McDermid, and Ibrahim Habli. "Guidance on the Safety Assurance of Autonomous Systems in Complex Environments (SACE)." https://www.york.ac.uk/media/assuring-autonomy/documents/SACEv1.pdf
47. "How to assure the safety of your machine learning component." Assuring Autonomy International Programme. https://www.assuringautonomy.com/amlas/tool
48. User Guide AdvoCATE, Product Model 2.0. https://ntrs.nasa.gov/api/citations/20220009664/downloads/AdvoCATE%20User%20Guide_1.4.pdf

The manufacturer's authorised representative in the EU is Springer
Nature Customer Service Centre GmbH, Europaplatz 3, 69115 Heidelberg,
Germany. If you have any concerns regarding our products, please
contact ProductSafety@springernature.com

Printed and bound by CPI Group (UK) Ltd, Croydon, CR0 4YY
21/09/2026
02220010-0013